Chemical Reactions in Clusters

TOPICS IN PHYSICAL CHEMISTRY
A Series of Advanced Textbooks and Monographs

Series Editor: Donald G. Truhlar

F. Iachello and R. D. Levine, *Algebraic Theory of Molecules*
P. Bernath, *Spectra of Atoms and Molecules*
J. Simons and J. Nichols, *Quantum Mechanics in Chemistry*
J. Cioslowski, *Electronic Structure Calculations on Fullerenes and Their
    Derivatives*
E. R. Bernstein, *Chemical Reactions in Clusters*

Chemical Reactions in Clusters

Edited by
Elliot R. Bernstein
Department of Chemistry
Colorado State University

New York   Oxford
OXFORD UNIVERSITY PRESS
1996

Oxford University Press

Oxford   New York
Athens   Auckland   Bangkok   Bombay
Calcutta   Cape Town   Dar es Salaam   Delhi
Florence   Hong Kong   Istanbul   Karachi
Kuala Lumpur   Madras   Madrid   Melbourne
Mexico City   Nairobi   Paris   Singapore
Taipei   Tokyo   Toronto

and associated companies in
Berlin   Ibadan

Library of Congress Cataloging-in-Publication Data
Chemical reactions in clusters / edited by Elliot R. Bernstein.
    p.   cm.—(Topics in physical chemistry)
Includes index.
ISBN 0-19-509004-7
1. Microclusters.   2. Molecular dynamics,   3. Chemical reaction.
Conditions and laws of.   I. Bernstein. E. R. (Elliot R.)
II. Series: Topics in physical chemistry series.
QD461.C4225   1996
541.39'4—dc20   95-10991

987654321

Printed in the United States of America
on acid-free paper

Preface

The study of van der Waals clusters, complexes, and molecules is one of the most dynamic and rapidly changing areas of physical chemistry research that I have encountered. In just the past 6 years or so, the focus of the research on van der Waals systems has shifted from the elucidation of energy levels and structures to the investigation of energy dynamics and chemical reactions. In the latter difficult area, the study of van der Waals systems seems to make its most valuable contribution to date. We are finally getting a glimpse of what minimum set of conditions is required for a dynamical event or chemical reaction to occur. The pace of the field is spectacular, and the number of systems being investigated is simply overwhelming.

I trust that this book gives the flavor of the pace, excitement, and accomplishments of the last few years of cluster research. For me, the most surprising and important feature of this volume is the breadth that this new area of physical chemistry demonstrates. The various experimental chapters cover ionic chemistry, hot atom chemistry, photochemistry, neutral molecule chemistry, electron and proton transfer chemistry, chemistry of radicals and other transient species, and vibrational dynamics and cluster dissociation. Of at least equal importance is that theoretical potential energy surface studies are not accessible for cluster systems and are being pursued. All of us associated with this project have tried to convey the fresh insights and contributions that van der Waals cluster research has brought to physical chemistry.

I would like to thank Bob Rogers at Oxford University Press for allowing us the time and freedom to bring this volume together and Jean Gilbert for helping me put together the various contributions from the other authors, as well as my own.

Fort Collins, Colorado   E.R.B.
November 1994

Contents

Contributors

Elliot R. Bernstein
Colorado State University

A. W. Castleman Jr.
Pennsylvania State University

James F. Garvey
State University of New York at Buffalo

Angels González-Lafont
Universidad Autonoma de Barcelona

C. Jouvet
Université Paris Sud

Roger E. Miller
University of North Carolina at Chapel Hill

D. Solgadi
Université Paris Sud

D. G. Truhlar
University of Minnesota

Gopalakrishnan Vaidyanathan
State University of New York at Buffalo

Shiqing Wei
Pennsylvania State University

Curt Wittig
University of Southern California

Ahmed H. Zewail
California Institute of Technology

Chemical Reactions in Clusters

1

Theoretical Approaches to the Reaction Dynamics of Clusters

ANGELS GONZÁLEZ-LAFONT AND DONALD G. TRUHLAR

1.1. INTRODUCTION

The theoretical treatment of cluster kinetics borrows most of its concepts and techniques from studies of smaller and larger systems. Some of the methods used for such smaller and larger systems are more useful than others for application to cluster kinetics and dynamics, however. This chapter is a review of specific approaches that have found fruitful use in theoretical and computational studies of cluster dynamics to date. The review includes some discussion of methodology; it also discusses examples of what has been learned from the various approaches, and it compares theory to experiment. A special emphasis is on microsolvated reactions—that is, reactions where one or a few solvent molecules are clustered onto gas-phase reactants and hence typically onto the transition state as well.

Both analytic theory and computer simulations are included, and we note that the latter play an especially important role in understanding cluster reactions. Simulations not only provide quantitative results, but they provide insight into the dominant causes of observed behavior, and they can provide likelihood estimates for assessing qualitatively distinct mechanisms that can be used to explain the same experimental data. Simulations can also lead to a greater understanding of dynamical processes occurring in clusters by calculating details which cannot be observed experimentally.

One interesting challenge that reactions in van der Waals and hydrogen-bonded clusters offer is the possibility of studying specifically how weak interactions or microsolvation bonds affect a chemical reaction or dissociation process. In that sense, theoretical studies of weakly bound clusters have proved to be useful in learning about the "crossover" in behavior from that of an isolated nonsolvated molecule in the gas phase to that for a molecule in a liquid or solid solvent.

It is very common to begin reviews with a disclaimer as to completeness. Such a disclaimer is, we hope, not required for this chapter because it is not a comprehensive review but a limited-scope discussion of selected work that illustrates some issues that we perceive to be especially important.

The chapter is divided into three parts. Section 1.2 discusses collisional and statistical theories for cluster reactions. This section is mainly (though not entirely) concerned with ion kinetics. It is well known that the chemistry of ions is strongly influenced by solvation. Thus, the reactivity of cluster ions is particularly interesting in defining the influence and role of solvent molecules on the chemistry of ions. The study of the reactivity of ions can be especially illuminating with respect to the molecular origin of solvation effects for which condensed phase studies only show the collective effect of solvent. With the development of experimental techniques for studying cluster ion dynamics in the gas phase, it has become possible to quantitatively explore the transition in the kinetics of ion–molecule reactions from their solvent-free behavior to their behavior in solution; thus, this kind of study is one of the purest examples of using cluster chemistry to bridge the gap between the gas phase and condensed matter (for representative examples, see Bohme and Mackay 1981; Bohme and Raksit 1984; Bohme and Young 1970; Castleman and Keesee 1986a,b; Hierl et al. 1986a,b; Jortner 1992; Leutwyler and Bösiger 1990; Syage 1994).

Section 1.2 has a particular emphasis on capture rate coefficients for exchange and association reactions, and it begins a discussion of the related issue of unimolecular dissociation of the association complexes. This provides a bridge into section 1.3 which considers the question of energy transfer in clusters from a more general point of view. Section 1.4 returns to the subject of reaction rates emphasized in section 1.2, but now considering cases where the theoretical model involves detailed consideration of the short-range forces in the vicinity of a tight dynamical bottleneck.

Most bimolecular cluster reactions can proceed through complexes. If A and B denote reactants and C and D denote products, at low pressure one has only direct reaction,

$$A + B \xrightarrow{k_0} C + D, \tag{1}$$

because there are no collisions to stabilize the intermediate, where k_0 is the rate coefficient. At higher pressures one has

$$A + B \underset{k^*_{-1}(E)}{\overset{k_1}{\rightleftarrows}} I^*(E) \begin{cases} \xrightarrow{M} I^*(E') \begin{cases} \xrightarrow{k^*_{-1}(E')} A + B \\ \xrightarrow{k^*_2(E')} C + D \end{cases} \\ \xrightarrow{M} I \begin{cases} \xrightarrow{k_{-1}} A + B \\ \xrightarrow{k_2(T)} C + D \end{cases} \\ \xrightarrow{k^*_2(E)} C + D \end{cases} \tag{2}$$

where k_1 is called either the collision rate coefficient or the capture rate coefficient, $I^*(E)$ and $I^*(E')$ are activated intermediates with energy E and E' (or energy distributions centered at E or E'), and I is a thermalized intermediate. Either I^* or I can convert to product by a unimolecular reaction, which is assumed to pass through a tight transition state. Assuming that every collision of $I^*(E)$ with M produces I eliminates the need to consider $I^*(E')$. This is called the strong collision approximation. A somewhat weaker approximation is to set the rate coefficient for $I^*(E) + M \rightarrow I + M$ equal to an empirical constant (usually called the collision efficiency or β) times the collision rate coefficient but again to neglect $I^*(E')$ [or, equivalently, to take $k_2^*(E) = k_2^*(E')$ and $k_{-1}^*(E) = k_{-1}^*(E')$]. With or without these additional assumptions, in the high pressure limit, mechanism (2) yields an overall rate coefficient

$$k_0 = k_1 k_2 / k_{-1}. \tag{1-1}$$

At low pressure it would yield,

$$k_0 = k_1 \frac{k_2^*(E)}{k_{-1}^*(E) + k_2^*(E)}; \tag{1-2}$$

however, this low pressure limit has meaning only for molecules that internally randomize their energy well enough to serve as their own heat bath; otherwise $k_2^*(E)$ and $k_{-1}^*(E)$ are not physically meaningful as kinetic constants. A more appropriate formulation of the low pressure limit involves only a single rate coefficient for converting reactants to products. We call this k_0 in eq. (1-1).

Transition state theory plays an important conceptual role in discussing such mechanisms. The transition state assumption is that there exists a perfect dynamical bottleneck somewhere along the reaction path. A perfect dynamical bottleneck is a hypersurface in coordinate space or phase space that separates reactants R from products P and has the property that any trajectory through it in the R → P direction originated on the R side and will proceed to products without recrossing it. In terms of the reaction we have been discussing, the low pressure k_0 would be expected to equal the capture rate coefficient k_1 of the mechanism shown in reaction (2) if the transition state assumption is exact for a dynamical bottleneck that occurs early in the collision, whereas it would equal the k_0 of eq. (1-1) if transition state theory at a tight dynamical bottleneck is exact. In general, the transition state assumption might not be perfect for either of these choices, and then one must consider recrossing effects at one or both imperfect dynamical bottlenecks. One (approximate) way of doing this is the unified statistical model (Garrett and Truhlar 1979c, 1982; Miller 1976; Truhlar et al. 1985b), which is based on the branching analysis of Hirschfelder and Wigner (1939). The only completely reliable way to estimate recrossing effects is to compute the full dynamics and compare to quantized transition state theory calculations. A popular approximation is to carry out a dynamical simulation using classical trajectories.

The situation is slightly different for an association reaction,

$$A^+ + B + M \rightarrow AB^+ + M. \tag{3}$$

Then, making the strong collision approximation, one considers the mechanism

$$A^+ + B \underset{k_{-1}}{\overset{k_1}{\rightleftharpoons}} (AB^+)^* \xrightarrow{k_2[M]} AB^+, \tag{2'}$$

where again an asterisk denotes an unequilibrated (hot, excited) species.

In terms of these mechanisms, section 1.2 is primarily concerned with theoretical models of k_1, k_{-1}, k_2^*, and k_2 that do not require a full potential energy surface. Section 1.3 is concerned with $I^*(E) \xrightarrow{M} I^*(E')$ or I or with energy transfer within $I^*(E)$ itself. Section 1.4 is primarily concerned with the calculation of k_0 from potential energy surfaces.

1.2. COLLISIONAL AND STATISTICAL THEORIES FOR CLUSTER REACTION RATES

1.2.1. Bimolecular Rate Coefficients

a. *Theory*

In the early development of chemical kinetics, it was often assumed that bimolecular reactions occur upon every encounter between a pair of reactants. Then the encounter becomes a capture event, and the reaction rate is the capture rate. Although it is well known that this assumption is usually not true, the concept of a "collision" or "capture" rate is useful for establishing an approximate upper bound on measured reaction rates. This approximate bound is particularly useful in ion-molecule chemistry where strong long-range forces dominate the early stages of the reaction dynamics. The capture approximation can be applied to calculate approximate rate coefficients for exothermic reactions which contain no potential barriers and which are dominated by attractive, long-range intermolecular forces. For example, if the long-range potential is dominated by a charge-induced dipole interaction, then the capture rate coefficient is given by the familiar Langevin (1905)/Gioumousis–Stevenson (1958) expression. According to this model, the reactants are treated as point particles, and the rate coefficient is that for passing the maximum in the effective potential consisting of a positive centrifugal potential added to negative ion-induced dipole potential. This is an "orbiting transition state" (McDaniel 1964). The rate coefficient then depends on the electric properties of the ion (charge and polarizability), is inversely proportional to the square root of the colliding pair's reduced mass, and is independent of temperature.

For systems with anisotropic potentials, such as the reactions of ions with molecules having permanent dipole moments, the application of capture theories is not as simple as for collision partners interacting by central potentials, since the rotational motion of the molecules becomes hindered by the presence of the ion as it approaches, and this strong perturbation of the rotational motion has to be included in any realistic theory (Moran and Hamill 1963). A variety of theoretical approaches have been developed to simplify this problem. The first of these approximations that could be generally applied was the average dipole orientation (ADO) theory of Su and Bowers (Bass et al. 1975; Su and Bowers 1973a,b, 1975). This theory used statistical methods to calculate the average orientation of the polar molecule in the ion field and a Langevin procedure to calculate the rate of passage over the resulting entrance channel effective barrier. Ridge and coworkers (Barker and Ridge 1976; Celli et al. 1980) treated the competition between free rotation and collision-induced calculated alignment differently; they calculated an average interaction energy between the ion and the dipole and then again used the Langevin procedure to calculate the rate coefficient. The major assumption in the original formulation of the ADO theory is that there is no angular momentum transfer between the rotating molecule and the ion–molecule orbital motion. While this assumption may be quite good at large ion–molecule separations it becomes less valid as the separation distance decreases. Conservation of angular momentum was considered in the formulation of an improved theory called angular-momentum-conserved ADO theory or AADO (Su et al. 1978).

A further advance occurred when Chesnavich et al. (1980) applied variational transition state theory (Chesnavich and Bowers 1982; Garrett and Truhlar 1979a,b,c,d; Horiuti 1938; Keck 1967; Wigner 1937) to calculate the thermal rate coefficient for capture in a noncentral field. Under the assumptions that a classical mechanical treatment is valid and that the reactants are in equilibrium, this treatment provides an upper bound to the true rate coefficient. The upper bound was then compared to calculations by the classical trajectory method (Bunker 1971; Porter and Raff 1976; Raff and Thompson 1985; Truhlar and Muckerman 1979) of the true thermal rate coefficient for capture on the ion–dipole potential energy surface and to experimental data (Bohme 1979) on thermal ion-polar molecule rate coefficients. The results showed that the variational bound, the trajectory results, and the "experimental upper bound" were all in excellent agreement. Some time later, Su and Chesnavich (Su 1985; Su and Chesnavich 1982) parameterized the collision rate coefficient by using trajectory calculations.

At low temperature the classical approximation fails, but a quantum generalization of the long-range-force-law collision theories has been provided by Clary (1984, 1985, 1990). His capture-rate approximation (called adiabatic capture centrifugal sudden approximation or ACCSA) is closely related to the statistical adiabatic channel model of Quack and Troe (1975). Both theories calculate the capture rate from vibrationally and rotationally adiabatic potentials, but these are obtained by interpolation in the earlier work (Quack and Troe 1975) and by quantum mechanical sudden approximations in the later work (Clary 1984, 1985).

The abundant experimental data on ionic clusters reacting with neutral molecules has been used to test some of these collision theories. In the next subsection, we briefly review several papers where comparisons between measured and theoretical rate coefficients have been made, and we summarize some of the important conclusions concerning the reactivity of clusters.

b. *Comparison to experiment*

i. *Exchange reactions.* In an early paper, Smith et al. (1981) studied the temperature dependence of the rate coefficients of the proton transfer and ligand exchange reactions $H^+(H_2O)_n + CH_3CN$ with $n = 1$–4 at $T = 200$–300 K. A slightly positive or zero temperature dependence was found; this result agrees well with collision theory calculations that predicted only a 1% rise in the rate coefficient with temperature. However, later improved theories do not agree so well. For example, the parameterized trajectory calculations of Su and Chesnavich (1982) predict a negative temperature dependence (approximately $T^{-1/2}$ as expected from the analytic models) of the collision rate coefficients for an ion–molecule reaction in which the neutral reactant has a permanent dipole moment. Viggiano et al. (1988a) studied the reactions $H^+(H_2O)_n + CH_3CN$, NH_3, CH_3OH, CH_3COCH_3 and C_5H_5N with $n = 2$–11. They found that the rate coefficients displayed a stronger temperature dependence, varying as T^{-1}, than the just-mentioned theoretical prediction of $T^{-1/2}$. The authors explained the observed discrepancy between experiment and theory as being due to a failure of the theory to account for the dipole-induced dipole interaction and the nonuniform distribution of charge in the clusters at low temperatures. The former interaction is more important for cluster ions than for bare ones because of the large polarizability of the ion. In later work (Viggiano et al. 1988b), the rate coefficients for proton transfer from $H^+(NH_3)_m \cdot (H_2O)_n$ with $m + n \leq 5$ were found to vary as $T^{-\lambda}$ with $\lambda = 0.3-1.7$. In more recent work, Yang and Castleman (1991a) analyzed the kinetics of $H^+(H_2O)_n$ + acetone, acetonitrile, and methyl acetate with $n = 1-60$, both at room temperature and at $T = 130$ K. The measured rate coefficients were found to agree within experimental error with values calculated using the Su–Chesnavich (1982) method for the entire range of cluster sizes and at both temperatures. In a following paper, Yang and Castleman (1991b) reported detailed experimental studies and theoretical calculations of the temperature and cluster size dependence of $H^+(H_2O)_n$ with CH_3CN, with $n = 1-30$, for temperatures in the range 130–300 K. Very good agreement was found between the experiments and Su–Chesnavich theory for both proton transfer and switching reactions for all the accessible cluster sizes at room temperature. The agreement of theory and experiment was also found to be very good for the dependence of the rates on temperature and cluster size. The same kind of good agreement was found for larger clusters, $H^+(H_2O)_n$ with $n = 4-45$, in their association reactions with CH_3CN (Yang and Castleman 1989).

Hierl et al. (1986a) studied the proton transfer ($X^- + HY \to HX + Y^-$) between $OH^-(H_2O)_n$ and HF with $n = 0-3$ as a function of hydration number and temperature in the range 200–500 K. Their experimental data agree within experimental error with theoretical predictions for collision rate coefficients derived using the ACCSA method introduced above. They included only ion–dipole and ion-induced dipole interactions and omitted the dipole–dipole interactions, which are estimated to raise the rate coefficient in the case analyzed by up to 30%. The agreement with other theoretical predictions (Su and Chesnavich 1982) was about 20%. From these comparisons, Hierl et al. (1988) concluded that inter-molecular proton transfer was occurring on essentially every collision throughout the ranges of hydration and temperature studied, and that the product tends to be hydrated. The former observation is consistent with other works—that is, proton transfer is usually fast (see, e.g., Viggiano et al. 1988a). The latter observation was explained by postulating a transition state structure of the form $HOH \cdots OH^- \cdots HF$, such that the formal transfer of a proton to the left and a water to the right is accomplished in actually by the transfer of an OH^- to the right. One may consider the proton as coming not from the donor but from the acceptor's solvation shell (kinetic participation of the solvation shell), and the reaction pathway is favorable (can occur at low energy) because the polar solvent molecule can stay close to the center of charge.

Hierl et al. (1986b) also studied the nucleophilic displacement reactions ($X^- + CH_3Y \to CH_3X + Y^-$) between $OD^-(D_2O)_n$ and CH_3Cl with $n = 0-2$ as a function of hydration number and temperature at 200–500 K. The reaction efficiencies (the reaction efficiency is defined as the ratio of the experimental rate coefficient to the theoretical collision rate coefficient) evaluated using ACCSA collisional rates showed that, in contrast to the proton transfer discussed above, nucleophilic displacement does not occur on every collision, and efficiencies decrease with increasing hydration and temperature. In fact, three water molecules were found to be sufficient to quench the reaction altogether, which is consistent with the much higher activation energies observed for S_N2 reactions in the condensed phase. The results were interpreted in terms of the model discussed by Olmstead and Brauman (1977), which views S_N2 reactions as proceeding along a potential energy profile with a double minimum; one minimum corresponding to the reactant ion–dipole complex and one to the product ion–dipole complex. These may be called the precursor and successor complexes. The efficiency of a reaction in the Olmstead–Brauman model results from the trade-off between two effects in the reactivity of the precursor ion–dipole complex: (1) differences in entropies of activation for the dynamical bottleneck between reactants and the precursor complex and the dynamical bottleneck near the central barrier between the minima, and (2) the magnitude of the central barrier resulting from the differential solvation of the reactants and the transition state. The central transition state involves Walden inversion. Solvent transfer was considered energetically unfavorable because it was assumed to require an energetically and entropically unfavorable transition state. We will return to this question in section 1.4. Related work (Bohme and Mackay 1981; Bohme and Raksit 1985; Henchman et al. 1983, 1985, 1987; Hierl et al. 1988) has been discussed in terms of the relative energetics of unsolvated and microsolvated species. Some experimentally observed micro-solvation effects may be understood quantitatively in terms of the attractive idea that cluster-ion studies in the gas phase bridge the gap between unsolvated gas phase reactions on one hand and the condensed phase reactions on the other hand. However, this is not always true (Bohme and Raksit 1984; Henchman et al. 1983, 1988; Hierl et al. 1988).

In early studies, Fehsenfeld and Ferguson (1974) determined the room-temperature rate coefficients for the reactions of CO_2 and other molecules with $OH^-(H_2O)_n$, $n = 0$ and 2–4, and with $O_3^-(H_2O)_n$, $n = 0$–2. In a later paper, Fahey et al. (1982) examined the reaction of CO_2 and other molecules $O_2^-(H_2O)_n$ with $n = 1$–4. Hierl and Paulson (1984) analyzed the energy dependence of the cross sections for the reactions of $OH^-(H_2O)_n$ with $n = 0$–3. More recently, Viggiano et al. (1990) have studied the temperature dependence of rate coefficients for reaction of $O^-(H_2O)_n + H_2$ or D_2 with $n = 0$–2. All these investigations deal with small hydrated clusters, and the reaction paths are those expected for gas phase species. For example, Hierl and Paulson (1984) found that CO_2 replaces water molecules in the hydrated cluster $OH^-(H_2O)_{n=1-2}$ to form $HCO_3^-(H_2O)_n$ with a rate coefficient nearly equal to the gas phase collision limit as evaluated with the AADO formalism introduced above. Interestingly, when $n = 3$ the measured rate coefficient was reported to be significantly lower than the calculated value (the reason was not discussed in the paper). In more recent work (Yang and Castleman 1991c), the reactions of large clusters $X^-(H_2O)_{n=0-59}$, $X = OH$, O, O_2 and O_3 with CO_2 were studied by Yang and Castleman. For the smaller solvated cluster ions, the rate coefficients are very close to the Langevin collision limit, and they vary as the negative square root of the reduced mass of the collision complex, as predicted by theory. The rate coefficients for those reactions that proceed at near the gas phase collision limits do not display any temperature dependences, as predicted by Langevin theory for the case where the neutral (here CO_2) has no permanent dipole moment. The differences between experimentally measured rate coefficients and the theoretical calculations become larger as cluster size increases.

For O_2^- and O_3^- clusters, these are explained by a change in sign of the reaction enthalpy, whereas reactions of hydrated OH^- with CO_2 are exothermic for all degrees of solvation, and the large discrepancy between the experimentally measured rate coefficients and the theoretically calculated values, also attributable to solvation, was explained by an association mechanism in which the unimolecular dissociation rate coefficient of the reaction intermediate increases and the rate coefficient for conversion to product decreases for progressively larger cluster sizes. (We will review theoretical treatments of association reactions in the next subsection.) Yang et al. (1991) made a similar comparison between experimental and theoretical collision rate coefficients (in this case evaluated by the Su–Chesnavich method) and showed that $OH^-(H_2O)$ with $n = 0$ or 1 reacts with CH_3CN via proton transfer and ligand switching reactions at nearly the collision rate. Further hydration greatly reduces the reactivity of $OH^-(H_2O)$ with $n > 1$, in disagreement with the collision theory. On the contrary, for all the cluster sizes studied, $O^-(H_2O)_n$ reacts with CH_3CN at nearly the collision rate via hydrogen transfer from acetonitrile to the anionic clusters. Hierl and Paulson (1984) had found their measured rate coefficients for the reactions between $OH^-(H_2O)_n$ and SO_2 to be comparable to those predicted by the AADO theory. The authors explained that SO_2 reacts more rapidly than CO_2 according to that theoretical formalism because SO_2 possesses a permanent dipole moment. The collision theory of Su and Chesnavich was also shown (Yang and Castleman 1991d) to correctly predict the rates of $X^-(H_2O)_n$ with $n = 1\text{–}3$ and $X = OH$, O, O_2 and O_3, with SO_2; these reactions proceed via ligand switching at room temperature. For larger clusters at low temperature, where association dominates the reaction mechanism, the measured rate coefficients are also very close to the collision limit, showing very little dependence on pressure, temperature, and cluster size, as predicted by the collision theory. In another study, Yang and Castleman (1990) analyzed the switching reactions $NaX_n^+ + L \rightarrow NaX_{n-1}L^+ + X$ with $X = H_2O$, NH_3, and CH_3OH, $n = 1\text{–}3$, and $L = NH_3$ or various organic molecules at room temperature. All of the measured rates are very fast, proceeding at or near the collision rate predicted by the parameterized trajectory calculations of Su and Chesnavich. Furthermore, the rate coefficients do not show a pressure dependence, and the type of ligand bound to the sodium ion has little effect on the reaction rate. These features agree well with expectations (Castleman and Keesee 1986b) since all the reactions are exothermic and barrierless, and the parent ions can be treated as point charges due to their small physical size compared with the distance at which the maximum of the centrifugal barrier in the Langevin model occurs.

ii. *Association reactions.* Ion–molecule association reactions have received an increasing amount of attention over the years. Initially, the primary emphasis was thermochemical (Hogg et al. 1966), and later interest turned to the association rate coefficients. Simple clustering reactions provide good systems for testing theoretical models. Theoretical developments have been made concurrently with the experimental work.

Our discussion is based on the overall reaction (3) and the mechanism in reaction (2′) presented above. The second step of reaction (2′) has a bimolecular rate coefficient of β times the $(AB^+)^* + M$ collision rate coefficient, where β is the stabilization efficiency, usually assumed to be temperature independent, with a value between 0 and 1 depending on the nature of the third body, M, and the reacting system, $A^+ + B$.

Although the general mechanism of cluster formation is well understood, the redistribution of energy in the intermediate excited complex $(AB^+)^*$ and the lifetime against dissociation back to the original reactants are major questions requiring further work. Therefore, theory can contribute by providing a better understanding of the unimolecular dissociation in terms of the statistical redistribution of energy within the excited intermediate. Although Rice–Ramsperger (1927)/Kassel (1928) (RRK) theory is sometimes used, the more sophisticated Rice–Ramsperger–Kassel–Marcus (RRKM) formulations (Forst 1983; Marcus 1952; Marcus and Rice 1951; Robinson and Holbrook 1972) and phase space theory (Bass et al. 1979; Bass and Jennings 1984; Caralp et al. 1988; Chesnavich and Bowers 1977; Light 1967; Light and Lin 1965; Nikitin 1965; Pechukas and Light 1965; Truhlar and Kuppermann 1969) give more insight because of their closer connection to the true molecular dynamics. In particular, RRKM theory is equivalent to transition state theory (Kreevoy and Truhlar 1986; Magee 1952; Rosenstock et al. 1952), and it allows an arbitrarily detailed description of the transition state. Phase space theory, in contrast, assumes that the collisional rate coefficient k_1 and the rate coefficient k_{-1} for dissociation of the complex are governed by an orbiting or other type of loose transition state (requiring less information but sometimes introducing error when the assumption is invalid), but—unlike the usual formulation of RRKM theory—it rigorously conserves angular momentum. Especially interesting fundamental questions are related to the effectiveness of collisions and radiation in removing energy from complexes, leading to stable clusters. Other interesting questions are the effect of competing reaction channels on clustering, and the pressure and temperature dependences of association reactions. These questions have been discussed in the literature (Castleman and Keesee 1986b; Viggiano 1986).

Some of the initial work dealt with the formation of proton-bound dimers in simple amines. Those systems were chosen because the only reaction that occurs is clustering. A simple energy transfer mechanism was proposed by Moet-Ner and Field (1975), and RRKM calculations performed by Olmstead et al. (1977) and Jasinski et al. (1979) seemed to fit the data well. Later, phase space theory was applied (Bass et al. 1979). In applying phase space theory, it is usually assumed that the energy transfer mechanism of reaction (2') is valid and that the collisional rate coefficients k_1 and k_{-1} can be calculated from Langevin or ADO theory and equilibrium constants.

Bass et al. (1981) published phase space theory models of the reaction $CH_3^+ + HCN \rightarrow (CH_3^+ \cdot HCN) + h\nu$, analyzing, in particular, radiative stabilization of the complex. Important work on radiative stabilization has also been published by Dunbar (1975), Herbst (1976) and Woodin and Beauchamp (1979).

In more recent work, Bass et al. (1983) applied the statistical phase space theory to clustering reactions of $CH_3OH_2^+$, $(CH_3)_2OH^+$, and $(CH_3OH)_2H^+$ with CH_3OH. Generally good agreement was found between the experimental and the theoretical rate coefficients. The authors also modeled molecular elimination, back dissociation, collisional stabilization, and sequential clustering reactions.

The capture theories are most directly useful for exothermic reactions whose reverse reaction is also bimolecular. For association reactions, the reverse reaction is unimolecular. Equating the association rate coefficient to the capture one is only valid in the high pressure limit where all complexes are stabilized by third-body collisions. If the association reaction is treated as bimolecular, the apparent second-order rate coefficient becomes independent of pressure only in this limit. This problem has been widely studied for the reverse dissociation reactions, and specialized techniques have been developed (Troe 1977b, 1979) for theoretical treatment of the "falloff" regime between the high pressure second-order and low pressure third-order limits. Chang and Golden (1981) discussed this issue using Troe's simplified model in which the requisite information for the low pressure limit is the collision efficiency β and the density of states of the association complex. The low pressure model is equivalent to calculating the bimolecular dissociation rate coefficient and combining it with the equilibrium constant. The results are similar to those obtaned by the somewhat more complicated RRKM theory of dissociation reactions.

The falloff region was treated for cluster reactions by Lau et al. (1982), who considered $H^+(H_2O)_{n-1} + H_2O \rightarrow H^+(H_2O)_n$ with $n = 2$–6 and by Bass et al. (1983), who treated $H^+(CH_3OH)_{n-1} + CH_3OH \rightarrow H^+(CH_3OH)_n$ with $n = 2$ and 3.

The association reactions $CF_3 + O_2$ and $CCl_3 + O_2$, although not cluster reactions, may be used to illustrate the issues. Ryan and Plumb (1982) and Danis et al. (1991) studied the kinetics of these reactions, the former in helium at 1.6×10^{16}–2.7×10^{17} molecules cm^{-3} and the latter in nitrogen in the 1–12 torr pressure range, as well as at 760 torr. Both groups carried out RRKM calculations for modeling the experimental results, and their results seem to be in reasonable agreement once the third-body efficiencies are taken into account. However, Danis et al. reported a strong temperature dependence for the rate coefficients at low pressure that could not be easily described using RRKM calculations. More recently, Fenter et al. (1993a) published new results on the same association reaction and they fit the experimental data by means of an RRKM calculation. This calculation was carried out using the strong collision hypothesis ($\beta = 1$, Troe 1977a), and a modified Gorin model (Davies and Pilling 1989; Garrett and Truhlar 1979d; Gorin 1938; Smith and Golden 1978) was used to represent the activated complex. The modified Gorin model is a phenomenological surrogate for variational transition state theory (see, e.g., Rai and Truhlar 1983) that does not require realistic potential functions. The analyzed experimental data were collected in the falloff region of the association reaction. Comparison of extrapolations with low and high pressure limiting rate coefficients from data taken in this region illustrates the state of the art of this kind of treatment. The same kind of calculations were reported by Caralp et al. (1988) for the association of peroxy radicals with NO_2 and by Fenter et al. (1993b) for the association reactions of $CHCl_2$ and CH_2Cl with molecular oxygen.

A more realistic treatment of the low pressure limit of the association rate coefficient requires a more complete treatment of energy transfer collisions, going beyond the assumption that all such effects can be subsumed under the guise of a constant collision efficiency. These more sophisticated treatments of energy transfer are discussed in section 1.3.

Further discussion of ion–molecule association reactions and cluster formation is provided in chapter 6 in this volume by Wei and Castleman.

1.2.2. Unimolecular Dynamics

One approach to examining the dynamics of reactive bimolecular collisions that proceed through a complex is to study the unimolecular dissociation of a species that corresponds to the reaction intermediate. Clearly, in considering the mechanism of reaction (2), the unimolecular rate coefficients k_{-1} and k_2 are just as essential to a complete picture as is the association rate coefficient k_1. These unimolecular rate coefficients are sometimes amenable to direct study. For instance, a stable intermediate for a gas phase S_N2 reaction was isolated and photolyzed by Wilbur and Brauman (1991), and product was formed with significantly higher efficiency in the photolysis than in bimolecular kinetics studies. Because of large-impact-parameter collisions, the bimolecular reaction proceeds with larger average angular momentum than the species observed in the photolysis experiments, and statistical models were used to determine whether the higher angular momentum in the bimolecular reaction could account for its low efficiency. The orbital angular momentum in the bimolecular reaction raises the average effective barrier by 2.5 kcal mol^{-1} when a fixed value of 8.6 Å is used for the distance between the centers of mass of the reactants at the association transition state. A variational transition state theory calculation of the transition state for association predicts that angular momentum raises the average effective barrier by approximately 1.5 kcal mol^{-1}, resulting in an efficiency change which accounts for about 30% of the effects seen. Calculations indicate that angular momentum also plays a significant role in the efficiency of product formation and lead one to expect differences in product energy distributions. In the energization of an intermediate, there is an energy regime in which an activated species has enough energy to cross the barrier to products but not enough energy to access the entrance channel. For species in this regime, formation of products has unit efficiency. For a low pressure bimolecular reaction, the reactants have energy at or above both channels of decay of the complex. Thus, the intermediate energy range is not accessed, and the efficiency is reduced. In related work, Graul and Bowers (1991) showed that the dissociation dynamics of $Cl^-(CH_3Br)$ is nonstatistical. Comparison of the experimental kinetic energy release distribution for metastable dissociation of the $Cl^-(CH_3Br)$ species with the distribution predicted by phase space theory revealed significant deviations, attributed to vibrational excitation of the CH_3Cl product.

Monomer evaporation from clusters has been studied extensively by Lifshitz and coworkers and interpreted in terms of transition state theory (Lifshitz 1993).

Sunner et al. (1989) used a semiempirical treatment to theoretically evaluate the rate coefficients of hydride transfer reaction sec-$C_3H_7^+ + iso$-$C_4H_{10} \rightarrow C_3H_8 + tert$-$C_4H_9^+$. Their kinetic scheme is based on a loose and excited chemically activated complex $(C_3H_7^+ \cdot C_4H_{10})^*$ formed at the Langevin rate. The complex can decompose back to reactants or form the products of the hydride transfer process [following the association mechanism of reaction (2)]. However, Hartree–Fock calculations with the STO-3G basis set and modified neglect of differential overlap (MNDO) semiempirical molecular orbital calculations indicate that the potential surface for this hydride transfer reaction does not have a central barrier—that is, is not of the double-minimum type.

Unimolecular dynamics of smaller clusters has also been studied. The HF dimer provides a particularly interesting system because it involves a highly quantal degenerate rearrangement consisting of a concerted double hydrogen-bond switch (Quack and Suhm 1991; Truhlar 1990).

1.3. ENERGY TRANSFER PATHWAYS IN CLUSTERS

Most of the models of ion–molecule association reviewed here do not consider the energy transfer process involved in stabilizing the intermediate of reaction mechanism (2′). Instead, the association rate coefficient is simply equated to that for ion–molecule capture, which is assumed to occur if the system passes the entrance-channel centrifugal barrier or entrance-channel vibrational transition state. However, there are two important dynamical steps in the mechanism of reaction (2′). One is the initial ion–molecule capture step, and the second is transfer of the reagent relative translational energy to vibrational and/or rotational modes of the complex. This energy transfer is necessary for formation of the excited complex (AB^+). Similar energy transfer issues occur in photodissociation (both direct photofragmentation and predissociation from a photoexcited resonance state), in cage effects, and in exchange reactions; and all these issues are discussed in this section.

1.3.1. Energy Transfer in Association

We begin by returning to the question of the low-pressure third-order rate coefficient for association reactions. A steady-state treatment of reaction mechanism (2′) leads to a bimolecular rate coefficient

$$k = k_1 \frac{k_2[M]}{k_{-1} + k_2[M]}, \qquad (1\text{-}3)$$

which at low pressure becomes

$$k_0 = \left(\frac{k_1}{k_{-1}}\right) k_2[M]. \qquad (1\text{-}4)$$

A considerable amount of work (Adams and Smith 1981, 1983; Bass and Jennings 1984; Bates 1979a,b; Böhringer and Arnold 1982; Böhringer et al. 1983; Headley et al. 1982; Herbst 1979, 1980, 1981; Jennings et al. 1982; Liu et al. 1985; Moet-Ner 1979; Moet-Ner and Field 1975; Nielson et al. 1978; Patrick and Golden 1985; van Koppen et al. 1984; Viggiano 1984; Viggiano et al. 1985) has been addressed to the evaluation of this low pressure limit—that is, the termolecular rate coefficient

$$k_{ter} = (k_1/k_{-1})k_2 \qquad (1\text{-}5)$$

and especially its temperature dependence. The theoretical treatments differ mainly in how the states of $(AB^+)^*$ are counted in calculating the equilibrium constant (k_1/k_{-1}).

Typically, experimental data is analyzed by assuming $k_{ter} = AT^{-n}$, and the value of n is found to be small (e.g., 2–4) for diatomics and triatomics but considerably larger (e.g., 10) for polyatomics (Adams and Smith 1981; Bass et al. 1983; Bass and Jennings 1984; Böhringer et al. 1983; Headley et al. 1982; Jennings et al. 1982; Moet-Ner 1979; Neilson et al. 1978; van Koppen et al. 1984; Viggiano 1984).

The "thermal" and "modified thermal" models of Herbst (1979, 1980, 1981) and Bates (1979a,b) predict a small value of n and appear to be consistent only with the data for small molecules. Phase space theory calculations seem to be more successful (Bass et al. 1979, 1983), both in predicting larger n and in reproducing the curvature of the $\log k$ vs. $\log T$ plots. Phase space modeling has shown that the large temperature dependence and the nonlinear shape of $\log k$ vs. $\log T$ is principally due to vibrational excitation of the reactants, in agreement with the conclusions of Bass and Jennings (1984) for smaller systems. Viggiano (1984) found that the log–log slopes for the clustering reactions $NO_3^-(HNO_3)_n + HCl$, with $n = 1$ and 2 deviated considerably from the value of 2.5 predicted by Bates' theory, and these results were attributed to the low energy vibrations and internal rotations of the clusters. Similar conclusions were obtained for the association reactions $HSO_4^-(HNO_3)_n + HCl$ with $n = 1$ and 2 (Viggiano et al. 1985).

A modification to the theories of Herbst and Bates for ion–molecule association rate coefficients was proposed by Viggiano (1986). In the low pressure limit, the modified theory of Viggiano is similar to phase space theory and predicts a similar temperature dependence. This theory was applied to several systems, and good agreement with experiment was obtained (Morris et al. 1991). The steepness of the negative temperature dependences of association reactions, increasing with increasing complexity of the system, can be correlated with the increasing number of active vibrational degrees of freedom as either the temperature or the size of the cluster is increased (Adams and Smith 1981, Viggiano et al. 1985).

A potentially important question in association reactions is the temperature dependence of the collisional stabilization step. While this dependence is usually small, it is not always negligible. The primary evidence for this temperature dependence is that results obtained with different buffers show appreciably different temperature dependences. This problem has received considerable theoretical attention (Bates 1979a; Böhringer et al. 1983; Herbst 1982; Moet-Ner 1979; Patrick and Golden 1985; Smith et al. 1984; Viggiano 1986; Viggiano et al. 1985).

Bates (1984) interpreted the temperature dependence of $O_2^+ + 2O_2 \rightarrow O_4^+ + O_2$ in terms of the energy randomization rate in the complex.

Ion–molecule association is seemingly well suited for the application of the quasiclassical trajectory (QCT) method (Porter and Raff 1976; Raff and Thompson 1985; Truhlar and Muckerman 1979). Since there is no potential barrier and the centrifugal potential is broad, quantum mechanical tunneling is typically unimportant. Energy transfer from relative translational to vibrational and/or rotational motions of the complex should be reasonably classical because of the large density of states involved. Furthermore, since the variational transition state has an early location along the reaction path, quantization of reactant vibrational motions should result in a reasonably correct treatment of these motions at the transition state; however, we will return to this issue a few paragraphs later. (And we will return to the energy transfer issue in subsection 1.3.2.)

The earliest use of classical trajectory studies to provide insight into the dynamics of possible energy transfer processes in association reactions was by Dugan et al. (Dugan and Canright 1971; Dugan and Palmer 1972; Dugan et al. 1969). Their study involved rigid molecules and so energy transfer occurred by a translation-to-rotation (T–R) mechanism. Later, Brass and Schlier (1988) studied the exchange of energy between relative translation and reagent vibration (T–V). In other other studies, Schelling and Castleman (1984) and Babcock and Thompson (1983a,b) used trajectories to study enhancement of T–R energy transfer and successful association events by anisotropy in the long-range portion of the ion–molecule interaction potential and by reagent internal energy. The most extensive series of trajectory studies of association reactions was carried out by Hase and coworkers, and some of these studies are discussed next.

Classical trajectory studies of the association reactions $M^+ + H_2O$ and $M^+ + D_2O$ with M = Li, Na, K (Hase et al. 1992; Hase and Feng 1981; Swamy and Hase 1982, 1984), $Li^+(H_2O) + H_2O$ (Swamy and Hase 1984), $Li^+ + (CH_3)_2O$ (Swamy and Hase 1984; Vande Linde and Hase 1988), and $Cl^- + CH_3Cl$ (Vande Linde and Hase 1990a,b) are particularly relevant to cluster dynamics. In these studies, the occurrence of multiple inner turning points in the time dependence of the association radial coordinate was taken as the criterion for complex formation. A critical issue (Herbst 1982) is whether the collisions transfer enough energy from translation to internal motions to result in association. Comparison of association probabilities from various studies leads to the conclusion that "softer" and/or "floppier" ions and molecules that have low frequency vibrations typically recombine the most efficiently. Thus, it has been found that $Li^+ + (CH_3)_2O$ association is more likely than $Li^+ + H_2O$ association, and similarly H_2O association with $Li(H_2O)^+$ is more likely than with the bare cation Li^+. The authors found a nonmonotonic dependence of association probability on the assumed H_2O bend frequency and also a dependence on the impact parameter, the rotational temperature, and the orientation of the H_2O dipole during the collision.

Classical trajectory calculations do not include quantum effects such as tunneling, interference, and zero point energy (ZPE), although in quasiclassical trajectory calculations ZPE effects are included approximately by quantizing reactant energies at the start of the trajectory (see, e.g., Agrawal et al. 1988; Truhlar and Muckerman 1979). Hase and coworkers have discussed the sensitivity of trajectory results to the treatment of zero point energy. In particular, comparisions (Clark and Collins 1990; Gomez Llorente et al. 1990; Lu and Hase 1989) of trajectory simulations with experiments and quantum dynamics, as well as arguments (Torres-Vega and Frederick 1990) based on classical quantum corre-spondence, indicate that it may be better to omit the reactant molecule's zero point energy or include only a small fraction of it in choosing trajectory initial conditions. This is because classical mechanics allows zero point energy to flow freely within the molecule, both incorrectly simulating an increased density of states and also allowing physically unrealistic processes to occur. As an example of the latter, trajectory studies on endothermic bimolecular reactions lead to incorrect threshold behavior (Gray et al. 1978, 1979a) and violations of detailed balance (Gray et al. 1979b). One approach for enforcing physically realistic quantization at dynamical bottlenecks is to find the dynamical bottleneck by variational transition state theory, quantize there, and run trajectories forwards and backwards from this point. When tunneling is included by a consistent transmission coefficient, this is called the unified dynamical model (Truhlar and Garrett 1987; Truhlar et al. 1982, 1985b); this model has been applied to reactions with tight transition states but not—so far—to loose dynamical bottlenecks.

Issues particularly germane to this chapter arise in the association of an ion, atom, or molecule with a cluster to form a larger cluster. One interesting issue is whether a cluster is large enough to act as its own heat sink. Another interesting issue is the mechansim by which a new molecule, added to a cluster, becomes "solvated" by moving to the interior. Cluster add-on and growth collisions have been studied by classical trajectory techniques with the aim of answering these and other questions (Alimi et al. 1990; Chartrand et al. 1991; de Pujo et al. 1993; Del Mistro and Stace 1992; Marks et al. 1991; Perera and Amar 1990). Reactive adsorption has also been studied (Adams 1990; Jellinek and Güvenç 1991; Raghavan et al. 1989). The kinetics of steady-state nucleation have been treated by Freeman and Doll (1988).

Hu and Hase (1992a,b, 1993) simulated association reactions of H in HAr_n microclusters with CH_3 to form CH_4 in order to study microscopic solvation dynamics. Classical trajectories and reaction path calculations were carried out for the $CH_3 + HAr_n$ systems with $n = 2, 4, 12$, and 13 and with H initially on the surface of the cluster. In addition, they studied collisions of CH_3 with an Ar_6HAr_6 cluster in which H is completely solvated in the interior of an Ar_{12} shell. Solvating the H atom with Ar_n was found to have three important effects on the association dynamics. (1) Caging of H by Ar_n attenuates the association probability by keeping H from coming into close contact with CH_3. Solvent shells isolate reactants, so that their opportunities for contact are reduced. (2) Since HAr_n has a larger polarizability than H, the long-range attractive forces are greater, and the collision cross section is increased at low temperature. (3) Trapping by CH_3 by Ar_n increases the probability of association. In collisions with low relative translational energy, the collision partners are trapped in a van der Waals well for long times. Physisorption of CH_3 on the surface of the Ar_6HAr_6 cluster provides sufficient time for the argon cage to relax, so that H and CH_3 can associate. If the H atom is initially attached to the surface of the Ar_n cluster this relaxation is not necessary, since reaction can occur by both direct (the CH_3 strips off the H in passing, or it initially hits the site where H is adsorbed on Ar_n) and hopping (the CH_3 is physisorbed elsewhere on the Ar_n surface and migrates to the H) mechanisms. (4) The solvation shell can serve as a chaperon. When a vibrationally and rotationally hot CH_4 is formed, Ar_n acts as an energy sink to stabilize the excited CH_4.

Schulte et al. (1993) calculated classical trajectories for the collision of thermal MO_5 clusters with Ne, Ar, and Xe. The simulation studied the dependence of the energy transfer rate on collision mass and atom-cluster interaction potential.

Kaplan et al. (1993) have carried out classical trajectory calculations of high energy collisions of He^+ and Li^+ with C_{60} in free space and on an iron substrate. The simulations demonstrate the implantation of He^+ to form endohedral He^+C_{60} at various energies, but Li^+ collisions with C_{60} do not form Li^+C_{60}. For Li^+C_{60}, the authors found insertion and fragmentation to form Li^+C_{54} and Li^+C_{56}. The authors studied the yields as functions of incident energy, incident angle, point of impact, and whether the C_{60} is on the substrate or in free space.

Many of the issues that arise in cluster growth collisions are well known from the study of accommodation and sticking coefficients for gaseous molecules colliding with, or adsorbing at, solid surfaces (see, e.g., Adamson 1982; Kiselev and Krylov 1985; Weinberg 1991; Zangwill 1988). A related new area of study, intermediate between cluster growth and bulk solution, is the phenomenon of molecular scattering by the surface of a liquid. For example a recent study of the collision of a D_2O molecule with the surface of liquid H_2SO_4 (Govoni and Nathanson 1994) involves the same competition of impulsive recoil, energy accommodation, trapping, desorption, and dissolution in the interior of the liquid as occurs when a molecule strikes a large cluster.

1.3.2. Energy Transfer in Dissociation

For a statistical approach to be valid for dissociation reactions, internal energy exchanges between modes should occur on time scales that are short compared with the dissociation time (Hase 1981; Truhlar et al. 1983). However, when the vibrational frequency spectrum has a gap, the high frequency modes may exchange energy slowly with the low frequency ones. Energy-gap (Beswick and Jortner 1981) and momentum-gap (Ewing 1979) "laws" have been proposed to quantify this effect. In one study of a system exhibiting such an effect, Desfrancois and Schermann (1991) investigated the classical dynamics of energy exchanges in a tetraatomic van der Waals cluster and found very long time scales for relaxation of high frequency degrees of freedom. Tardiff et al. (1990) studied energy transfer from the CH and OH stretch modes of $CF_3H(H_2O)_3$ into the dissociative mode leading to $CF_3H + (H_2O)_3$ and found that the OH mode leads to more rapid dissociation, probably because the water trimer is structurally coupled ot the C–F bonds but not to the C–H bond.

A particularly important difficulty for classical theories arises in cases where the large ZPE from high frequency (stiff) vibrational modes (or part of it) is illegally transferred to other, soft modes. Consider a classical simulation of a van der Waals (vdW) cluster containing a diatomic molecule with a high frequency stretch. Not putting ZPE in the diatomic stretch may result in an incorrect description of various system properties, since the effective coupling among vibrational modes is changed, but giving the diatomic an energy equal to the quantum ZPE will allow part of the energy to flow to the weak bonds leading to unphysical dissociation of the cluster. There have been several theoretical attempts to devise methods that force the classical calculation to retain at least zero point energy in each mode (Bowman et al. 1989; Miller et al. 1989; Varandas and Marques 1994), but these methods may lead to unphysical chaotic behavior (Sewell et al. 1992). One method was proposed by Alimi et al. (1992) specifically for molecules held in weakly bound clusters. This method treats the high frequency modes by semiclassical Gaussian wavepackets and the soft modes by classical dynamics using the time dependent self-consistent field approach (Alimi et al. 1990; Barnet et al. 1988) to couple the classical and the semiclassical modes. The resulting algorithm is very similar in form to a classical trajectory calculation, is stable, and appears to be free of unphysical effects. The method was illustrated by test applications to models of the van der Waals clusters I_2He and $(HBr)_2$ in their ground states, which dissociate at the expense of their ZPE in classical trajectory calculations but that remain stable in the new method.

Quantum mechanical methods do not suffer these difficulties, and they have been widely applied to predissociative vibrational energy transfer in dimers (see, e.g., Balint-Kurti 1990; Chu 1984; Clary 1989, 1991; Hutson 1990; LeRoy 1984; Tucker and Truhlar 1988; Zhang and Zhang 1993), but they become impractical for larger ones.

Intracluster reactions are often induced by photons, as discussed in several chapters in this volume. One reason for the strong interest in such processes is that the relative geometries of the reagents are dictated by the cluster geometry so that one can, in principle, control the stereochemistry between them as if carrying out bimolecular collisions with oriented reactants (Wittig et al. 1988).

The process of predissociation in a cluster involves breaking a weak inter-molecular bond following a high energy excitation of one of the molecular constituents of the complex. A central conclusion emerging from theoretical studies of this process is that statistical unimolecular rate theories are inapplicable to the predissociation of small van der Waals clusters. The most widely studied example is $X\cdots BC$, where X is a rare gas atom and BC is a chemically bonded diatomic molecule. In this process, which has been studied in hundreds of papers, part of the internal energy of BC is transferred to the van der Waals bond, causing its dissociation (Beswick and Jortner 1981). Larger clusters, such as $X\cdots BC\cdots Y$, with Y also being a rare gas atom, X_nBC with $n = 2$, and B_2X_n with $n = 17\text{--}71$, have also been studied (Bačić et al. 1992; Delgado-Barrio et al. 1987; García-Vela et al. 1990a,b, 1991b; Le Quéré and Gray 1993; Potter et al. 1992; Schatz et al. 1983; Shin 1988; Villareal et al. 1989). The presence of at least two substituents weakly bonded to BC leads to complex dynamical behavior, and, as n increases, a transition to statistical liquid-like behavior may occur (García-Vela et al. 1990b, 1991b).

1.3.3. Cage Effect

A very fundamental difference between reactions in condensed matter and isolated molecular processes in the gas phase is the cage effect: when a reaction or excitation process occurs in a cluster or in a condensed phase, the surrounding solvent molecules may prevent the separation of the reaction products or excited interacting species or delay such separation, confining the nascent species to the initial "cage" for an extended period of time. As in the work reviewed in subsection 1.3.2, this involves the interplay of dissociation and energy transfer. There, the emphasis was on transfer of energy into a dissociative mode; here, it is on transfer out of such a mode. If one can study the dynamics as a function of the size of the clusters, one can ask several interesting questions. For example, how does the probability for caging and the associated energy relaxation dynamics depend on cluster size, structure, and temperature? At what size cluster is bulk-like behavior observed?

Experimental evidence for caging has been found with even a single solvent atom. This effect has been the subject of several theoretical studies which have appeared in the literature (Beswick et al. 1987; García-Vela et al. 1991a, 1992a,b, 1993; Noorbatcha et al. 1984; Segall et al. 1993). For example, in a classical trajectory study of the photodissociation dynamics of HCl in the Ar $\cdots$ HCl cluster, García-Vela and coworkers have shown that a significant cage effect appears in the presence of even one solvent atom. The light atom has a high probability of colliding up to several times with the two heavy atoms that form the walls of the cage. There is substantial probability of tranferring up to 25% of its initial kinetic energy. The cage effect leads to broad kinetic energy distributions and isotropic angular distributions of the fragments. A quasiclassical treatment led to the same qualitative results and conclusions as the fully classical simulation, although there were some significant differences between them (García-Vela et al. 1991a). In a later study, García-Vela et al. (1992a,b) used an approach that treats the hydrogen atom quantum mechanically while the heavy atoms are described classically. The time dependent self-consistent field approximation (Gerber et al. 1982a,b, 1986) was used to couple the quantum and classical modes. On the whole, qualitatively good agreement was found between the results of the (partly quantum) hybrid method and the purely classical ones of the earlier paper, despite the light mass of H. However, quantum diffraction oscillations were found in the angular distribution of the hydrogen fragment, and interference effects were found in its kinetic energy distribution. The peaks in the kinetic energy distribution are directly related to the resonance levels, which are not, however, seen in the absorption spectrum, which is structureless.

McCoy et al. (1993) extended this kind of study to $(HCl)_2$. In this case, they again used a quasiclassical method in which quantal initial conditions were combined with classical trajectories; now, however, the quantal initial conditions were simulated by a Wigner transform of an anharmonic quantal ground state vibrational probability density obtained by the diffusion Monte-Carlo method. This initial condition was lifted to the repulsive excited state for one of the monomers, and trajectories were calculated. Although caging was observed, there were, at most, two internal collisions of H between HCl and Cl, as compared with up to six for H between Ar and Cl. The difference is due to the possibility of H + HCl reaction in the $(HCl)_2$ system. McCoy et al. observed that trajectories reaching the reactive part of the potential energy surface led to very efficient energy transfer, a well known effect (Thompson 1976) in noncluster dynamics.

Caging has also been studied in larger clusters. For example, Amar and Berne (1984) simulated caging of photoexcited Br_2 in neutral clusters of between 8 and 70 argon atoms. Caging was found to be particularly effective when the cluster structure had solvent atoms along the diatomic axis. The transition from a single shell of cluster atoms around the chromophore (Br_2Ar_{20}) to a two-shell cluster (Br_2Ar_{70}) gave little difference in caging behavior, which already approximates bulk behavior, but two shells were required for relaxation behavior to become similar to that of the condensed phase, where the diatomic relaxes without significant dissociation. Scharf et al. (1986, 1988) observed a similar phenomenon in a classical dynamical study of the excimer dynamics of rare gas clusters. In particular, they found reactive molecular-type behavior in small clusters and nonreactive solid-state-type behavior in a large cluster. One can generalize that reactive vibrational predissociation is characteristic of vibrational energy flow in small clusters, while nonreactive vibrational energy redistribution is typical in condensed phases. (We will focus on vibrational relaxation pathways in the following paragraphs.)

Similar results were found in other studies. For example, Amar (1987) modeled the dynamics of $Br_2^- \cdots Ar_n$ clusters following photodissociation of Br_2^-, and he found a correlation between structure and recombination mechanism similar to what had been observed (Amar and Berne 1984) for the corresponding neutral cluster; in particular, argon atoms approximately collinear with the diatomic axis were especially effective in promoting recombination. Alimi and Gerber (1990) used the hybrid quantum–classical method mentioned above to study the dynamics of the cage effect, corresponding to H chattering between heavy atoms, in the photodissociation of HI in clusters of the type $Xe_n HI$ with $n = 1$–12. They found that a cage effect exists for all the clusters, including $n = 1$. For $n = 1$, these resonances are short lived, about 40 fs, and relatively unimportant, but, for $n = 5$, they are long lived, about 0.5 ps, and they dominate the process. As in the studies discussed in the previous paragraph, the larger cluster behavior already shows strong quantitative similarity to the corresponding condensed-matter reaction (Gerber and Alimi 1990).

Perera and Amar (1989) found more detailed support for the structural control of caging in classical dynamics calculations on a model of Br_2^- in large clusters of Ar and CO_2. The dissociation channel was found to become closed, as a function of cluster size, between 11 and 12 CO_2 molecules in the $Br_2^-(CO_2)_n$ clusters, correlating with the appearance of double-capped minimum energy structures. This correlation was found in the $Br_2^- Ar_n$ clusters as well. Collisions between a vibrating diatomic molecule in a cluster and the solvent particles may cause V–T energy transfer and rapid evaporation of the cluster.

Amar and Perera (1991) have also performed classical trajectory simulations of the photodissociation dynamics of I_2^- in $I_2^-(CO_2)_n$. Papanikolas et al. (1991) interpreted their experimental results on the same system in terms of these simulations. Their calculations showed that CO_2 molecules first cluster around the I_2^- waist to form a solvent cylinder. Within this cylinder, the products of I_2^- photodissociation can undergo large-amplitude motion along the internuclear axis. This type of large-amplitude motion was also observed in the classical trajectories for $Br_2^-(CO_2)_n$ (Perera and Amar 1989). During this motion, one or more of the CO_2 molecules could slip between the dissociating iodine atoms, thereby creating a solvent-separated pair and hindering recombination.

Structural effects have also been observed in $Ar_n HCl$ clusters with $n = 1$ and 2 (García-Vela et al. 1994). In this case, the effect of cluster size on the cage effect depends on the excitation energy of HI and on the specific region of the potential surface that it accesses.

1.3.4. Energy Transfer in Exchange Reactions

As discussed previously, most work studying reaction dynamics in clusters by means of theoretical simulations has been directed to photoinduced unimolecular reactions of molecules in van der Waals or hydrogen-bonded clusters. Fewer simulations have been carried out on the effect of van der Waals or hydrogen-bonded clustering on bimolecular reactions. However, Hurwitz et al. (1993) have recently published a classical trajectory simulation of the hydrogen transfer reactions between an $O(^3P)$ atom and a hydrocarbon molecule weakly bound to an argon atom. The authors concluded that the Ar is not boiled off early in the reaction process. Rather, the solvent atom remains attached to the R group throughout the reaction. The $O + H-R \cdots Ar$ reaction is not direct, and throughout the lifetime of a collision complex several hydrogen chattering events occur. Coupled by an anisotropic interaction to the transient H bending mode in the $O \cdots H-R$ subsystem, the Ar takes part of the collision energy, which leads to substantially colder energy distributions for the OH product than were found for the corresponding reaction with the free hydrocarbon. The complexing with Ar also leads to considerably longer collision complex lifetimes, and the OHR subsystem explores a larger portion of the phase space in the cluster reaction than in the unclustered one.

The initial capture steps of the exchange reactions $I + IAr_n \rightarrow I_2 + Ar_n$ with $n = 12$ and 54 have been studied by trajectories by Hu and Martens (1993).

Kaukonen et al. (1991) reported classical trajectory calculations for $[Na_4Cl_3]^+Ar_n + Cl^-$ with $n = 12$ and 32 and for $[Na_{14}Cl_{12}]^{+2}Ar_{30} + Cl^-$. Their results showed that it is possible to "tune" the relative probabilities for different product isomers by varying the initial vibrational temperature of the reactants and relative translations energy between collisional partners and/or the number of embedding Ar atoms.

1.4. MODELING CLUSTER REACTIONS WITH TIGHT TRANSITION STATES

In this section, we discuss studies where specific cluster reactions with tight transition states are modeled in terms of multidimensional potential energy surfaces or force fields (which, for practical purposes, we take here to be synonymous). In the last few years, there has been a considerable growth in the number of applications of potential energy surfaces to the study of chemical reactivity (Casavecchia 1990; Dunning and Harding 1985; Duran and Bertrán 1990; Gianturco 1989; Kaufman 1987; Kuntz 1985; Schatz 1989; Slanina 1986; Stone 1990; Truhlar 1981; Truhlar et al. 1985a, 1986, 1987; Truhlar and Gordon 1990). This now includes a number of applications of potential surfaces to reactive dynamics in and on clusters. As for other sections of this chapter, we will discuss selected examples from the literature. Examples chosen for discussion are especially concentrated on cluster reactions related to solvent effects.

Potential energy surfaces can be built starting from experimental data (e.g., bond strengths, geometries, infrared and fluoresence spectra, molecular beam scattering cross sections, viscosity, diffusion coefficients, line broadening parameters, ultrasonic dispersion, or data on chemical reaction rates, cross sections, activation energies, threshold energies, kinetic isotope effects, or product energy distributions). They can also be built starting from theoretical calculations of the electronic structure of the system of interest, since the Born–Oppenheimer electronic energy plus the nuclear coulomb repulsion is the potential energy surface for interatomic motion. The pure theoretical approach, in which no experimental data are used, is called *ab initio*. *Ab initio* electronic structure calculations can be carried out at the self-consistent field molecular orbital level or, more accurately, including electron correlation. Various basis sets, differing greatly in size and quality, may also be employed. The empirical and theoretical approaches can be combined into a mixed approach, yielding semiempirical methods which combine experimental and theoretical data to construct potential energy surfaces. Within the semiempirical realm, we can distinguish general parameterizations (Dewar et al. 1985; Pople and Segal 1966; Stewart 1989), based on fitting the parameters to data for many molecules (but not usually to data for reaction rates), and specific models, designed to represent a single range of compounds or even only a single reaction; and in the latter case, often involving data for that specific reaction—for example, its activation energy. Examples of all these types of potential energy applied to the study of reactivity of clusters are given in the following.

In modern work, the simplest level of examination of a potential energy surface by electronic structure theory consists in at least optimizing the geometries and calculating the energies of the reactant and transition state stationary points (the former being a minimum, the latter a first-order saddle point). Successively more complete studies include the product stationary point, hessians (second derivatives of the potential with respect to geometry) and vibrational frequencies at stationary points, steepest-descent reaction paths, vibrational frequencies along such paths, and, finally, a semiglobal (valid only near the reaction path) or even completely global analytic representation of the potential energy surface.

Theoretical interpretation of bimolecular nucleophilic substitution (S_N2) reactions has a long history (de la Mare et al. 1955; Dostrovsky et al. 1946; Shaik et al. 1992) Anion-neutral S_N2 reactions that are characterized by activation energies in the range 15–30 kcal mol^{-1} in solution often proceed with no or small barriers in the gas phase. Because this large solvation effect is very interesting, and because of their simplicity, S_N2 processes have become the prototype for recent work on solvent effects. We have previously mentioned (in the Introduction and in subsection 1.2.1.b.i) some of the experimental work designed to follow the transition in the kinetics due to the stepwise hydration of the nucleophile or proton acceptor, bridging the gap between the gas phase and solution. Theoretical studies have also explored this transition, as discussed next. We center attention on anion-neutral S_N2 reactions exhibiting Walden inversion.

In particularly thorough examples of the traditional physical organic approach, Parker (1969) and Abraham (1974) interpreted solvent effects on Walden inversion reactions by using thermodynamic transfer functions. However, in order to explain the reaction rate decrease upon solvation from a microscopic point of view, quantum mechanical electronic structure calculations must be carried out. Microsolvated S_N2 reactions were initially studied in this way, with the CNDO/2 semiempirical molecular orbital (MO) method, by using the supermolecule approach, which was applied to $F^-(H_2O)_n + CH_3F(H_2O)_m$ with $n = 4, 6$, and 8 and $m = 4$ and 7 (Cremaschi et al. 1972). The authors carried out partial optimization of the stationary points and found that solvation stabilizes the transition state much less than the reactants. In the absence of solvent, some of the transition states were calculated to have negative energy barriers (preceded and followed by wells, so that the barrier is a local maximum). In other work, Morokuma (1982) tried to explain the gas phase data obtained by Bohme and Mackay (1981) by performing *ab initio* MO calculations with the 3-21G basis set. In particular, he studied the symmetric chloride exchange reaction $Cl^-(H_2O)_n + CH_3Cl$ with $n = 1$ and 2. The calculated energy profile reproduces the double-well potential postualted by Olmstead and Brauman (1977), as discussed in subsection 1.2.1.b.i. A noteworthy aspect of Morokuma's work is the analysis of several reaction paths. For $n = 2$, for instance, Morokuma found that the most favorable path is the initial migration of one H_2O, with little or no barrier to form an intermediate complex, followed by a transition state corresponding to the CH_3 inversion, and finally followed by migration of the other H_2O molecule. For the paths studied, the energy barrier was found to increase gradually when n changes from 0 to 2. Ohta and Morokuma (1985) carried out *ab initio* MO calculations with a flexible basis set to examine the potential energy surfaces for the S_N2 reactions $OH^-(H_2O)_n + CH_3Cl(H_2O)_m \rightarrow HOCH_3 + Cl^- + (n + m)H_2O$, where the reactants are complexed with up to two water molecules. Although the water transfer process was found to be a part of the rate determining step in the symmetric reaction studied previously (Morokuma 1982), in these reactions it was found to take place after the transition state of the rate determining step. In general, in highly exothermic reactions such as $OH^-(H_2O)_n + CH_3Cl(H_2O)_m \rightarrow HOCH_3 + Cl^- + (n + m)H_2O$, water transfer is not likely to be involved at the dynamical bottleneck of the rate determining step, because the transition state for exothermic reactions is early. Thus, CH_3 inversion precedes water transfer. In many cases, for exothermic reactions, the water never transfers (Hierl et al. 1986b), a feature not only of the S_N2 mechanism, but also of cluster reactions involving proton transfer, where a propensity rule was proposed (Hierl et al. 1986a): "The most efficient channel is the least exothermic, yielding the ionic product with the minimum number of solvate molecules."

According to Magnera and Kerbale (1984), S_N2 reactions with transition states that are more stable than reactants may exhibit a negative temperature dependence. However, adding even one or two water molecules of microsolvation may be enough to create a positive barrier. For example, Hirao and Kebarle (1989) found by *ab initio* electronic structure calculations that the gas phase energy profile along the reaction coordinate for the reaction $Cl^-(H_2O)_2 + CH_3Br$ is very much closer to that for aequeous solution than to that for the unhydrated gas phase case. In contrast, Kong and Jhon (1986) determined, with a simpler potential energy surface, that about 60 water molecules are needed for representing solution-like behavior in the chloride exchange reaction and the S_N2 process $F^- + CH_3Cl$. Thus, convergence toward the bulk limit may be considered fast or slow, depending on one's point of view.

Marcus (1956) and Levich (1970) have shown that nonequilibrium solvent relaxation plays an important role in homogeneous outer-sphere electron transfer reactions. Marcus' equation can also be applied to the interpretation of nucleophilic substitution reactions in the gas phase and in solution (Albery 1980; Albery and Kreevoy 1978; Lewis and Slater 1980; Pellerite and Brauman 1980, 1982; Wolfe et al. 1981), which is not surprising since this equation is a quadratic free energy relation that may be derived under fairly general assumptions (Kreevoy and Truhlar 1986), involving either equilibrium or nonequalibrium solvation. Thus, it is hard to infer from phenomenological application of the Marcus equation how important nonequilibrium solvation effects are. Analyses based on reorientation of water dipoles indicate that bulk electric polarization of the solvent may sometimes be significantly out of equilibrium with the reaction coordinate in S_N2 reactions in bulk water (Lee and Hynes 1988, Truhlar et al. 1993). What about the microsolvated case? Several studies have been carried out to investigate this question.

Jaume et al. (1984) studied the contribution of solvent relaxation to the reaction coordinate of the $F^-(H_2O) + CH_3F(H_2O)S_N2$ reaction. Potential energy calculations were performed using the *ab initio* MO method with the 3-21G basis set. The authors found large variation of the solvation parameters along the reaction path and concluded that solvent coordinates are an important part of the reaction coordinate. They showed that the solvent acts not only as a medium for the reaction but also as a rectant. Thus, the solvent does not adjust its position to the changing chemical system but rather takes part in it.

The Menshutkin reaction (Abboud et al. 1993) is a special case of the S_N2 reaction where the reactants are uncharged but the products are charged, in contrast to the more usual S_N2 reactions where both reactants and products are charged. The difference in charge states translates into an opposite effect of the solvent on these two types of reactions. Usual S_N2 reactions are slowed down by solvent because charge is delocalized at the transition state, whereas the Menshutkin reaction is favored by the presence of the solvent because charge separation is created. Microsolvation effects on the reaction between ammonia and methyl bromide, which was taken as a prototype for the Menshutkin reaction, were analyzed by Solà et al. (1991). In this study, two water molecules were considered, one solvating the bromine atom in CH_3Br and the other solvating the ammonia. The effect of including the two water molecules is twofold: first, there is a decrease in the energy barrier; second, the transition state occurs earlier along the reaction path. As the system advances along the reaction path in an H_2O solvent, the N–O and Br–H distances shorten so there is a contraction of the solvation shell around both fragments. This is opposite to what happens in typical S_N2 reactions, where one solvent shell is contracted and another one is expanded. The coordinates of the two solvent distances make significant contributions to the transition vector, which is the normal-mode eigenvector associated with the imaginary frequency at the saddle point, and this again indicates participation of the solvent in the reaction coordinate.

Theoretical studies of the microsolvation effect on S_N2 reactions have also been reported by our coworkers and ourselves (González-Lafont et al. 1991; Truhlar et al. 1992; Tucker and Truhlar 1990; Zhao et al. 1991b, 1992). Two approaches were used for interfacing electronic structure calculations with variational transition state theory (VST) and tunneling calculations. We analyzed both the detailed dynamics of microsolvation and also its macroscopic consequences (rate coefficient values and kinetic isotope effects and their temperature dependences). The first approach was applied to the study of the microhydrated S_N2 reaction of a chloride ion with methyl chloride, where the microsolvation consisted of one or two water molecules, in both cases solvating the anion (Tucker and Truhlar 1990). A semiglobal analytical potential energy function was created for the unsolvated reaction, with a barrier height chosen to fit an experimental rate coefficient value and with a shape based on correlated *ab initio* calculations for the gas phase system. The potential function has reaction-coordinate-dependent force constants and partial charges for all atoms of the solute. The interaction potential for the water molecules was added by molecular mechanics. All degrees of freedom were included, and the resulting analytic potential energy surface is a function of 36 coordinates. With the addition of just two water molecules, a definite trend toward the solution phase reaction profile was observed. For example, the barrier height, relative to reactants as zero of energy, increases from a value of 3.1 kcal mol^{-1} in the gas phase, to 5.4 kcal mol^{-1} for the monohydrated reaction, and to 10.7 kcal mol^{-1} for the dihydrated reaction, as compared with the accepted value of the solution phase barrier of about 26–27 kcal mol^{-1}. To go beyond classical transition state theory, the potential surface was used (Tucker and Truhlar 1990) to calculate, for both $n = 1$ and 2, the minimum energy path (MEP) in iso-inertial coordinates (Shavitt 1969; Truhlar et al. 1985b; Truhlar and Kuppermann 1970, 1971), also called the intrinsic reaction coordinate (IRC) (Fukui 1981; Morokuma and Kato 1981), or, more suitably, the intrinsic reaction path (IRP). In contrast with the simpler approaches to defining a reaction coordinate used in previous cluster modeling studies (Jaume *et al.* 1984; Morokuma 1982; Ohta and Morokuma 1985), the MEP is an intrinsic property of the system, independent of the theoretician's model or choice of isoinertial coordinate system. For both the monohydrated and the dihydrated reactions it was found that, as the reaction proceeds, the water molecules migrate from the approaching chloride to the leaving chloride. Because the reaction under study is thermoneutral, the unsolvated ion is less likely to be a significant product than it is in the exothermic reactions typically studied experimentally. Rate coefficients were calculated by canonical variational transition state theory (CVT) (Lu et al. 1992; Truhlar et al. 1985b) for a tight dynamical bottleneck with a quantum tunneling correction based on a semiclassical approximation (Skodje et al. 1981), assuming small curvature of the reaction path. The rate coefficients decrease with increasing hydration of the system, as expected from the increasing barrier height that results from better solvation of the reactant than the transition state.

Since these calculations treat all 36 degrees of freedom on an equal footing, they include nonequilibrium solvation to the extent it is present. For comparison, the rate coefficients for the monohydrated reaction were also evaluated under the equilibrium solvation approximation. The extent of nonequilibrium solvation was evaluated by comparing calculations in which the degrees of freedom of the water molecule participate in the reaction coordinate to those in which they do not. Two different methods for defining the generalized transition state theory dividing surface under the equilibrium solvation approximation yielded quite different values for the rate coefficient. The most appropriate approximation, as shown by the variational transition state criterion, was found to give an increase of only 10% compared with the nonequilibrium approach. From this result, one may conclude that strong solute–solvent coupling does not necessarily mean that nonequilibrium solvation effects are important. The different estimate of the nonequilibrium effects observed with the less accurate equilibrium approach indicates that one must distinguish the failure of the equilibrium solvation approximation from the failure of a less than optimum way of implementing it.

A second approach that we have used (González-Lafont et al. 1991) to interface electronic structure calculations with VTST and tunneling calculations is "direct dynamics" (Baldridge et al. 1989; González-Lafont et al. 1991; Liu et al. 1993), by which we mean basing the dynamics calculations directly on the output of the electronic structure calculations without explicit fitting or interpolation. This methodology was applied to the $Cl^-(H_2O)_n + CH_3Cl \rightarrow CH_3Cl + Cl^-(H_2O)_n$ reaction with $n = 0$, 1, and 2. Instead of using an analytical potential energy function, the energy and gradient were calculated whenever needed by neglect-of-diatomic-differential-overlap (NDDO) semiempirical MO theory with parameters adjusted, starting from the general AM1 (Dewar et al. 1985; Zoebisch and Dewar 1988) parameterization. The method was labeled NDDO-SRP to denote NDDO with specific reaction parameters. Thus, rather than use one of the general parameterizations available for NDDO semiempirical molecular orbital theory as a completely predictive theoretical tool, this study used the NDDO molecular orbital framework as a partly predictive and partly fitting tool, resulting in an implicit potential energy function for a specific reaction. The results were compared, in detail, with previous calculations (Tucker and Truhlar 1990; Zhao et al. 1991b) based on the explicit (analytic) multidimensional semiglobal analytic potential function. Comparison with the results obtained with the analytical surface revealed some important differences that could be tested by future work; for instance, the structures of the saddle points for both $n = 1$ and $n = 2$ are rather different on the two surfaces. Perhaps more significantly, in many respects the two quite different approaches agree remarkably well. From an energetic point of view, the potential energy barriers on the implicit surface increase with stepwise hydration, as found earlier with the explicit surfaces.

The correspondence between the two sets of calculated α-deuterium secondary kinetic isotope effects (KIEs), as well as heavy water microsolvent kinetic isotope effects, and their interpretation in terms of specific modes is very encouraging. For instance, both approaches predict an inverse secondary α-deuterium KIE for CD_3Cl, approximately independent of the extent of solvation of Cl^-. [In general, vibrational frequencies and zero point energies are larger for lighter masses, because light particles are more quantum mechanical and have more widely spaced energy levels. Typically, the vibrations of a system become looser (i.e., have lower frequencies and hence lower zero point energies) as a system passes from reactants to the transition state, because force constants usually go down (since half bonds of transition states have weaker force constants than whole bonds of reactants). Thus, typically zero point energies go down in passing to the transition state, which would release energy *into* the reaction coordinate. Since zero point effects are, as just discussed, larger for lighter masses, this release typically is larger for lighter systems, and thus isotopically lighter systems typically react faster. This result is called a normal KIE. In the case under discussion, the effect of isotropic substitution is dominated by modes that increase in frequency so we get an inverse effect.] The effect on the vibrational contribution to the CD_3KIE of adding a single water molecule is traceable almost entirely to a single transition state mode—the CH_3 or CD_3 internal rotation around the Cl–Cl axis. The differences between the two approaches are more significant in comparing microsolvent kinetic isotope effects—that is, the effect on the rate coefficient of changing D_2O to H_2O in the microsolvent. In particular, the NDDO-SRP surface predicts an inverse microsolvent KIE for $n = 1$ and $n = 2$, while the microsolvent KIEs calculated from the earlier analytic surfaces are normal in both cases. However, while in the analytic approach the vibrational contribution to the global solvent KIE was also found to be normal for $n = 2$, the vibrational factor was inverse for $n = 1$ (the normal global KIE resulted from the rotational contribution). In order to achieve a more detailed understanding, the vibrational contribution to the global KIE was further factored into contributions from individual vibrational modes. From this analysis, it was concluded that mid-frequency modes are not bystanders but have an important normal contribution, and the inverse microsolvent kinetic isotope effects are primarily due to the low frequency vibrations for this reaction.

The calculations carried out with the two approaches mentioned above showed the sensitivity of solvent KIEs to the low frequency vibrations associated with the coupling of the solute to the solvent. To treat this coupling more accurately, a new potential energy function for $Cl(H_2O)^-$ was calibrated (Zhao et al. 1991a). The parameters of the new potential were determined to improve agreement with experiment for the dipole moment of water, with new-extended-basis-set correlated electronic calculations for the dissociation energy, geometry, and frequencies of the complex, with 369 *ab initio* interaction energies (Dacre 1984), and with one calculation for a geometry close to the $Cl(H_2O)^-$ configuration at the previously calculated saddle point of the reaction $^*Cl(H_2O)^- + CH_3Cl \rightarrow CH_3^*Cl + Cl(H_2O)^-$. We also attempted to converge the individual and total vibrational contributions to the equilibrium isotope effect for $Cl(H_2O)^- + D_2O^- \rightarrow Cl(D_2O) + H_2O$. New calculations of rate coefficients and secondary kinetic isotope effects for the microsolvated S_N2 reaction $Cl(H_2O)^- + CH_3Cl$ were then carried out (Zhao et al. 1992) based on the new chloride–water potential energy function combined with a more accurate internal potential function for the water molecule and the original analytic solute potential. At all temperatures, the microsolvent KIE is inverse with the new potential in contrast to the normal KIE calculated with the original potential, but is in agreement with the result obtained with the NDDO-SRP direct dynamics approach. The change in the direction of the predicted microsolvent kinetic isotope effect can be understood in terms of the effects of the solvent–solute interaction on coordinates other than the reaction coordinate; comparison of the vibrational contributions to the microsolvated KIEs indicates that the stronger interaction of the solvent and solute enhances the inverse tendency. The dominant effects on this microsolvent KIE (μSKIE) come from the low frequency and high frequency modes (in particular, the symmetric stretch mode), and it is their inverse contributions that convert the normal microsolvated KIE to an inverse one.

The contributions to the microsolvent KIE from various classes of mode are illustrated in Table 1-1. These factors indicate that the primary error in the original study was to miss the effect of about 10% in the high frequency modes, with a

Table 1-1. Factors Contributing to the Predicted Rate Coefficient Ratio μSKIE = $k[Cl^-(H_2O) + CH_3Cl]/k[Cl^-(D_2O) + CH_3Cl]$

			Modes			
Potential function	Translation	Rotation	Low frequency vibrations	Mid-frequency vibrations	High frequency vibrations	μSKIE
1 (Tucker and Truhlar 1990)	1.03	1.41	0.43	1.67	1.00	1.04
2 (González-Lafont et al. 1991)	1.03	1.43	0.46	1.61	0.85	0.93
3 (Zhao et al. 1992)	1.03	1.41	0.40	1.67	0.90	0.87

difference in the same direction of about 7% in the low frequency contribution. The final conclusion is that the inverse microsolvent kinetic isotope effect (μSKIE) for this reaction is primarily due to a very significant contribution from low frequency modes, almost, but not quite, compensated by "normal" contributions from rotation and mid-frequency modes, and helped by a small, but significant, contribution from high frequencies.

In light of these new results, it is interesting to review the available theories for bulk solvent kinetic isotope effects. Two fundamentally different models have been put forth to explain bulk solvent kinetic isotope effects. Swain and Bader (1960) proposed a model in which the libration frequencies of solvent molecules are responsible for solvent isotope effects. Each librational frequency in pure water is controlled by its coordination to four surrounding water molecules which make up the solvent "cage." The cage is the source of the structure of liquid water. The change in librational frequencies due to introduction of a solute is associated with the change imposed on the water librations in the four-coordinate cell. This theory, originally developed for equilibrium isotope effects (Swain and Bader 1960), was later applied to solvent KIEs (Swain et al. 1960; Swain and Thornton 1961a,b). In this case, it was assumed that transition states that are larger in size cause more breakdown in the structure of water, but effects due to "electrical" differences (changes in partial charges) of the solute at the transition state and in its reactant state were neglected (Swain et al. 1960).

Arnett and McKelvey (1969) presented a review of thermodynamic properties of highly dilute solutions in H_2O and D_2O and strongly emphasized the issues of structure breaking and structure making. Laughton and Robertson (1969) reviewed work on equilibrium and kinetic solvent isotope effects in terms of structural reorganization associated with librational degrees of freedom and the structural stability of hydrogen-bonded structures adjacent to the solute, and they concluded that solvent KIEs in S_N2 solvolyses are related to "solvent reorganization about the developing anion." Thornton and Thornton (1971) reviewed both equilibrium and kinetic solvent isotope effects and differentiated among several factors, including changes in internal vibrations of solvent molecules strongly coupled to the solute, changes in water structure (especially for small ions), changes in liberational frequencies, and exchange effects.

The second model of solvent isotope effects, due to Bunton and Shiner (1961a,b) concentrates on the hydrogen stretching frequencies of internal O–H vibrations in hydrogen-bonded solvent molecules and ignores the contributions of bending modes and librations. This model does not require any considerations of water structure, and it would be expected to be applicable to microsolvation without substantial modification. Bigeleisen (1960), Newton and Friedman (1985), Friedman and Newton (1986), and Kneifel et al. (1994) also emphasized the significant change in the O–L stretch frequency of ion–L_2O complexes upon the substitution of L = D for L = H. For cations, Newton and coworkers (in the latter three references) found no support for the libration-dominant Swain–Bader model as far as the first-shell hydration is concerned, but, for bulk solvation, they concluded that libration appears to be more important.

Mitton et al. (1969) discussed both models in the context of methanolysis rates in CH_3OH vs. CH_3OD and found the argument involving solvent librations provided a more understandable explanation of the magnitudes of the effects than changes in the internal vibrational frequencies of the solvent molecules, but the argument involving hydrogen bonding explained the direction of the substituent effects.

Gold and Grist (1972) concluded that strongly basic anions perturb the water molecules that are hydrogen bonded to them, thereby contributing to inverse solvent kinetic isotope effects, but they did not provide evidence for a molecular explanation.

Schowen (1972) used the Swain–Bader theory to explain solvent kinetic isotope effects on S_N2 reactions involving halides, in terms of the numer of water molecules solvating a halide ion in the transition state (assumed to be three) vs. the number solvating it in the bulk (four). The contribution of a single halide–water hydrogen bond was also taken to depend on the partial charge on the halide, which could be consistent with either theory.

In this light, the single-water microsolvent KIE is very illuminating. The inverse character of this effect is consistent with the direction predicted by the solvent–structure argument, but for only one water it is clear that water structure breaking is impossible, and this indicates that the cell model of water structure breaking is probably not the sole reason for the inverse KIEs observed in bulk solent KIEs. Water molecule O–H vibrations and water molecule librations are both involved in the microsolent KIE, and they clearly must both be assessed for understanding cluster and bulk solvent KIEs. We must keep in mind, though, that bulk solvation may lead to a change in the structure of the reactant and/or transition state hydrogen bonds, causing a different effect to dominate the solvent KIE in the gas phase and in aqueous solution.

The prediction of an inverse μSKIE, even for a single water molecule, has recently been strikingly confirmed by experimental studies of the reaction $F^-(D_2O)$ with several methyl halides (O'Hair et al. 1994). One of these, $F^-(H_2O) + CH_3Cl$, has also been studied theoretically (Hu and Truhlar 1994) using transition state theory for a tight dynamical bottleneck whose properties were estimated using high level electronic structure calculations (electron correlation by Møller–Plesset second-order perturbation theory and an aug-cc-pVDZ basis set). The theoretical μSKIE is 0.65, and the calculated value is 0.65 as well. The need for high level theory for quantitative results was demonstrated by repeating the calculation with PM3 semiempirical molecular orbital theory, which yielded μSKIE = 0.78. Factorization of the high level calculated isotope effect yields 1.05, 1.35, 0.74, 1.14, and 0.55, respectively, for the same classification of mode types as in Table 1-1. Thus, for this reaction, the high frequency modes provide the dominant effect. The isotope effect is dominated by the O–H stretching mode of the hydrogen bond between F^- and H_2O, in accord with the seminal analysis of Bunton and Shiner (1961a,b). In passing from reactants to the transition state, the zero point energy requirement of this mode increases by 1.2 kcal in the $F^-(H_2O)$ case but by only 0.9 kcal mol^{-1} in the $F^-(D_2O)$ case. In either case, this energy becomes unavailable for overcoming the barrier. Since 0.3 kcal mol^{-1} more energy is tied up in this vibration at the transition state for the $F^-(H_2O)$ case than for the $F^-(D_2O)$ case, less energy is available for crossing the barrier in the former case. Since less energy is available for crossing the barrier in the $F^-(H_2O)$ case, its rate is slower.

Why does the critical frequency go up? The answer is like a double negative. The hydrogen bond to F^- weakens the internal O–H bond of water. The hydrogen bond is stronger in reactants than in the transition state (because there is a full charge on F^-, rather than a partial charge) so the O–H bond is weakened more in reactants than in the transition state—that is, the O–H bond is stronger in the transition state, so it has a higher force constant, a higher frequency, and a greater zero point energy requirement at the transition state than at reactants.

It would be interesting, in future work, to examine the effect of anharmonicity on the calculated results.

At the transition state for the $F^-(H_2O) + CH_3Cl$ reaction, the hydrogen bond from water is entirely to the F—that is, there is no evidence for water transfer at the methyl transfer transition state (see Figure 1-1), which is consistent with the propensity rule mentioned above, in that the water does not migrate. However, it is also consistent with the possibility that water migrates after the transition state is passed.

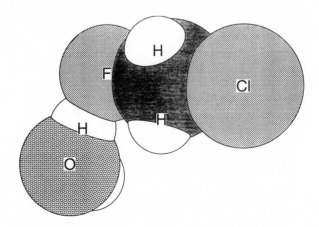

Figure 1-1. Space-filling view of the transition state structure for the $F^-(H_2O) + CH_3Cl$ S_N2 reaction.

Several other studies have been concerned with cases where the solvent molecules intervene in specific ways in chemical reactions, such as acting as bifunctional catalysts (e.g., Bertrán 1989; Buckingham et al. 1986; Field et al. 1984; Lledós and Bertrán 1984, 1985a,b, 1986; Lledós et al. 1986; Moreno et al. 1987; Nguyen et al. 1991; Nguyen and Ha 1984; Nguyen and Hegarty 1984; Nguyen and Ruelle 1987; Ohta and Morokuma 1987; Peeters and Leroy 1991; Ruelle 1986, 1987a,b; Ruelle et al. 1985, 1986; Ventura et al. 1987; Weiner et al. 1985; Williams 1987; Williams et al. 1983; Yamabe et al. 1984). It has been shown that the potential barrier of the 2-hydroxipyridine → pyridone process decreases noticeably when one or two water molecules intervene in the hydrogen transfer process (Field et al. 1984; Lledós and Bertrán 1986). Similar results have been obtained for *cis*-hydroxyimine → formamide (Lledós and Bertrán 1984), and keto-enol tautomerization (Ventura et al. 1987; Weiner et al. 1985). The barrier to decarboxylation of a series of organic acids is decreased noticeably by intervention of a water molecule as a bifunctional catalyst (Lledós et al. 1986; Nguyen and Ruelle 1987; Ruelle 1986, 1987a,b; Ruelle et al. 1985). The presence of a second solvent molecule, which acts as a bifunctional catalyst, clearly favors numerous hydrolysis processes (Buckingham et al. 1986; Nguyen and Ha 1984; Nguyen and Hegarty 1984; Nguyen and Ruelle 1987; Williams 1987; Williams et al. 1983).

To illustrate these specific effects, consider the vinyl alcohol → acetaldehyde tautomerization process. In this reaction, a chain containing two water molecules is needed to connect the hydrogen to be transferred and the nonsubstituted carbon. As compared with the gas phase transfer process, the barrier in the two-water cluster (Lledos et al. 1986; Ventura et al. 1987) is dramatically lowered, in good agreement with experimental results. However, the barrier is still sufficiently high to cast some doubt on the feasibility of such a mechanism. To understand how the energy is decreased by the two-water chain, one can look at the transition state: the hydroxylic hydrogen has been transferred to the first water molecule, whereas this molecule simultaneously begins to transfer one hydrogen to the second water molecule, and this, in turn, is hydrogen bonded to carbon. Thus, the hydrogen is transferred in a concerted way with the two-water group acting as a bifunctional catalyst, since it simultaneously accepts and releases a proton. A similar behavior was reported by Williams et al. (1983) on the assistance of a second water molecule to the addition of water to formaldehyde (Williams et al. 1980) and by Williams (1987) for the effect of one or two waters on the addition of ammonia to formaldehyde. In contrast, in a molecular orbital calculation carried out by Bouchoux and Hoppilliard (1990), the acid-catalyzed dehydration of ethanol via solvated complexes was found to be increasingly unfavored as the number of polar solvent molecules increases.

A question of great interest in the kinetics of reactions with cluster wells along the reaction path is the accuracy of transition state theory. Some classical trajectory calculations have indicated the possibility of large breakdowns (Cho et al. 1992), but these potential errors may be overestimated by the classical mechanical method employed, as discussed elsewhere (Truhlar et al. 1992; Wladkowski et al. 1992).

Tunneling effects on the microsolvated S_N2 reactions studied so far are small. Processes dominated by tunneling are excited state proton transfer in

$C_{10}H_9OH \cdot NH_3$ and $C_6H_5OH(NH_3)_5$ clusters, and these processes have been treated theoretically by Hineman et al. (1992) and by Syage (1993), respectively. Syage compared two models, one employing the Marcus theory of condensed phase reactions (Cohen and Marcus 1968; Marcus 1964, 1968) and one (Hineman et al. 1992) corresponding to an isolated solute, and, considering both his own data and those of Hineman et al. (1992), found better agreement with the latter. It would be interesting to repeat this kind of analysis using a full potential energy surface. Further discussion of proton transfer in the naphthol/ammonia and phenol/ammonia systems is provided in chapter 5 in this volume by Bernstein.

1.5. CONCLUDING REMARKS

A wide variety of dynamical approximations have been applied to cluster dynamics and kinetics. Most calculations to date are based on simplified potentials and classical mechanics or statistical methods. In the near future, we can expect to see more work with detailed potential energy surfaces (both analytic and implicitly defined by electronic structure calculations) and progress in sorting out quantum effects and treating them more accurately.

ACKNOWLEDGMENTS: The authors are grateful to Wei-Ping Hu for assistance with the figure and to Richard Schowen for helpful comments on solvent isotope effects. This work was supported in part by the U.S. Department of Energy, Office of Basic Energy Sciences.

REFERENCES

Abboud, J.-L.; Notario, R.; Bertrán, J.; Solà, M. 1993 Adv. Phys. Org. Chem. 19:1

Abraham, M. H. 1974 Prog. Phys. Org. Chem. 11:1

Adams, J. E. 1990 J. Chem. Phys. 92:1849

Adams, N. G.; Smith, D. 1981 Chem. Phys. Lett. 79:563

Adams, N. G.; Smith D. 1983 In Reactions of Small Transient Species, Kinetics and Energetics, A. Fontijn and M. A. A. Clyne, eds., Academic Press, New York, p. 311

Adamson, A. W. 1982 Physical Chemistry of Surfaces, 4th Ed., John Wiley & Sons, New York

Agrawal, P. M.; Thompson, D. L.; Raff, L. M. 1988 J. Chem. Phys. 88:5948

Albery, W. J. 1980 Annu. Rev. Phys. Chem. 31:227

Albery, W. J.; Kreevoy, M. M. 1978 Adv. Phys. Org. Chem. 16:87

Alimi, R.; García-Vela, A.; Gerber, R. B. 1992 J. Chem. Phys. 96:2034

Alimi, R.; Gerber, R. B. 1990 Phys. Rev. Lett. 64:1453

Alimi, R.; Gerber, R. B.; Hammerich, A. D.; Kosloff, R.; Ratner, M. A. 1990 J. Chem. Phys. 93:6484

Amar, F. G. 1987 In The Chemistry and Physics of Small Clusters, P. Jena, S. Khanna, and B. Rao, eds., Plenum Press, New York, p. 207

Amar, F. G.; Berne, B. J. 1984 J. Phys. Chem. 88:6720

Amar, F. G.; Perera, L. 1991 Z. Phys. D 20:173

Arnett, E. M.; McKelvey, D. R. 1969 In Solute–Solvent Interactions, J. F. Coetzee and C. D. Ritchie, eds., Marcel Dekker, New York, p. 344

Babcock, L. M.; Thompson, D. L. 1983a J. Chem. Phys. 78:2394

Babcock, L. M.; Thompson, D. L. 1983b J. Chem. Phys. 79:4193

Bačić, Z.; Kennedy-Mandsiuk, M.; Moscowitz, J. W.; Schmidt, K. E. 1992 J. Chem. Phys. 97:6472

Baldridge, K. K.; Gordon, M. S.; Steckler, R.; Truhlar, D. G. 1989 J. Phys. Chem. 93:5107

Balint-Kurti, G. G. 1990 In Dynamics of Polyatomic van der Waals Complexes, N. Halberstadt and K. C. Janda, eds., Plenum Press, New York, p. 59

Barker, R. A.; Ridge, D. P. 1976 J. Chem. Phys. 64:4411

Barnet, R. N.; Landman, U.; Nitzan, A 1988 J. Chem. Phys. 89:2242

Bass, L. M.; Cates, R. D.; Jarrold, M. F.; Kirchner, N. J.; Bowers, M. T. 1983 J. Amer. Chem. Soc. 105:7024

Bass, L.; Chesnavich, W. J.; Bowers, M. T. 1975 Chem. Phys. Lett. 34:119

Bass, L.; Chesnavich, W. J.; Bowers, M. T. 1979 J. Amer. Chem. Soc. 101:5493

Bass, L. M.; Kemper, P. R.; Anicich, V. G.; Bowers, M. T. 1981 J. Amer. Chem. Soc. 103:5283

Bass, L.; Jennings, K. R. 1984 Int. J. Mass Spectrom. Ion. Proc. 58:307

Bates, D. R. 1979a J. Phys. B 12:4135

Bates, D. R. 1979b J. Chem. Phys. 71:2318

Bates, D. R. 1984 J. Chem. Phys. 81:298

Bertrán, J. 1989 In *New Theoretical Concepts for Understanding Organic Reactions*, J. Bertrán and I. G. Csizmadia, eds., Kluwer, Dordrecht, p. 231

Beswick, J. A.; Jortner, J. 1981 Adv. Chem. Phys. 47 (P. 1):363

Beswick, J. A.; Monot, R.; Philippoz, J. M.; van den Bergh, H. 1987 J. Chem. Phys. 86:3965

Bigeleisen, J. 1960 J. Chem. Phys. 32:1583

Böhringer, H.; Arnold, A. 1982 J. Chem. Phys. 77:5534

Böhringer, H.; Arnold, F.; Smith, D.; Adams, N. G. 1983 Int. J. Mass Spectrom. Ion Phys. 52:25

Bohme, D. K. 1979 In *Interaction Between Ions and Molecules*, P. Ausloos, ed., Plenum Press, New York, p. 489

Bohme, D. K.; Mackay, G. I. 1981 J. Amer. Chem. Soc. 103:978

Bohme, D. K.; Raksit, A. B. 1984 J. Amer. Chem. Soc. 106:3447

Bohme, D. K.; Raksit, A. B. 1985 Can. J. Chem. 63:3007

Bohme, D. K.; Young, L. B. 1970 J. Amer. Chem. Soc. 92:7354

Bouchoux, G.; Hoppilliard, Y. 1990 J. Amer. Chem. Soc. 112:9110

Bowman, J. M.; Gadzy, B.; Sun, Q. 1989 J. Chem. Phys. 91:2859

Brass, O.; Schlier, Ch. 1988 J. Chem. Phys. 88:936

Buckingham, A. D.; Handy, N. C.; Rice, J. E.; Somasundram, K.; Dijugraff, C. 1986 J. Comp. Chem. 7:283

Bunker, D. L. 1971 Methods Comp. Phys. 10:287

Bunton, C. A.; Shiner, V. J., Jr. 1961a J. Amer. Chem. Soc. 83:42

Bunton, C. A.; Shiner, V. J., Jr. 1961b J. Amer. Chem. Soc. 83:3207

Caralp, F.; Lesclaux, R.; Rayez, M. T.; Rayez, J. C.; Forst, W. 1988 J. Chem. Soc. Faraday Trans. 2 84:569

Casavecchia, P. 1990 In *Dynamics of Polyatomic van der Waals Complexes*, N. Halberstadt and K. C. Janda, eds., Plenum Press, New York, p. 123

Castleman, A. W., Jr.; Keesee, R. G. 1986a Annu. Rev. Phys. Chem. 37:525

Castleman, A. W., Jr.; Keesee, R. G. 1986b Chem. Rev. 86:589

Celli, F.; Weddle, G.; Ridge, D. P. 1980 J. Chem. Phys. 73:801

Chang, J. S.; Golden, D. M. 1981 J. Amer. Soc. 103:496

Chartrand, D. J.; Shelley, J. C.; LeRoy, R. J. 1991 J. Phys. Chem. 95:8310

Chesnavich, W. J.; Bowers, M. T. 1977 J. Chem. Phys. 66:2306

Chesnavich, W. J.; Bowers, M. T. 1982 Progr. Reaction Kinet. 11:137

Chesnavich, W. J.; Su, T.; Bowers, M. T. 1980 J. Chem. Phys. 72:2641

Cho, Y. J.; Vande Linde, S. R.; Zhu, L.; Hase, W. L. 1992 J. Chem. Phys. 96:8275

Chu, S.-I. 1984 ACS Symp. Ser. 263:263

Clark, D. L.; Collins, M. A. 1990 J. Chem. Phys. 92:5602

Clary, D. C. 1984 Mol. Phys. 53:1

Clary, D. C. 1985 Mol. Phys. 54:605

Clary, D. C. 1989 In *Supercomputer Algorithms for Reactivity, Dynamics, and Kinetics of Small Molecules*, A. Laganà, ed., Kluwer, Dordrecht, p. 295

Clary, D. C. 1990 Annu. Rev. Phys. Chem. 41:61

Clary, D. C. 1991 J. Chem. Phys. 96:90

Cohen, A. O.; Marcus, R. A. 1968 J. Phys. Chem. 72:4249

Cremaschi, P.; Gamba, A.; Simonetta, M. 1972 Theor. Chim. Acta 25:237

Dacre, P. D. 1984 Mol. Phys. 51:633

Danis, F.; Caralp, F.; Rayez, M. T.; Lesclaux, R. 1991 J. Phys. Chem. 95:7300

Davies, J. W.; Pilling, M. J. 1989 In *Bimolecular Collisions: Advances in Gas Phase Photochemistry and Kinetics*, M. N. R. Ashford and J. E. Baggott, eds., Royal Society of Chemistry, London, p. 105

de la Mare, P. D. B.; Fowden, L.; Hughes, E. D.; Ingold, C. J.; Mackie, J. D. H. 1955 J. Chem. Soc. 3200

de Pujo, P.; Mestdagh, J.-M.; Visticot, J.-P.; Cuvellier, J.; Meynadier, P.; Sublemontier, O.; Lallement, A.; Berlande, J. 1993 Z. Phys. D 25:357

Del Mistro, G.; Stace, A. J. 1992 Chem. Phys. Lett. 196:67

Delgado-Barrio, G.; Villarreal, P. Varadé, A.; Martín, N.; García, A. 1987 In *Structure and Dynamics of Weakly Bound Molecular Complexes*, A. Weber, eds., D. Reidel, Dordrecht, p. 573

Desfrancois, C.; Schermann, J. P. 1991 Z. Phys. D 20:185

Dewar, M. J. S.; Zoebisch, E. G.; Healy, E. F.; Stewart, J. J. P. 1985 J. Amer. Chem. Soc. 107:3902

Dostrovsky, I.; Hughes, E. D.; Ingold, C. K. 1946 J. Chem. Soc. 173

Dugan, J. V., Jr.; Canright, R. B., Jr. 1971 Chem. Phys. Lett. 8:253

Dugan, J. V., Jr.; Palmer, R. W. 1972 Chem. Phys. Lett. 13:144

Dugan, J. V., Jr.; Rice, J. H.; Magee, J. L. 1969 Chem. Phys. Lett. 3:323

Dunbar, R. 1975 Spectrochim. Acta 31A:797

Dunning, T. H., Jr.; Harding, L. B. 1985 In *Theory of Chemical Reaction Dynamics*, Vol. 1, M. Baer, ed., CRC Press: Boca Raton, p. 1

Duran, M.; Bertrán, J. 1990 Rep. Mol. Theor. 1:57

Ewing, G. E. 1979 J. Chem. Phys. 71:3143

Fahey, D. W.; Böhringer, H.; Fehsenfeld, F. C.; Ferguson, E. E. 1982 J. Chem. Phys. 76:1799

Fehsenfeld, F. C.; Ferguson, E. E. 1974 J. Chem. Phys. 61:3181

Fenter, F. F.; Lightfoot, P. D.; Caralp, F.; Lesclaux, R.; Niiranen, J. T.; Gutman, D. 1993a J. Phys. Chem. 97:4695

Fenter, F. F.; Lightfoot, P. D.; Niiranen, J. T.; Gutman, D. 1993b J. Phys. Chem. 97:5313

Field, M. J.; Hillier, I. H.; Guest M. F. 1984 J. Chem. Soc. Chem. Commun.: 1310

Forst, W. 1983 *Theory of Unimolecular Reactions*, Academic Press, New York

Freeman, D. L.; Doll, J. D. 1988 Adv. Chem. Phys. 70:139

Friedman, H. L.; Newton, M. D. 1986 J. Electroanal. Chem. 204:21

Fukui, K. 1981 Acc. Chem. Res. 14:363

García-Vela, A.; Gerber, R. B.; Buck, U. 1994 J. Phys. Chem. 98:3518

García-Vela, A.; Gerber, R. B.; Imre, D. G. 1992a J. Chem. Phys. 97:7242

García-Vela, A.; Gerber, R. B.; Imre, D. G.; Valentini, J. J. 1993 Chem. Phys. Lett. 202:473

García-Vela, A.; Gerber, R. B.; Valentini, J. J. 1991a Chem. Phys. Lett. 186:223

García-Vela, A.; Gerber, R. B.; Valentini, J. J. 1992b J. Chem. Phys. 97:3297

García-Vela, A.; Villareal, P.; Delgado-Barrio, G. 1990a J. Chem. Phys. 92:496

García-Vela, A.; Villareal, P.; Delgado-Barrio, G. 1990b J. Chem. Phys. 92:6504

García-Vela, A.; Villareal, P.; Delgado-Barrio, G. 1991b J. Chem. Phys. 94:7868

Garrett, B. C.; Truhlar, D. G. 1979a J. Chem. Phys. 70:1593

Garrett, B. C.; Truhlar, D. G. 1979b J. Chem. Phys. 83:1052

Garrett, B. C.; Truhlar, D. G. 1979c J. Chem. Phys. 83:1079

Garrett, B. C.; Truhlar, D. G. 1979d J. Amer. Chem. Soc. 101:4534

Garrett, B. C.; Truhlar, D. G. 1982 J. Chem. Phys. 76:1853

Gerber, R. B.; Buch, V.; Ratner, M. A. 1982a In *Intramolecular Dynamics*, J. Jortner and B. Pullman, eds., D. Reidel, Dordrecht, p. 171

Gerber, R. B.; Buch, V.; Ratner, M. A. 1982b J. Chem. Phys. 77:3022

Gerber, R. B.; Kosloff, R.; Berman, M. 1986 Comp. Phys. Rep. 5:59

Gianturco, F. A. 1989 In *New Theoretical Concepts for Understanding Organic Reactivity*, J. Bertrán and I. G. Csizmadia, eds., Kluwer, Dordrecht, p. 257

Gioumousis, G.; Stevenson, D. P. 1958 J. Chem. Phys. 29:294

Gold, V.; Grist, S. 1972 J. Chem. Soc. Perkin II 1972:89

Gomez Llorente, J. M.; Hahn, O.; Taylor, H. S. 1990 J. Chem. Phys. 92:2762

González-Lafont, A.; Truong, T. N.; Truhlar, D. G. 1991 J. Phys. Chem. 95:4618

Gorin, E. 1938 Acta Physicochim. URSS 9:681

Govoni, S. T.; Nathanson, G. M. 1994 J. Amer. Chem. Soc. 116:779

Graul, S. T.; Bowers, M. T. 1991 J. Amer. Chem. Soc. 113:9696

Gray, J. C.; Fraser, G. A.; Truhlar, D. G. 1979a Chem. Phys. Lett, 68:359

Gray, J. C.; Garrett, B. C.; Truhlar, D. G. 1979b J. Chem. Phys. 70:5921

Gray, J. C.; Truhlar, D. G.; Clemens, L.; Duff, J. W. 1978 J. Chem. Phys. 69:240

Hase, W. L. 1981 In *Potential Energy Surfaces and Dynamics Calculations*, D. G. Truhlar, ed., Plenum Press, New York, p. 1

Hase, W. L.; Darling, C. L.; Zhu, L. 1992 J. Chem. Phys. 96:8295

Hase, W. L.; Feng, D.-F. 1981 J. Chem. Phys. 75:738

Headley, J. V.; Mason, R. S.; Jennings, K. R. 1982 J. Chem. Soc. Faraday Trans. 2 78:933

Henchman, M. J.; Hierl, P. M.; Paulson, J. F. 1985 J. Amer. Chem. Soc. 107:2812

Henchman, M. J.; Hierl, P. M.; Paulson, J. F. 1987 ACS Adv. Chem. Ser. 215:83

Henchman, M. J.; Paulson, J. F.; Hierl, P. M. 1983 J. Amer. Chem. Soc. 105:5509

Henchman, M.; Paulson, J. F.; Smith, D.; Adams, N. G.; Lindinger, W. 1988 In *Rate Coefficients in Astronomy and Chemistry*, A. D. Williams and T. J. Miller, eds., D. Reidel, Dordrecht, p. 201

Herbst, E. 1976 Ap. J. 205:94

Herbst, E. 1979 J. Chem. Phys. 70:2201

Herbst, E. 1980 J. Chem. Phys. 72:5284

Herbst, E. 1981 J. Chem. Phys. 75:4413

Herbst, E. 1982 Chem. Phys. 78:323

Hierl, P. M.; Ahrens, A. F.; Henchman, M.; Viggiano, A. A.; Paulson, J. F.; Clary, D. C. 1986a J. Amer. Chem. Soc. 108:3140

Hierl, P. M.; Ahrens, A. F.; Henchman, M.; Viggiano, A. A.; Paulson, J. F.; Clary, D. C. 1986b J. Amer. Chem. Soc. 108:3142

Hierl, P. M.; Ahrens, A. F.; Henchman, M.; Viggiano, A. A.; Paulson, J. F.; Clary, D. C. 1988 Faraday Discuss. Chem. Soc. 85:37

Hierl, P. M.; Paulson, J. F. 1984 J. Chem. Phys. 80:4890

Hineman, M. F.; Bruckner, G. A.; Kelley, D. F.; Bernstein, E. R. 1992 J. Chem. Phys. 97:3341

Hirao, K.; Kebarle, P. 1989 Can. J. Chem. 67:1261

Hirschfelder, J. O.; Wigner, E. 1939 J. Chem. Phys. 7:616

Hogg, A. M.; Haynes, R. M.; Kebarle, P. 1966 J. Amer. Chem. Soc. 88:28

Horiuti, J. 1938 Bull. Chem. Soc. Japan 13:210

Hu, W.; Truhlar, D. G. 1994 J. Amer. Chem. Soc. 116:7797

Hu, X.; Hase, W. L. 1992a J. Phys. Chem. 96:7535

Hu, X.; Hase, W. L. 1922b Z. Phys. D 25:57

Hu, X.; Hase, W. L. 1993 J. Chem. Phys. 98:7826

Hu, X.; Martens, C. C. 1993 J. Chem. Phys. 99:9532

Hurwitz, Y.; Rudich, Y.; Naaman, R.; Gerber, R. B. 1993 J. Chem. Phys. 98:2941

Hutson, J. M. 1990 In *Dynamics of Polyatomic van der Waals Complexes*, N. Halberstadt and K. C. Janda, eds., Plenum Press, New York, p. 67

Jasinski, J. M.; Rosenfeld, R. N.; Golden, D. M.; Brauman, J. I. 1979 J. Amer. Chem. Soc. 101:2259

Jaume, J.; Lluch, J. M.; Oliva, A.; Bertrán, J. 1984 Chem. Phys. Lett. 106:232

Jellinek, L.; Güvenç, Z. B. 1991 Z. Phys. D 19:371

Jennings, K. R.; Headley, J. V.; Mason, R. S. 1982 Int. J. Mass Spectrom. Ion. Phys. 45:315

Jortner, J. 1992 Z. Phys. D 24:247

Kaplan, T.; Rasolt, M.; Karimi, M.; Mostoller, M. 1993 J. Phys. Chem. 97:6124

Kassel, L. S. 1928 J. Phys. Chem. 32:1065

Kaufman, J. J. 1987 Int. J. Quantum Chem. 29:179

Kaukonen, H.-P.; Landman, U.; Cleveland, C. L. 1991 J. Chem. Phys. 95:4997

Keck, J. C. 1967 Adv. Chem. Phys. 13:85

Kiselev, V. F.; Krylov, O. V. 1985 *Adsorption Processes on Semiconductor and Dielectric Surfaces*, Springer-Verlag, Berlin, pp. 51 and 223.

Kneifel, C. L.; Newton, M. D.; Friedman, H. L. 1994 J. Mol. Liq. 60:107

Kong, Y. S.; Jhon, M. S. 1986 Theor. Chim. Acta 70:123

Kreevoy, M. M.; Truhlar, D. G. 1986 In *Investigation of Rates and Mechanisms of Reactions*, Part I, 4th Ed., C. F. Bernasconi, ed., Wiley, New York, p. 13

Kuntz, P. J. 1985 In *Theory of Chemical Reaction Dynamics*, Vol. 1, M. Baer, eds., CRC Press, Boca Raton, p. 71

Langevin, P. 1905 Ann. Chim. Phys. 5:245

Lau, Y. K.: Ikuta, S.; Kebarle, P. 1982 J. Amer. Chem. Soc. 104:1462

Laughton, P. M.; Robertson, R. E. 1969 In *Solute–Solvent Interactions*, J. F. Coetzee and C. D. Ritchie, eds., Marcel Dekker, New York, p. 400

Le Quéré, F.; Gray, S. K. 1993 J. Chem. Phys. 98:5396

Lee, S.; Hynes, J. T. 1988 J. Chem. Phys. 88:6863

LeRoy, R. J. 1984 ACS Symp. Ser. 263:231

Leutwyler, S.; Bösiger, J. 1990 Chem. Rev. 90:489

Levich, V. G. 1970 In *Physical Chemistry: An Advanced Treatise*, Vol. 9B, H. Eyring, O. Henderson, and W. Jost, eds., Academic Press, New York, p. 985

Lewis, E. S.; Slater, C. D. 1980 J. Amer. Chem. Soc. 102:1619

Lifshitz, C. 1993 In *Cluster Ions*, C. Y. Ng, T. Baer, and I. Powis, eds., John Wiley, London, p. 121

Light, J. C. 1967 Discuss. Faraday Soc. 44:14

Light, J. C.; Lin, J. 1965 J. Chem. Phys. 43:3209

Liu, S.; Jarrold, M. F.; Bowers, M. T. 1985 J. Phys. Chem. 89:3127

Liu, Y.-P.; Lu, D.-H.; González-Lefont, A.; Truhlar, D. G.; Garrett, B. C. 1993 J. Amer. Chem. Soc. 115:7806

Lledós, A.; Bertrán, J. 1984 Theochem. 107:233

Lledós, A.; Bertrán, J. 1985a Theochem. 21:73

Lledós, A.; Bertrán, J. 1985b Theochem. 24:211

Lledós, A.; Bertrán, J. 1986 An. Quim. 82:194

Lledós, A.; Bertrán, J.; Ventura, O. N. 1986 Int. J. Quantum Chem. 30:467

Lu, D.-H.; Hase, W. L. 1989 J. Chem. Phys. 91:7490

Lu, D.-H.; Truong, T. N.; Melissas, V. S.; Lynch, G. C.; Liu, Y.-P.; Garrett, B. C.; Steckler, R.; Isaacson, A. D.; Rai, S. N.; Hancock, G. C.; Lauderdale, J. G.; Joseph, T.; Truhlar, D. G. 1992 Computer Phys. Commun. 71:235

Magee, J. L. 1952 Proc. Natl. Acad. Sci. U.S.A. 38:764

Magnera, T. F.; Kebarle, P. 1984 In *Ionic Processes in the Gas Phase*, M. A. Almoster, ed., D. Reidel, Dordrecht, p. 135

Marcus, R. A. 1952 J. Chem. Phys. 20:359

Marcus, R. A. 1956 J. Chem. Phys. 24:966

Marcus, R. A. 1964 Annu. Rev. Phys. Chem. 15:155

Marcus, R. A. 1968 J. Phys. Chem. 72:891

Marcus, R. A.; Rice, O. K. 1951 J. Phys. Colloid Chem. 55:894

Marks, A. J.; Murrell, J. N.; Stace, A. J. 1991 J. Chem. Phys. 94:3908

McCoy, A. B.; Hurwitz, Y.; Gerber, R. A. 1993 J. Phys. Chem. 97:12516

McDaniel, E. W. 1964 *Collision Phenomena in Ionized Gases*, Wiley, New York, p. 701

Miller, W. H. 1976 J. Chem. Phys. 65:2216

Miller, W. H.; Hase, W. L.; Darling, C. L. 1989 J. Chem. Phys. 9:2863

Mitton, C. G.; Gresser, M.; Schowen, R. L. 1969 J. Amer. Chem. Soc. 91:2045

Moet-Ner, M. 1979 In *Gas Phase Ion Chemistry*, Vol. I, M. T. Bowers, ed., Academic Press, New York, p. 198

Moet-Ner, M.; Field, F. H. 1975 J. Amer. Chem. Soc. 97:5339

Moran, T. F.; Hamill, W. H. 1963 J. Chem. Phys. 39:413

Moreno, M.; Lluch, J. M.; Oliva, A.; Bertrán, J. 1987 Can. J. Chem. 65:2774

Morokuma, K. 1982 J. Amer. Chem. Soc. 104:3732

Morokuma, K.; Kato, S. 1981 In *Potential Energy Surfaces for Dynamics Calculations*, D. G. Truhlar, ed., Plenum Press, New York, p. 243

Morris, R. A.; Viggiano, A. A.; Paulson, J. F.; Henchman, M. J. 1991 J. Amer. Chem. Soc. 113:5932

Neilson, P. V.; Bowers, M. T.; Chau, M.; Davidson, W. R.; Aue, D. H. 1978 J. Amer. Chem. Soc. 100:3649

Newton, M. D.; Friedman, H. L. 1985 J. Chem. Phys. 83:5210

Nguyen, K. A.; Gordon, M. S.; Truhlar, D. G. 1991 J. Amer. Chem. Soc. 113:1596

Nguyen, M. T.; Ha, T.-K. 1984 J. Amer. Chem. Soc. 106:599

Nguyen, M. T.; Hegarty, A. T. 1984 J. Amer. Chem. Soc. 106:1552

Nguyen, M. T.; Ruelle, P. 1987 Chem. Phys. Lett. 138:486

Nikitin, E. E. 1965 Teor. Eksperim. Khim. 1:135 [English transl.; 1965 Theor. Exp. Chem. 1:831]

Noorbatcha, I.; Raff, L. M.; Thompson, D. L. 1984 J. Chem. Phys. 81:5658

O'Hair, R. A. J.; Davico, G. E.; Hacaloglu, J.; Dang, T. T.; DePuy, C. H.; Bierbaum, V. M. 1994 J. Amer. Chem. Soc. 116:3609

Ohta, K.; Morokuma, K. 1985 J. Phys. Chem. 89:5845

Ohta, K.; Morokuma, K. 1987 J. Phys. Chem. 91:401

Olmstead, W. N.; Brauman, J. I. 1977 J. Amer. Chem. Soc. 99:4219

Olmstead, W. N.; Lev-On, M.; Golden, D. M.; Brauman, J. I. 1977 J. Amer. Chem. Soc. 99:992

Papanikolas, J. M.; Gord, J. R.; Levinger, N. E.; Ray, D.; Vorsa, V.; Lineberger, W. C. 1991 J. Phys. Chem. 95:8028

Parker, A. J. 1969 Chem. Rev. 69:1

Patrick, R.; Golden, D. M. 1985 J. Chem. Phys. 82:75

Pechukas, P.; Light, J. C. 1965 J. Chem. Phys. 42:3281

Peeters, D.; Leroy, G. 1991 J. Canad. Chem. 69:1376

Pellerite, M. J.; Brauman, J. I. 1980 J. Amer. Chem. Soc. 102:5993

Pellerite, M. J.; Brauman, J. I. 1982 ACS Symp. Ser. 198:81

Perera, L.; Amar, F. G. 1989 J. Chem. Phys. 90:7354

Perera, L.; Amar, F. G. 1990 J. Chem. Phys. 93:4884

Pople, J. A.; Segal, G. A. 1966 J. Chem. Phys. 44:3289

Porter, R. N.; Raff, L. M. 1976 In *Dynamics of Molecular Collisions*, Part B, W. H. Miller, ed., Plenum Press, New York, p. 1

Potter, E. D.; Liu, Q.; Zewail, A. H. 1992 Chem. Phys. Lett. 200:605

Quack, M.; Suhm, M. A. 1991 Chem. Phys. 183:187

Quack, M.; Troe, J. 1975 Ber. Bunsenges, Phys. Chem. 79:170

Raff, L. M.; Thompson, D. L. 1985 In *Theory of Chemical Reaction Dynamics*, Vol. 3, M. Baer, ed., CRC Press, Boca Raton, p. 1

Raghavan, K.; Stave, M. S.; DePristo, A. E. 1989 J. Chem. Phys. 91:1904

Rai, S. N.; Truhlar, D. G. 1983 J. Chem. Phys. 79:6046

Rice, O. K.; Ramsperger, H. C. 1927 J. Amer. Chem. Soc. 49:1617

Robinson, T. J.; Holbrook, K. H. 1972 *Unimolecular Reactions*, Wiley, New York

Rosenstock, H. M.; Wallenstein, M. B.; Wahrhaftig, A. L.; Eyring, H. 1952 Proc. Natl. Acad. Sci. U.S.A. 38:667

Ruelle, P. 1986 Chem. Phys. 110:263

Ruelle, P. 1987a J. Comput. Chem. 8:158

Ruelle, P. 1987b J. Amer. Chem. Soc. 109:1722

Ruelle, P.; Kesselring, U. W.; Nam-Tran, H. J. 1985 Theochem. 25:41

Ruelle, P.; Kesselring, U. W.; Nam-Tran, H. 1986 J. Amer. Chem. Soc. 108:371

Ryan, K. R.; Plumb, I. C. 1982 J. Phys. Chem. 86:4678

Scharf, D.; Jortner, J.; Landman, U. 1986 Chem. Phys. Lett. 84:2783

Scharf, D.; Jortner, J.; Landman, U. 1988 J. Chem. Phys. 88:4273

Schatz, G. C. 1989 Rev. Mod. Phys. 61:669

Schatz, G. C.; Buch, V.; Ratner, M. A.; Gerber, R. B. 1983 J. Chem. Phys. 79:1808

Schelling, F. W.; Castleman, A. W., Jr. 1984 Chem. Phys. Lett. 111:47

Schowen, R. L. 1972 Prog. Phys. Org. Chem. 9:275

Schulte, J.; Lucchese, R. R.; Marlow, W. H. 1993 J. Chem. Phys. 99:1178

Segall, J.; Wen, Y.; Singer, R.; Wittig, C.; García-Vela, A.; Gerber, R. B. 1993 Chem. Phys. Lett. 207:504

Sewell, T. D.; Thompson, D. J.; Gezelter, J. D.; Miller, W. H. 1992 Chem. Phys. Lett. 193:512

Shaik, S. S.; Schlegel, H. B.; Wolfe, S. 1992 *Theoretical Aspects of Physical Organic Chemistry: The S_N2 Mechanism*, Wiley, New York

Shavitt, I. 1968 J. Chem. Phys. 49:4048

Shin, H. K. 1988 J. Chem. Phys. 89:2943

Skodje, R. T.; Truhlar, D. G.; Garrett, B. C. 1981 J. Phys. Chem. 85:3019

Slanina, Z. 1986 *Contemporary Theory of Chemical Isomerism*, D. Reidel, Dordrecht

Smith, D.; Adams, N. G.; Alge, E. 1981 Planet. Space Sci. 29:449

Smith, D.; Adams, N. G.; Alge, E. 1984 Chem. Phys. Lett. 105:317

Smith, G. P.; Golden, D. M. 1978 Int. J. Chem. Kinet. 10:489

Solà, M.; Lledós, A.; Duran, M.; Bertran, J.; Abboud, J. L. M. 1991 J. Amer. Chem. Soc. 113:2873

Stewart, J. J. P. 1989 J. Comp. Chem. 10:221

Stone, A. J. 1990 In *Dynamics of Polyatomic Van der Waals Complexes*, N. Halberstadt and K. C. Janda, eds., Plenum Press, New York, p. 329

Su, T. 1985 J. Chem. Phys. 82:2164

Su, T.; Bowers, M. T. 1973a J. Chem. Phys. 58:3027

Su, T.; Bowers, M. T. 1973b Int. J. Mass Spectrum. Ion Phys. 12:347

Su, T.; Bowers, M. T. 1975 Int. J. Mass Spectrom. Ion Phys. 17:211

Su, T.; Chesnavich, W. J. 1982 J. Chem. Phys. 76:5183

Su, T.; Su E. C. F.; Bowers, M. T. 1978 J. Chem. Phys. 69:2243

Sunner, J. A.; Hirao, K.; Kebarle, P. 1989 J. Phys. Chem. 93:4010

Swain, C. G., Bader, R. F. W. 1960 Tetrahedron 10:182

Swain, C. G.; Bader, R. F. W.; Thornston, E. R. 1960 Tetrahedron 10:200

Swain, C. G.; Thornton, E. R. 1961a J. Amer. Chem. Soc. 83:3884

Swain, C. G.; Thornton, E. R. 1961b J. Amer. Chem. Soc. 83:3790

Swamy, K. N.; Hase, W. L. 1982 J. Chem. Phys. 77:3011

Swamy, K. N.; Hase, W. L. 1984 J. Amer. Chem. Soc. 106:4071

Syage, J. A. 1993 J. Phys. Chem. 97:12523

Syage, J. A. 1994 In *Ultrafast Dynamics of Chemical Systems*, J. D. Simon, ed., Kluwer, Dordrecht, p. 289

Tardiff, J.; Deal, R. M.; Hase, W. L.; Lu, D.-H. 1990 J. Cluster Sci. 1:335

Thompson, D. G. 1976 Acc. Chem. Res. 9:338

Thornton, E. K.; Thornton, E. R. 1971 In *Isotope Effects in Chemical Reactions*, C. J. Collins and N. S. Bowman, eds., Van Nostrand Reinhold, New York, p. 213

Torres-Vega, G.; Frederick, J. H. 1990 J. Chem. Phys. 93:8862

Troe, J. 1977a J. Chem. Phys. 66:4745

Troe, J. 1977b J. Chem. Phys. 66:4758

Troe, J. 1979 J. Phys Chem. 83:114

Truhlar, D. G. (ed.) 1981 *Potential Energy Surfaces and Dynamics Calculations*, Plenum Press, New York, 1981

Truhlar, D. G. 1990 In *Dynamics of Polyatomic Van der Waals Complexes*, N. Halberstadt and K. C. Janda, eds., Plenum Press, New York, 1990, p. 159

Truhlar, D. G.; Brown, F. B.; Schwenke, D. W.; Steckler, R. 1985a In *Comparision of Ab Initio Quantum Chemistry with Experiment for Small Molecules*, R. J. Bartlett, ed., D. Reidel, Dordrecht, p. 95

Truhlar, D. G.; Brown, F. B.; Steckler, R.; Isaacson, A. D. 1986 In *Theory of Chemical Reaction Dynamics*, D. C. Clary, ed., D. Reidel, Dordrecht, 1986, p. 285

Truhlar, D. G.; Garrett, B. C. 1987 Faraday Discuss. Chem. Soc. 84:465

Truhlar, D. G.; Gordon, M. S. 1990 Science 249:491

Truhlar, D. G.; Hase, W. L.; Hynes, J. T. 1983 J. Phys. Chem. 87:2664, 5523(E)

Truhlar, D. G.; Isaacson, A. D.; Garrett, B. C. 1985b In *Theory of Chemical Reaction Dynamics*, Vol. 4, M. Baer, ed., CRC Press, Boca Raton, p. 65

Truhlar, D. G.; Isaacson, A. D.; Skodje, R. T.; Garrett, B. C. 1982 J. Phys. Chem. 86:2252

Truhlar, D. G.; Kuppermann, A. 1969 J. Phys. Chem. 73:1722

Truhlar, D. G.; Kuppermann, A. 1970 J. Chem. Phys. 52:4480

Truhlar, D. G.; Kupperman, A. 1971 J. Amer. Chem. Soc. 93:1840

Truhlar, D. G.; Lu, D.-H.; Tucker, S. C.; Zhao, X. G.; González-Lafont, A.; Truong, T. N.; Maurice, D.; Liu, Y.-P.; Lynch, G. C. 1992 ACS Symp. Ser. 502:16

Truhlar, D. G.; Muckerman, J. T. 1979 In *Atom–Molecule Collision Theory*, R. B. Bernstein, ed., Plenum Press, New York, p. 505

Truhlar, D. G.; Schenter, G. K.; Garrett, B. C. 1993 J. Chem. Phys. 98:5756

Truhlar, D. G.; Steckler, R.; Gordon, M. S. 1987, 87:217

Tucker, S. C.; Truhlar, D. G. 1988 J. Chem. Phys. 88:3667

Tucker, S. C.; Truhlar, D. G. 1990 J. Amer. Chem. Soc. 112:3347

van Koppen, P. A. M.; Jarrold, M. F.; Bowers, M. T.; Bass, L. M.; Jennings, K. R. 1984 J. Chem. Phys. 81:288

Vande Linde, S. R.; Hase, W. L. 1988 Comp. Phys. Commun. 51:17

Vande Linde, S. R.; Hase, W. L. 1990a J. Phys. Chem. 94:6148

Vande Linde, S. R.; Hase, W. L. 1990b J. Chem. Phys. 93:7962

Varandas, A. J. C.; Marques, J. M. C. 1994 J. Chem. Phys. 100:1908

Ventura, O. N.; Lledós, A.; Bonaccorsi, R.; Bertrán, J.; Tomasi, J. 1987 Theor. Chim. Acta 72:175

Viggiano, A. 1984 J. Chem. Phys. 81:2639

Viggiano, A. A. 1986 J. Chem. Phys. 84:244

Viggiano, A. A.; Dale, F.; Paulson, J. F. 1985 J. Geophys. Res. 90:7977

Viggiano, A. A.; Dale, F.; Paulson, J. F. 1988a J. Chem. Phys. 88:2469

Viggiano, A. A.; Morris, R. A.; Dale, F.; Paulson, J. F. 1988b J. Geophys. Res. 93:9534

Viggiano, A. A.; Morris, R. A.; Deakyne, C. A.; Dale, F.; Paulson, J. F. 1990 J. Phys. Chem. 94:8193

Villareal, P.; Varadé, A.; Delgado-Barrio, G. 1989 J. Chem. Phys. 90:2684

Weinberg, W. H. 1991 In *Dynamics of Gas–Surface Interactions*, C. T. Rettner and M. N. R. Ashfold, eds., Royal Society of Chemistry, Cambridge, p. 171

Weiner, S. J. U.; Singh, U. C.; Kollman, P. A. 1985 J. Amer. Chem. Soc. 107:2219

Wigner, E. 1937 J. Chem. Phys. 5:720

Wilbur, J. L.; Brauman, J. I. 1991 J. Amer. Chem. Soc. 113:9699

Williams, I. H. 1987 J. Amer. Chem. Soc. 109:6299

Williams, I. H.; Maggiora, G. M.; Schowen, R. L. 1980 J. Amer. Chem. Soc. 102:7831

Williams, I. H.; Splanger, D.; Femec, D. A.; Maggiora, G. M.; Schowen, R. L. 1983 J. Amer. Chem. Soc. 105:31

Wittig, C.; Sharpe, S.; Beaudet, R. A. 1988 Acc. Chem. Res. 21:341

Wladkowski, D. B.; Lim, K. F.; Allen, W. D.; Brauman, J. I. 1992 J. Amer. Chem. Soc. 114:9136

Wolfe, S.; Mitchell, D. J.; Schlegel, H. B. 1981 J. Amer. Chem. Soc. 103:7694

Woodin, R. L.; Beauchamp, J. L. 1979 Chem. Phys. 41:1

Yamabe, T.; Yamashita, K.; Kaminoyama, M.; Koizumi, M.; Tachibana, A.; Fukui, K. 1984 J. Phys. Chem. 88:1459

Yang, X.; Castleman, A. W., Jr. 1989 J. Amer. Chem. Soc. 111:6845

Yang, X.; Castleman, A. W., Jr. 1990 J. Chem. Phys. 93:2045

Yang, X.; Castleman, A. W., Jr. 1991a Int. J. Mass Spectrom. Ion Proc. 109:339

Yang, X.; Castleman, A. W., Jr. 1991b J. Chem. Phys. 95:130

Yang, X.; Castleman, A. W., Jr. 1991c J. Amer. Chem. Soc. 113:6766

Yang, X.; Castleman, A. W., Jr. 1991d J. Phys. Chem. 95:6182

Yang, X.; Zhang, X.; Castleman, A. W. Jr. 1991 J. Phys. Chem. 95:8520

Zangwill, A. 1988 Physics at Surfaces, Cambridge University Press, Cambridge, p. 363

Zhang, D. H.; Zhang, J. Z. 1993 J. Chem. Phys. 99:6624

Zhao, X. G.; González-Lafont, A.; Truhlar, D. G.; Steckler, R. 1991a J. Chem. Phys. 94:5544

Zhao, X. G.; Lu, D.-H.; Lu, Y.-P.; Lynch, G. C.; Truhlar, D. G. 1992 J. Chem. Phys. 97:6369

Zhao, X. G.; Tucker, S. C.; Truhlar, D. G. 1991b J. Amer. Chem. Soc. 113:826

Zoebisch, E. G.; Dewar, M. J. S. 1988 Theochem. 180:1

2

Weakly Bound Molecular Complexes as Model Systems for Understanding Chemical Reactions

R. E. MILLER

2.1. INTRODUCTION

An issue of central importance in chemical reaction dynamics is the nature of the energy transfer processes within and between reactants and products. At a fundamental level, bond rupture and formation can be understood in terms of the transfer of energy into the reaction coordinate, causing a bond to break, and then relaxation of the energy away from the newly formed bonds in the product molecules to the other degrees of freedom of the system. By their very nature, these processes are highly anharmonic, making their detailed characterization a formidable challenge. In recent years, spectroscopists have taken on the challenge of trying to characterize the quantum states of a molecule at the high vibrational energies corresponding to the chemically interesting regime. At these energies, the density of states becomes extremely high and the coupling between the states very strong, the result being that the vibrations can no longer be characterized in terms of simple isolated local or normal modes. In the extreme limit, where RRKM theory (Wardlaw and Marcus 1987) applies, there is rapid energy redistribution that, at least approximately, samples the available states statistically, allowing us to overlook many of the fine details. Although we are still far from having a complete understanding of the quantum state dynamics of systems in this regime, the recent progress that has been made in both experiment (Felker and Zewail, 1985; Go et al., 1990; Parameter 1982, 1983; Smalley 1982) and theory (Stuchebrukhov and Marcus 1993; Uzer 1991) is helping to better define the important processes. Ultimately, the detailed characterization of all the intramolecular couplings in a molecule would provide us with a basis for understanding the chemistry at a fundamental level, in both the statistical and nonstatistical regimes. After all, energy transfer from one vibrational mode of a molecule to another is determined by the intermode couplings, which, in the ground electronic states of molecules, are predominantly due to anharmonic and/or coriolis effects. Of course, the problem becomes even more challenging when one moves from the realm of isolated molecules to solvated systems. As this volume is meant to illustrate, finite sized molecular clusters can provide us with new ways of approaching many of these complex problems.

The often overwhelming complexity (Pique et al. 1989) which arises in the spectroscopy of molecules at high energies is due to two factors. First, there is simply the high density of states corresponding to the large number of ways the energy can be distributed among the available vibrational degrees of freedom. Second, there is a diverse range of coupling mechanisms between these many vibrational modes. Fundamentally, it is really the coupling of the energy between the vibrational modes of the molecule which is of greatest interest, and ultimately can provide us with an understanding of molecular energy transfer. It therefore seems appropriate to look for systems which have all of the strong anharmonic and coriolis couling effects, only at much lower energies where the density of states is low. The hope is that by isolating just a few interacting states, a more detailed characterization of the important coupling terms and dynamical mechanisms can be obtained.

Weakly bound molecular complexes represent such a class of photochemical systems, The specific dynamics of interest here is vibrational predissociation, which involves excitation of a high frequency vibration associated with one of the monomer units within the complex, and subsequent energy transfer into the weak bond. Due to the presence of this fragile intermolecular bond, the energy threshold for dissociation of these systems is often very low. This fact, when combined with the relatively weak coupling between the initially excited intramolecular vibration and the intermolecular modes, often gives rise to dynamics which is in the highly nonstatistical regime (Butz et al. 1986; Dayton et al 1989; Dayton and Miller, 1988, 1989; Fraser and Pine 1989a; Jucks and Miller 1988a; Lovejoy et al. 1987; Lovejoy and Nesbitt 1990; Nesbitt et al. 1989b; Pine and Fraser 1988). Studies of these systems can thus provide important insights into the nature of the associated dynamics since the deviations from statistical behavior can reveal the "preferences" the system has for particular photodissociation channels. The weakness and floppiness of the bond between the constituent molecules (< 5 kcal/ mol) provides the anharmonic and coriolis coupling that gives rise to the vibrational dynamics of interest. Single quantum vibrational excitation of one of the constituent monomer units are usually sufficient for dissociation to occur.

This vibrational predissociation process has been the subject of intense experimental (Bohac et al. 1992a,b Bohac and Miller 1993a; Brouard et al. 1991; Burak et al. 1987; Butz et al. 1986; Cline et al. 1988; Dayton et al. 1989) and theoretical (Clary 1992; Evard et al. 1988; Halberstadt et al 1986, 1987; Villarreal et al. 1991; Zhang et al. 1992a,b) study. The experimental studies, rather artificially, are stratified into two categories; namely, those carried out in the ground and electronically excited states. Unfortunately, there has been very little overlap in the molecules being studied or the vibrations excited by these two classes of experimental methods. The studies on electronically excited states have focused either on diatomic rare gas systems (Alfano et al 1992; Bernstein, 1994; Bernstein et al. 1984; Levy, 1983; Osborn et al. 1992; Outhouse et al. 1993) or large aromatic molecules (Levy 1983; Bernstein et al. 1984). In the latter case, excitation of the low frequency vibrations accessible via electronic transitions often show intramolecular vibrational energy redistribution (IVR) rates that are competitive with, and even faster than, vibrational predissociation. On the other hand, the high frequency vibrations (H–X) that are normally studied by infrared spectroscopy in the ground electronic state appear to be in the weak coupling regime (Miller 1989). There is still considerable discussion about whether this is due to the fact that the coupling is fundamentally weaker in the ground electronic state or if the high frequency vibrations are simply more decoupled from the low frequency bath, thus displaying more nonstatistical dynamics. One of the future directions for this field must therefore be to find common ground between the ground and electronically excited state studies in order to resolve some of these issues. The focus of the present article is on the experiments designed to study vibrational predissociation on the ground electronic state surface using near infrared spectroscopy as a probe.

To obtain a thorough understanding of the vibrational dynamics of these systems we must address two important issues. The first is the characterization of the parent complex; namely, its structure and ultimately the associated inter-molecular potential energy surface and the perturbations it affords to the intramolecular vibrational degrees of freedom of the constituent molecules. Through the combined efforts of microwave, and near and far infrared spectro-scopies, considerable progress has recently been made in determining such poten-tials with unprecedented detail. The basic approach is to identify as many bound states of the van der Waals complex as possible (Bemish et al 1993; Dvorak et al. 1991; Farrel et al. n.d.; Lovejoy and Nesbitt 1989) and to compare these with the results of multidimensional quantum calculations based upon an assumed potential energy surface, with the eventual goal of fitting the parameterized potential to the experimental data (Bemish et al. 1993; Hutson 1992; Nesbitt et al. 1989a). In many cases, *ab initio* calculations can be useful in constraining the potential surface in regions that are not sampled by the available spectroscopy. The second issue deals with the determination of the rate of photodissociation of the complex and the final state distribution of the resulting fragments. The experimental and theoretical characterization of the final state distribution of the fragments can provide us with direct information on the nature of the energy transfer processes leading up to dissociation, and the energy transfer mechanisms that lead to excitation of the various degrees of freedom of the fragments. The focus of this chapter is on the experimental approaches that have been taken in order to address the latter of these two issues. In the following sections, we will discuss a number of experimental methods that have been used to obtain final state information on the photo-fragments and illustrate some of these with particular systems we have studied in our laboratory. This will serve to illustrate the diversity of behaviour that can be seen in these systems. The final section reviews where we are and projects where future developments in this field might take us.

2.2. EXPERIMENTAL APPROACHES

Recent developments in the field of infrared laser molecular beam spectroscopy have lead to a wealth of information on both the structure (Jucks et al. 1988; Klemperer 1978, 1987; Lovejoy and Nesbitt 1987) and potential energy surfaces (Cohen and Saykally 1991a,b, 1992; Hutson 1988, 1990) associated with these weakly bound molecular complexes in the ground electronic state (Fraser and Pine 1989b; Gough et al. 1977; Jucks et al., 1988; Kleiner et al. 1991; McIlroy et al. 1991; Miller 1988; Nesbitt and Lovejoy 1988, 1990; Pine and Fraser 1988). For semirigid complexes, the ro-vibrational structure observed in an infrared spectrum yields accurate rotational constants, which can often be used to unambiguously determine the structure of the complex. For more floppy systems, the rotational/vibrational/tunneling spectroscopy can provide direct information on the associated potential energy surface, as discussed previously. By far the vast majority of these spectroscopic studies are based upon the use of free jet expansion sources to both form the species of interest and to overcome Doppler and pressure broadening effects. The information that comes from these high resolution spectroscopic experiments is extremely important as a foundation for understanding the dynamical processes that are occurring on the corresponding potential energy surfaces.

Although these spectroscopic studies are an integral part of the problem being addressed here, they are somewhat outside the scope of this chapter, which is intended to be focused on the dissociation dynamics which follows vibrational excitation. Nevertheless, it is important to point out that, until recently, much of what we have learned about the dissociated dynamics has come from the interpretation of the homogeneous linebroadening observed in these spectra in terms of the vibrational predissociation lifetimes or rates (Dayton et al. 1991; Dayton and Miller 1988; Huang et al. 1986a; Jucks and Miller 1987b; Pine and Fraser 1988). Unfortunately, experiments of this type always leave one to speculate concerning: (1) the final outcome of the photochemical event, and (2) the detailed dynamical events leading up to dissociation. To address these questions, we clearly need more sophisticated experiments that allow us to determine the condition of the photofragments. The "ideal" for such an experiment is one where the initial parent state is fully defined in terms of its velocity, orientation, vibrational and rotational state, and where the internal and translational states of all the products are also measured. This includes not only the scalar properties, such as the product internal state distributions, but also various scalar and vector correlations which may exist (Hall et al. 1988; Houston 1987; Vasudev et al. 1984).

A number of methods have recently been developed and applied to the study of the infrared photodissociation of complexes (Bohac et al. 1992a,b; Casassa et al. 1988; Hetzler et al. 1991; Marshall et al. 1992). For example, pump–probe experiments have been carried out in free jet expansions using visible UV fluorescence and resonantly enhanced multiphoton ionization (REMPI) detection (Casassa et al. 1988; Hetzler et al. 1991) following pulsed infrared laser excitation. In cases where the resolution of the probe laser is sufficient to measure to Doppler profiles associated with the fragment transitions, it is possible to extract information concerning not only the rotational and vibrational distributions, but also the translation recoil energy. Vector correlations can also be obtained in some cases (Marshall et al. 1992; Shorter et al. 1992). The primary limitation with these methods has been their lack of generality. For both laser-induced fluorescence (LIF) and MPI, there are rather severe restrictions on the molecules that can be studied. This is problematic in several respects. First, since the theoretical calculations of these dynamical processes are extremely computationally intensive (Clary 1992; Zhang et al. 1992a,b,c), our best opportunity for making comparisons between experiment and theory will come as we develop experimental methods that are not restricted with regards to the molecules that can be studied. Second, it is important to realize that much of the progress that has been made in understanding the dynamics of these complexes has come from comparing the results for many different systems, establishing trends, and identifying special cases. Therefore, a general method is clearly needed for studying these processes at the state-to-state level. For this reason, we have been exploring the use of infrared spectroscopy as a probe of molecular dynamics, since most molecules possess at least one allowed vibrational band that can be used for both the pump and the probe steps.

Our first step in this direction was taken in 1988 when we reported the measurement of photofragment angular distributions for the HF dimer (Dayton et al. 1989) using the optothermal technique. This method is based upon the fact that infrared laser excitation can result in a significant change in the molecular beam energy, which can be detected by a liquid helium cooled bolometer (Gough et al. 1977; Miller 1988). Thus, for stable molecules, with infrared fluorescence lifetimes that are long with respect to the molecular flight time, the bolometer signal is due to the laser-induced increase in the molecular vibrational energy. For the weakly bound complexes of interest here, the vibrationally excited states are generally predissociative so that the fragments recoil out of the molecular beam, missing the bolometer detector, and giving a signal of opposite sign (or phase). A large number of complexes have been studied spectroscopically in this way (Miller 1986, 1988), namely by scanning the laser in frequency while using the bolometer to measure the resulting change in the molecular beam energy. To obtain information on the photofragment distributions, the bolometer can equally well be used to detect the fragments that recoil out of the molecular beam. The experimental apparatus used for this purpose is shown in Figure 2-1. The molecular beam source is rotated about the photolysis point so that the photofragment angular distribution can be measured. Given that the spectroscopy of

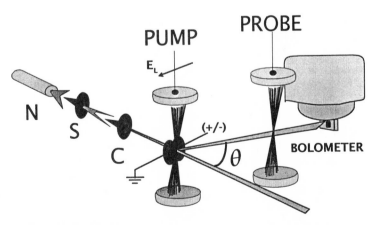

Figure 2-1. A schematic diagram of the apparatus used to record photofragment angular distributions of complexes. An F-center laser is used to pump transitions in the parent complex which leads to dissociation. A second F-center laser is used as a probe to state selectively detect the fragments. The electrodes are used to orient the parent molecules prior to excitation.

the complex is already assigned, the pump laser (a cw color-center laser) can be tuned into resonance with a single ro-vibrational transition. In this way, only complexes in a well-defined ro-vibrational state are dissociated, ensuring the selection of the intial state of the parent complex.

The detailed shape of the resulting angular distributions is determined by a large number of factors, including many dealing with the geometry of the apparatus. The important dynamical information is contained in the recoil energy of the photofragments, which determines to what angle in the laboratory frame the fragments can recoil. and the state-to-state probabilities for populating the various final states. In general, the photofragments are ejected over a range of angles in the center-of-mass frame so that the laboratory frame angular distribution is not a delta function, even for a single photofragment channel. As Figure 2-2 illustrates, for low recoil velocities with respect to the stream velocity of the parent molecule, there is a maximum angle to which the fragments can scatter in laboratory frame, corresponding to the recoil and stream velocities being perpendicular to one another. For the case of equal mass fragments, the laboratory frame angular distribution shows a single peak for each photofragment channel, as illustrated in Figure 2-2. The peak in the distribution comes from the fact that a range of center-of-mass angles contributes to the same laboratory angle when the detector is positioned tangent to the Newton sphere. The detailed shape of the observed angular distribution depends upon the shape of the center-of-mass photofragment angular distribution. A detailed discussion of how such information can be used to determine vectorial properties of the photodissociation is given elsewhere (Marshall et al. 1992) and will be discussed briefly in a later section of

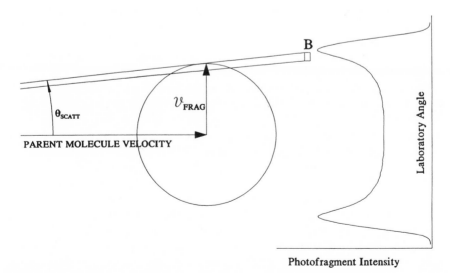

Photofragment Intensity

Figure 2-2. Photofragment angular distribution for a single channel. A Newton diagram showing the recoil of fragments out of the molecular beam. The angular distribution appearing to the right is calculated assuming monoenergetic and spherically symmetric recoil of the fragments.

this chapter. The calculation shown in Figure 2-2 is based upon a spherically symmetric dissociation in the center-of-mass frame.

As we will see in later sections of this chapter, when the photofragment state density is very low, which is to say that the energy spacing between levels is large, the resulting peaks in the laboratory frame angular distribution can be resolved. In such cases, their relative intensities provide direct information on the probabilities of populating these states, while their angular position determines the corresponding recoil kinetic energy.

The difficulty with the above approach is that as soon as the density of states becomes moderately high, the assignment of the peaks in the angular distribution becomes ambiguous and eventually the structure disappears entirely, giving a smoothly decaying angular distribution. Under these conditions, the only information that can be extracted from the distribution is the average kinetic energy released to the photofragments. To overcome this difficulty, we have added a second color-center laser to the apparatus in the manner shown in Figure 2-1. By directing the output of the second laser between the photolysis region and the detector, the fragments can be state selectively detected. This is done by tuning the probe laser through the various monomer ro-vibrational transitions. When the probe transition corresponds to a lower state that is populated by the dissociation process, the corresponding fragments are vibrationally excited by the probe laser giving rise to a bolometer signal that is proportional to the population in the probed level. In this way, the relative state-to-state probabilities can be determined without the need for resolving the peaks in the angular distribution.

For binary complexes formed from monomers of unequal mass, there is the added complication that, due to conservation of momentum, the two fragments scatter to different angles in the laboratory frame. As a result, each photodissociation channel give rise to two peaks in the angular distribution. This added complexity makes it even more difficult to obtain a unique assignment of the angular distribution for such cases. To overcome this difficulty, we have introduced a set of electrodes so that a DC electric field can be applied to the photolysis region. In this way, the parent molecules can be oriented (Block et al. 1992; Friedrich and Herschbach 1991; Loesch and Remscheid 1990; Rost et al. 1992) in the manner shown in Figure 2-3 for the N_2–HF complex. As the figure indicates, only moderate electric fields are required to modify the rotational wavefunctions to the point where they become one sided, corresponding to the molecules undergoing pendular-type motion in the field. Under these conditions, there is essentially perfect left–right orientation imposed on the system, which is to say that the probability of a molecule being antioriented with the field is essentially zero. Assuming the dissociation occurs along the intermolecular axis of the complex, the two fragments will preferentially recoil in opposite directions in the laboratory frame. In this way, we can record the angular distribution for the two fragments separately. As discussed below, the photofragment angular distributions resulting from molecules that have been oriented in this way can also provide information on the molecule fixed frame differential cross sections that cannot be achieved by normal laser polarization alignment methods (Gericke et al. 1986; Hall et al. 1986; Houston 1987; Vasudev et al. 1984).

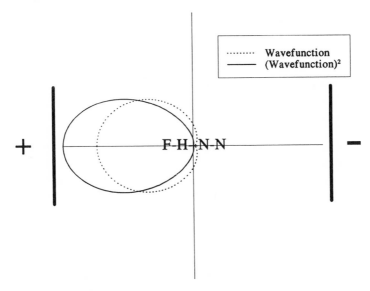

Figure 2-3. An illustration of how well molecules can be oriented in electric fields of moderate strength. The wavefunction shown in the figure is for the $M_j = 0$ pendular state for N_2–HF at $16.6 \, \text{kV cm}^{-1}$.

2.3. SEVERAL ILLUSTRATIONS: WHERE THEORY AND EXPERIMENT MEET

2.3.1. Vibrational (V) to Rotational (R) Energy Transfer: Hydrogen Fluoride Dimer

The hydrogen fluoride (HF) dimer has become a prototype in the study of hydrogen-bonded molecular complexes, from both the experimental (Barton and Howard 1982; DeLeon and Muenter 1984; Dyke 1984; Dyke et al. 1972; Gutowsky et al. 1985; Himes and Wiggens 1971; Howard et al. 1984; Huang et al. 1986b; Lafferty et al. 1987; Lisy et al. 1981; Pine and Fraser 1988; Pine and Howard 1986; Pine and Lafferty 1983; Pine et al. 1984; Puttkamer and Quack 1986; 1987a,b, 1989; Puttkamer et al. 1989, 1991; Vernon et al. 1982) and theoretical (Alexander and DePristo 1976; Aoyama and Yamakawa 1979; Barton and Howard 1982; Bunker et al. 1988; Ewing 1980; Halberstadt et al. 1986; Hancock et al. 1988; Hougen and Ohashi 1985; Jensen et al. 1990; Kofranek et al. 1988; Lischka 1979; Rybak et al. 1991; Sun and Watts 1990; Yarkony et al. 1974) points of view. Both the structure of the dimer and its tunneling dynamics were established nearly two decades ago by Klemperer and coworkers (Dyke et al. 1972) using the molecular beam electric resonance method. A large number of subsequent microwave, near and far infrared, and overtone studies have mapped out a great deal of the spectroscopy of this system, including work on all of the deuterated isotopomers. The near infrared and overtone studies have shown that the vibrational predissociation lifetime is strongly dependent upon the nature of the initial intramolecular vibrational mode excited, as determined from the associated homogeneous linewidths. In particular, excitation of the fundamental "free" HF vibration results in a vibrational predissociation lifetime which is approximately 20 times longer than that associated with excitation of the hydrogen-bonded HF stretch. This can be understood if one considers that the hydrogen-bonded HF stretch is much more strongly coupled to the intermolecular hydrogen-bonding coordinate than is the free HF stretch. For the present purposes, this system is interesting since only a single vibrational channel is energetically accessible, namely quenching of the H–F vibrational state from $v = 1$ to $v = 0$.

The theoretical work on the HF dimer includes studies of the potential energy surface (Alexander and DePristo 1976; Aoyama and Yamakawa 1979; Barton and Howard 1982; Bunker et al. 1988; Ewing 1980; Halberstadt et al. 1986; Hancock et al. 1988; Hougen and Ohashi 1985; Kofranek et al. 1988; Lischka 1979; Puttkamer and Quack 1989; Rybak et al. 1991; Sun and Watts 1990; Yarkony et al. 1974), as well as calculations of the lifetimes and state-to-state photodissociation cross sections (Halberstadt et al. 1996; Zhang and Zhang 1993). The first calculation of the final state distribution for the HF dimer (Halberstadt et al. 1986) was based upon a pseudo atom–diatom approximation, which predicted that the proton donor HF would be produced in the highest energetically accessible rotational state. This is consistent with an impulsive dissociation in which the proton donor "pushes off" the partner (which in this calculation is an atomic approximation to the other HF molecule). Since the associated force acts upon the hydrogen atom of the proton donor, which is far from its center of mass, this fragment is strongly torqued and appears with large rotational excitation.

In a number of publications, we have reported angular distributions for several different initial states of the dimer (Bohac et al. 1992c; Bohac and Miller, 1992; Dayton et al. 1989; Marshall et al. 1992). Due to the low density of photofragment states, the angular distributions are easily assigned and the state-to-state probabilities can be obtained by fitting. It is important to point out that the basis of the assignment is conservation of energy, where the translational recoil energy is the primary measurement. As a result, the experiment can be used to determine the intermolecular correlations between the rotational (and, more generally, the internal) states of the two fragments. The determination of the (j_1, j_2) correlations is crucial to understanding the dynamics. Examination of the fitted probabilities reveals a preference for producing one highly rotationally excited fragment in coincidence with a low j fragment. This is in qualitative agreement with the simple pseudodiatomic picture discussed earlier. Nevertheless, the agreement is far from quantitative, owing to the fact that the proton acceptor can be produced in a range of rotational angular momentum states. The result is a distribution of j states for the proton donor which is much broader than that obtained from this simple treatment. Recently, Zhang and Zhang (1993) carried out a diatom–diatom calculation which gives a correlated distribution that is in better, but still only qualitative, agreement with the experiment. Although the calculation picks out the same subset of important final state channels as observed in the experiment, there are significant differences that depend on the potential energy surface that is used in the calculation (Bemish et al. 1994b; Zhang and Zhang 1993), as is evident from Figure 2-4. Since the photodissociation process is really a scattering event, it is likely that the state-to-state cross sections are sensitive to a different part of the potential than the spectroscopy of the bound states. This is perhaps the reason why the bound state results are reasonably well described by the available potentials, while the photofragment cross sections are not.

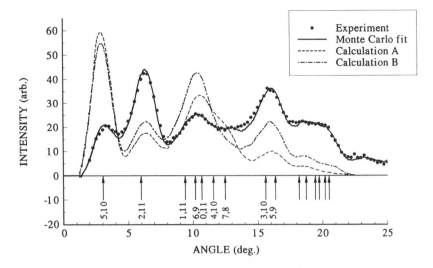

Figure 2-4. A comparison of the experimental HF dimer angular distribution and those calculated by Zhang and Zhang (1993) using the potential energy surfaces of Bunker et al. (1988, 1990) (calculation A) and Quack and Suhm (1991) (calculation B).

We have carried out a number of subsequent experiments on this system that are all consistent with an impulsive mechanism for the photodissociation. First, we have excited the combination band associated with both the F–F stretching and *trans*-bending vibrations of the dimer and find that the final state distributions are modified in a way that is consistent with the corresponding change in the initial state wavefunction, and, thus, the sampling of initial geometry of the complex (Bohac and Miller, 1992). Work that is in progress shows that for both the DF–HF and HF–DF systems, it is the proton donor that is perferentially excited to high rotational states. Finally, we have oriented the HF dimer using the electric field method discussed earlier, so that the fragments arising from the proton donor and acceptor positions can be distinguished in the experiment. These experiments show that it is indeed the proton donor that is the higher rotational energy fragment.

The HF dimer has provided us with an ideal system for studying V–R energy transfer. The system is complex enough, due to the correlations that exist between the two molecular fragments, to give rise to distributions that directly probe the dynamics, and yet simple enough to understand in straightforward terms. Despite the large effort that has been expended in the study of this system, a great deal remains to be done. The significant differences between theory and experiment suggest that more work on the full six-dimensional potential energy surface is needed, particularly in the repulsive region which is likely to dominate in the scattering processes considered here. More experimental work on both the spectroscopy and dynamics, with emphasis in higher intramolecular and inter-molecular vibrationally excited states, will be needed to test these potentials over as broad a range of configuration space as possible. Elastic and inelastic scattering data (Copeland et al. 1982; Haugen et al. 1984; Vohralik and Miller 1985) should also be included in future analyses to provide complementary information on the potential energy surface.

2.3.2. Intermolecular V–V Energy Transfer in D_2–HF: Comparisons with H_2–HF

The HF dimer represents the ultimate achievement, so far as our efforts to reduce the number of vibrational channels available to the system are concerned, since there is only one. However, as we are interested in understanding the nature of the interactions between vibrational states in molecules, it is clearly necessary to move on to systems where there are more possibilities. At the same time, we wish to keep the total number of available channels to a minimum to avoid the problem of being unable to make an assignment based upon the angularly resolved method discussed earlier. The D_2–HF system is ideal from this point of view since both the D_2 $v = 0$ and $v = 1$ channels are energetically accessible. In addition, there has been a great deal of spectroscopic and theoretical work done on this system that provides us with a starting point. The first spectroscopic work, exciting the H–F stretching vibration, was reported by Lovejoy et al. (1988). They found that the D_2–HF lifetimes are considerably shorter than those of H_2–HF (Lovejoy et al. 1987), which was interpreted as being due to the fact that the intramolecular V–V channel is open in the former case and closed in the latter. Clary (1992) has reported *ab initio* calculations of the potential energy surface, from which he has calculated both the energy levels needed to explain the spectroscopy and the dissociation dynamics. Calculations have also been carried out using the same potential surface for both D_2–HF (Zhang et al. 1992b,c) and H_2–HF (Zhang et al. 1992a). The calculations from both groups provide predisssociation lifetimes and final rotational state distributions, affording us another opportunity to make comparisons between theory and experiment.

Before proceeding to discuss the state-to-state dynamics of these systems, it is important to point out that the *ab initio* potential energy surface for this system gives calculated spectra that are in excellent agreement with experiment. We take this as evidence that, at least in the region of the well, the potential is quite realistic. Since the dominant electrostatic interaction in this system is that between the quadrupole of the H_2 and the dipole of the HF, the intramolecular coupling was included semiempirically in terms of the stretching dependence of these quantities (Clary 1992).

Figure 2-5(a) shows a comparison between the experimental angular distribution for one of the states of D_2–HF and one calculated based upon the probabilities reported by Clary (1992). It is important to point out that the calculated distribution also depends upon the dissociation energy (D_0) of the complex, which determines the angular position of the various channels. The calculation shown in the figure was carried out using the value for D_0 determined from the *ab initio* potential. It is clear from this comparison that both the dissociation energy of the complex and the calculated state-to-state probabilities are in excellent agreement with experiment. In addition, the lifetimes calculated for this system agree very well with those determined experimentally (Clary 1992; Lovejoy et al. 1988). As discussed in detail elsewhere Bohac and Miller 1993a),

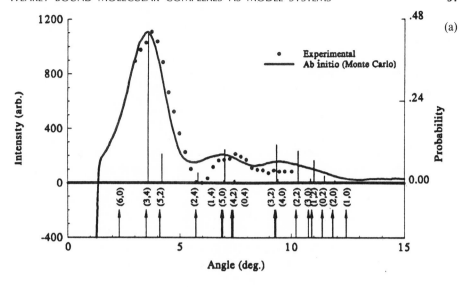

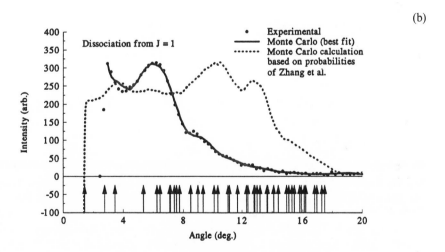

Figure 2-5. Comparisons between the experimental and calculated angular distributions for (a) *ortho*-D_2–HF and (b) *ortho*-H_2–HF. In both cases, the calculated distributions were based upon the dissociation energies and state-to-state probabilities from the *ab initio* potential.

there are differences between the experimental and theoretical state-to-state probabilities for some of the initial states of this system, which suggests that the coupling may need to be modified slightly to give quantitative agreement. Nevertheless, the agreement is good enough to convince us that the electrostatic quadrupole–dipole vibrational coupling responsible for the dissociation of this complex, via the V–V process, has been properly identified. It is interesting to note that for the D_2–HF system, the preferred channels are those in which $j_{D2} \approx j_{HF}$ (Bohac and Miller 1993a), in stark contrast with the HF dimer discussed earlier.

The situation is very different for the H_2–HF complex, as illustrated in Figure 2-5(b). Here, the experiment disagrees qualitatively with the calculations of Zhang et al. (1992a), based on the same potential surface. In particular, the calculations overemphasize the importance of the low j, high kinetic energy channels (which appear at large angles in the angular distribution) and, in addition, give a lifetime which is considerably longer than experiment (Jucks and Miller 1987a). We interpret this as being due to the fact that the only mechanism for dissociation in this case is V–R, T. As a result, the final state distribution is extremely sensitive to the anisotropy of the potential energy surface. The fact that the calculated lifetime is too long and the rotational distributions too cold suggests that the anisotropy of the *ab initio* potential is too weak. The large translation energy gaps associated with the low energy rotational channels inhibits dissociation, making the calculated lifetimes too long. Since the spectroscopic data are well reproduced by this potential, the problem is unlikely to be in the region of the attractive well. The more likely explanation is that the potential is too isotropic in the repulsive wall, which is understandable given that this part of the potential was not carefully explored in the original *ab initio* study (Clary and Knowles 1990). Here again, the interaction between theory and experiment has helped us to gain new insights into both the nature of the dynamical processes involved in these systems and the regions of the potential surface that require further attention. With further improvements in both the experimental and theoretical methods, there is clearly hope that similar cooperation could provide insights into the dynamics of larger systems that display much richer behavior.

2.3.3.   Intramolecular V–V Energy Transfer: Ar–CO$_2$

In this section, we consider a rather different dynamical process, namely that leading to a change in the vibrational state of a monomer subunit as a result of the presence of the intermolecular force field due to its partner. Spectroscopic evidence for the influence of intermolecular interactions on intramolecular coupling strengths already exists. For example, the Fermi resonance in acetylene is known to be quenched as a result of formation of a hydrogen bond with HF (Huang and Miller 1989). The very fact that these complexes dissociate in a manner that changes the vibrational state of the monomer subunits, is indicative of the coupling between the intramolecular and intermolecular degrees of freedom. Here, we consider the case of the photodissociation of Ar–CO$_2$ to illustrate the specific example, of intramolecular V–V energy transfer. These rare gas/molecular complexes are ideal for this purpose since all of the vibrational density of states is associated with CO$_2$ and the argon simply acts as a perturbation to the intramolecular force field of the molecule. The spectroscopic literature on this complex is quite extensive, beginning with the determination of its structure as T-shaped (Steed et al. 1979) using microwave spectroscopy. These results were later used by Hough and Howard (1987) to generate a number of potential energy surfaces which were capable of reproducing the available data. More recently, there have been a number of near infrared studies of this complex, the first being carried out in absorption by Randall et al. (1988) and by Sharpe et al. (1988) in the 4.25 μm spectral region (v_4 of the complex). Fraser et al. (1988) have since reported optothermal spectra in the 2.7 μm region, corresponding to excitation to both members of the $(101)/(02^01)$ Fermi diad in the CO_2 monomer. Most recently, is a direct absorption study by Sharpe et al. (1991), in which a combination band was observed, resulting from excitation of the low frequency in-plane bending vibration (v_b) along with the asymmetric stretch of the CO_2 monomer. Combining this result with the fundamental asymmetric stretch frequency, they were able to determine an accurate value for the intermolecular bending frequency. In all of these studies, the linewidths of the associated transitions were instrument limited, indicating that the lifetime of the complex is rather long (> 50 ns). All of this spectroscopic data has helped to define the potential energy surface for this system—a necessary prerequistite to a thorough understanding of the dynamics.

Even though the rotational constant of CO_2 is quite small and its vibrational density of states is moderately large, the angular distribution for this system shows assignable structure (Bohac et al. 1992a,b). This is only possible in view of the fact that there is no state density associated with the argon atom. Another factor that helps in this case is the fact that the complex can be excited to more than one initial state, owing to the $(10^01)/(02^01)$ Fermi resonance. This is helpful since both sets of data must be explained by the same value of the dissociation energy, in this case 166 cm^{-1}. Analysis of the data clearly shows that the dominant dissociation channels correspond to an intramolecular V–V energy transfer process, where the vast majority of the excess energy is accommodated in the CO_2 fragment vibration. The small rotational constant of the fragment ensures that only a small fraction of the energy appears in rotation. Upon excitation of the upper member of this Fermi diad, we find that dissociation occurs specifically into the highest open vibrational channel of the CO_2 monomer, namely the upper member of the $(21^10)/(13^10)/(05^10)$ triad shown in Figure 2-6. Dissociation into this state requires relatively few vibrational quantum number changes, which is well known from other studies (Ewing 1987) to be an important factor in controlling the strength of vibrational coupling. Excitation of the lower member of the Fermi diad again leads to specific dissociation into a single vibrational level: in this case, the middle member of the $(21^10)/(13^10)/(05^10)$ triad. This is despite the fact that there is another fragment vibrational level in closer resonance, namely the upper member of the $(13^30)/(05^30)$ diad. Nevertheless, dissociation into this state would require a large change in the vibrational quantum numbers of the CO_2, which is presumably not as favorable. Overall, the vibrational selectivity we observe in this system can be understood if one considers the interplay between a number of competing effects, namely: (1) energy gap considerations, which favor the most nearly resonant channel, (2) the magnitude of the vibrational quantum number change, which is preferentially small, and (3) conservation of angular momentum, which also tends to favor nearly resonant channels, since the rotational degree of freedom of the CO_2 fragment cannot accommodate much energy without accessing high angular momentum states. These are obviously very quantitative ideas and

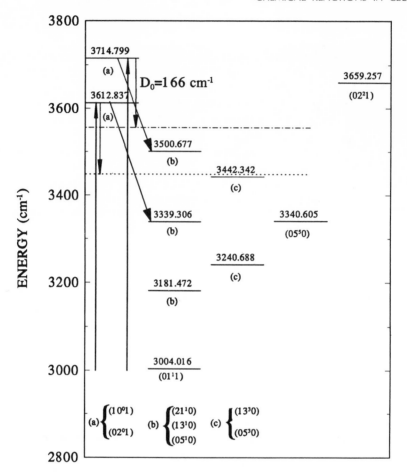

Figure 2-6. An energy level diagram showing the vibrational states of interest in the vibrational predissociation of Ar–CO_2. The experimental results show that the intramolecular V–V process, indicated by the diagonal arrows, is the dominant mechanism for dissociation. CO_2 vibrational energy levels designated (a), (b), and (c) are the Fermi diads and Fermi triad discussed in the text.

more progress into understanding the detailed vibrational coupling mechanisms involved in such cases will clearly require detailed dynamical calculations on realistic potential surfaces, which include the appropriate intramolecular couplings. For example, the role that the Fermi resonances play in the dynamics of this system is still far from being clear.

2.4. LARGER SYSTEMS WITH MIXED DYNAMICS

The systems we have discussed thus far clearly fall into the sparse density of states regime and display highly nonstatistical dynamics. Unfortunately, even modest increases in the complexity of the system lead to angular distributions that can

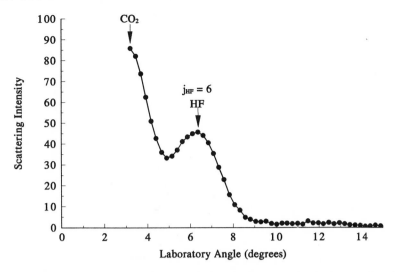

Figure 2-7. CO_2–HF angular distribution: H–F stretch $v = 1$, $J = 1$. A zero electric field angular distribution for the CO_2–HF complex. The two peaks observed in this distribution are associated with the two fragments. Pump–probe experiments reveal that the HF peak arises from dissociation into $j_{HF} = 6$.

no longer be assigned directly. As a result, if progress is to be made in studying systems with higher densities of states and more diverse dynamics, new methods for probing the final states must be developed. To illustrate the utility of the experimental methods introduced in an earlier section of this chapter, we consider the photodissociation of two systems, namely N_2–HF and CO_2–HF. The problem that are encountered can be appreciated by considering the angular distribution for CO_2–HF, shown in Figure 2-7. The large number of vibrational states accessible upon excitation of the H–F stretch, coupled with the small rotational constant of the CO_2 fragment, results in several hundred possible photofragment channels that could account for the quite limited structure observed in this angular distribution. In fact, the two peaks that are observed are simply due to the different masses of the two fragments, resulting in the HF fragment recoiling to a larger angle than the heavier CO_2. Clearly, the recoil kinetic energy alone is insufficient to make an assignment of the internal states of the fragments. The first approach we have taken to alleviate this problem is to use the pump–probe method illustrated in Figure 2-1. This involved setting the bolometer at 6° to detect the HF fragment produced by the pump laser photolysis and tuning the probe laser through the various HF monomer transitions to determine the rotational distribution of this fragment. Under these conditions, only the $j = 6$ state of the HF was observed. Given a reasonable estimate of the dissociation energy of this complex of 700 cm^{-1}, we estimate that the energy available to the CO_2 fragment is approximately equal to that of the $(00^0 1)$ vibrational state, namely the asymmetric stretch. The small rotational constant of the CO_2 fragment ensures that the amount of energy that can be accommodated in rotation is small, so we are left to conclude that this channel corresponds to the coincident production of

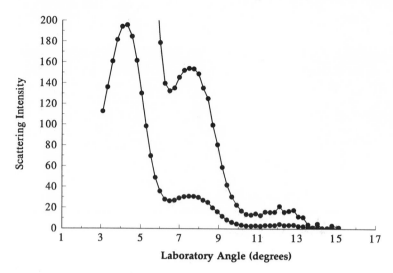

Figure 2-8. A pendular state ($M_j = 0$) photofragment CO_2–HF angular distribution corresponding to detection of the CO_2 fragment, showing at least three important dissociation channels. The expansion factor for the second trace is five.

HF ($j = 6$) and CO_2 ($00°1$). As discussed elsewhere (Bohac and Miller 1993b), this propensity may be due to strong dipole–dipole coupling between the initially excited H–F stretch and the asymmetric stretch of the CO_2, resulting in a very specific intermolecular V–V energy transfer process.

Further progress can be made by examining the pendular state angular distribution shown in Figure 2-8, corresponding to detection of the CO_2 fragment. To our surprise, this distribution shows three peaks, only one of which (namely the largest peak at small angles) is explained in terms of the dissociation channel discussed earlier. The other two were strongly overlapped in the zero field angular distribution by the HF fragment intensity. From this, we conclude that there are three vibrational–rotational fragment channels of importance in this system. Although, from the relative intensities of these three peaks, one might be tempted to conclude that the other two are relatively unimportant compared with HF ($j = 6$) and CO_2 (00^01), when these intensities are corrected for the falloff in detection efficiency with increasing angle, the probabilities for all three channels are significant. The pendular state data clearly provide us with new information that changes our overall understanding of this system. Without this data, we are misled into thinking that dissociation is completely dominated by a single V–V channel, which is clearly not the case. Work in progress is aimed at assigning the two extra channels to specific vibrational channels of the CO_2 fragment.

The N_2–HF system behaves in a somewhat similar manner, although the vibrational and rotational density of states is considerably lower, making the assignment of the final states somewhat easier. In this case, we find that (Bohac and Miller 1993b) dissociation occurs via both of the energetically accessible vibrational channels, namely N_2 $v = 0$ and $v = 1$, produced in coincidence with HF $j = 12$ and $j = 7$, respectively. In this case, there is sufficient structure in the angular distribution to also allow for the assignment of the rotational state distribution of the N_2 fragment (Bemish et al. 1994a). It is interesting that in both of these systems, more than one vibrational channel is important in the dissociation process. This is despite that fact that the coupling terms that give rise to these channels are quite different and there is no reason to expect that they would have comparable magnitude. For example, the N_2 $v = 0$ channel requires that the HF fragment experience a large torque, making this channel sensitive to the anisotropy of the potential (as in H_2–HF), while the V–V channel, which gives rise to $v = 1$, involves some form of coupling between the two intramolecular vibrational coordinates (as in D_2–HF). A possible explanation for why multiple vibrational channels are observed with comparable probability is that the transition state is of mixed vibrational character so that it is here that the final state vibrational distribution is determined. That is to say, the various vibrational channels are not produced by different dissociation mechanisms, but rather result from the evolution of a coherent superposition of vibrational states that define the transition state. Unfortunately, for these systems, the theory has not yet advanced to the point where detailed comparisons can be made.

The pendular state data obtained for N_2–HF was particularly helpful in determining the rotational distribution of the N_2 fragment, due to the reduced congestion that comes from the fact the the two fragments are detected separately. To a first approximation, the two pendular state angular distributions should sum to give the zero field result. However, careful examination of the results shows that there are differences between the zero field and pendular state angular distributions. One possibilty is that the dynamics are modified by the electric field, changing the state-to-state probabilities. Although this possibility has not yet been entirely ruled out, it seems unlikely for this system, given that the interaction energy with the electric field is only a fraction of a wavenumber. Rather, these differences most likely arise from the fact that the two sets of data are sensitive in different ways to the photofragment angular distributions in the molecule fixed frame (Wu et al. 1994). In contrast with experiments that rely upon the polarization of the excitation laser to impose an alignment on the excited state molecules (the zero field experiment), where only the P_0 and P_2 terms in the anisotropy of the photofragment angular distribution are determined (Greene and Zare 1982; Yang and Bersohn 1974; Zare 1982), the pendular state experiment depends upon all of the terms in the Legendre polynomial expansion, in a way that is determined by the magnitude of the electric field (Wu et al. 1994). As a result, the pendular state angular distribution depends in detail upon the functional form of the photofragment angular distribution in the molecule fixed frame $[f(\theta\phi)]$. Since the number of these distributions with the same value of β is infinite, the sum of the two pendular state distributions need not be equal to the zero field result. Preliminary interpretation of this data seems to suggest two things. First, there is some evidence that the $f(\theta, \phi)$ function is different for the two dissociation channels, namely V–V and V–R. Second, the results suggest that a small fraction of the dissociative events lead to the recoil of the fragments in the antiaxial direction, which is to say that the N_2 fragment recoils towards the positive electrode in Figure 2-3 and the HF to the negative. If confirmed by future investigations, these results would have important implications concerning the nature of the dissociative trajectories, namely that in a small fraction of the cases there is some form of orbiting resonance that allows the overall complex to effectively rotate before finally dissociating. Although these findings need to be confirmed by higher field studies, which will tend to accentuate the differences, the results do show that there is new dynamical information to be gained by studying the photodissociation of molecules that have been oriented in an electric field.

2.5. LARGER CLUSTERS: SELECTIVE DISSOCIATION OR PHOTOINDUCED EVAPORATION

Up to this point, our focus has been on the dissociation dynamics of binary complexes, where the methods are now capable of providing quite complete information. In contrast, the study of vibrationally induced dissociation in larger clusters is still in its infancy. It is generally assumed and expected that, in the large cluster limit, the vibrational excitation energy will be quickly thermalized, leading to a process that is best described as evaporation (Insepov and Karataev 1991). In view of the highly nonstatistical processes observed in binary complexes, however, it is not obvious at what cluster size this transition to thermal behavior will occur. Higher order clusters clearly present a formidable challenge for studies at the state-to-state level. As a result, this section is necessarily more qualitative and is intended to simply raise several important issues and to suggest possible courses of action for future studies. The experimental results and the conclusions we draw are by no means definitive.

Consider, for a moment, the vibrational excitation of a molecule within a cluster of moderate size. Assuming the energy is quickly randomized in the thermal bath of the cluster, the net result of this specific vibrational excitation is simply an increase in the temperature of the cluster. If this temperature jump is sufficient to make the cluster metastable, monomers will begin to evaporate. This process will cool the cluster, thus decreasing the evaporation rate for subsequent monomers (Insepov and Karataev 1991). Since the photon energy is quickly redistributed among the many degrees of freedom of the complex, one expects that the kinetic energy of the evaporating molecules will be quite low in comparison with the corresponding binary complex. Figure 2-9 shows a set of angular distributions obtained for $(HF)_2$, $(HF)_3$ and $(HF)_{n>3}$. It should be emphasized that the apparatus is configured in such a way that the detector views only the photolysis region. As a result, fragments are only detected from clusters that dissociate on a time scale faster than 1 μs, so that fragments originating from clusters undergoing sequential evaporation, on a time scale greater than this will not be detected. The angular distributions clearly show that the average photofragment kinetic energy decreases rapidly with increasing cluster size and that the density of states is already large enough in the case of the trimer to wash out the structure that is quite distinct for the dimer.

In the spirit of proposing directions for future study, it is interesting to consider several characteristics of these distributions—in particular, for the case of $(HF)_n$. Careful inspection of the latter distribution shows that at small angles there

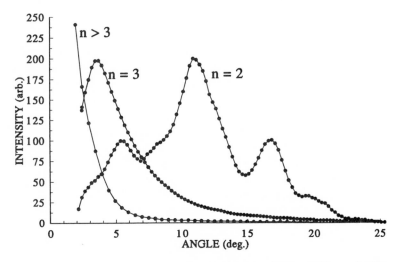

Figure 2-9. A comparison of the angular distributions for $(HF)_2$, $(HF)_3$ and $(HF)_{n>3}$.

is a very rapid falloff in the photofragment intensity, whereas at larger angles there is a rather broad plateau that corresponds to a high kinetic energy tail in the photofragment distribution. In fact, attempts at fitting this distribution to a Maxwellian velocity distribution at a single temperature were unsuccessful. At least two temperatures are needed to obtain a good fit to the distribution, 8.5 K to account for the small angle portion of the distribution and 100 K for the plateau. One possible explanation for this is that we are seeing sequential evaporative cooling of the cluster. Namely, within our 1 μs window more than one monomer evaporates, the first giving rise to the highest kinetic energies and the later ones being substantially cooler. Alternatively, the evaporation process might be highly nonthermal, proceeding by at least two mechanisms that give rise to both translationally "hot" and "cold" fragments. Although tantalizing, these data are clearly not sufficient to unambiguously determine the mechanisms involved. Therefore, part of the future in this area will be the experimental determination of the rotational state distributions of the evaporating species, which will provide us with independent information that might aid in distinguishing between the various possible dynamical processes.

Consider the use of the pendular state orientation method in the study of polar clusters. For example, the HCN trimer is known from previous spectroscopic measurements to have a linear polar isomer (HCN–HCN–HCN), in addition to a nonpolar cyclic form (Jucks and Miller 1988b). The large dipole moment and small rotational constant of the linear isomer make it ideal for pendular state orientation experiments (Block et al. 1992). This system has two nonequivalent hydrogen bonds and it is interesting to ask the question: Can excitation of different vibrational modes of this cluster provide us with a means of selectively dissociating one bond or the other? If so, this would clearly show that the "evaporation" process in this system is nonstatistical. For example, if the terminal C–H stretch is initially excited and the hydrogen bond between this molecule and the adjoining dimer is an effective bottleneck to energy flow, one might expect that this bond would break. On the other hand, if the energy is first randomized throughout the cluster and then the monomer units begin to "evaporate," either bond could break, presumably with statistical probabilities. Similar arguments could be made for exciting the hydrogen-bonded C–H stretches. By orienting the trimer, prior to excitation, it should be possible to distinguish between these mechanisms. Conservation of momentum dictates that the angular distribution will be more strongly forward peaked on the side to which the dimer fragment recoils, compared with the monomer side. Therefore, assuming that the process is selective, it should be possible to determine which bond has broken by noting which fragment recoils toward which electrode. There are clearly many interesting questions that remain to be answered in these larger systems. Fortunately, experimental methods are now becoming available that, in future, should provide us with data we need to at least begin.

2.6. CONCLUSIONS AND FUTURE PROSPECTS

In this chapter, we have reviewed work on the dissociation dynamics of a number of complexes, which illustrate the wide range of behavior these systems exhibit. The goal in studying these systems is to gain a better understanding of the nature of the vibrational coupling in molecules, at energies where dissociation can occur. We find that, for the small systems considered here, excitation of high frequency vibrations gives rise to highly specific and nonstatistical behavior. Clearly, our search for the onset of statistical behavior will need to continue into the realm of even more complex systems. Fortunately, the methods we have discussed in this chapter are quite robust and should carry us quite far in this search. The extension of this work to the higher state density regime will include both the study of complexes formed from larger monomer units and higher order clusters. It will also be important to understand the apparent differences between the behavior of complexes in electronically excited states and the ground state behavior observed in the systems considered here. The more rapid intramolecular vibrational redistribution generally observed in the former case could arise from differences in the nature of the vibrational coupling between the ground and electronically excited states, or simply because of the very different character of the vibrational energy initially excited in these systems. The focus in the ground electronic state, as the systems discussed earlier in this chapter illustrate, has been on high frequency H–X stretches, which one might expect to be rather isolated and thus behave nonstatistically. On the other hand, the electronically excited state studies have concentrated on IVR out of low frequency vibrations which are likely to be more strongly coupled to the bath of vibrational states. The extension of the experimental techniques considered here to larger molecules and lower frequency vibrations would obviously help in closing this gap.

As discussed in detail in other chapters of this volume, an important application of weakly bound complexes is in the steric control of chemical reactions (Shin et al. 1991a,b). In these studies, a molecular complex is used as a precursor to provide some control over the reactant geometry. The reaction is initiated by UV photolysis of one of the monomer constituents. In the case of H–X photolysis, a high energy hydrogen atom is launched at the partner, initiating the chemistry of interest. A possible future direction in this field is to introduce an near infrared step prior to photolysis. In this way, the initial state of the complex can be selected, giving even more control over the initial geometry of the parent molecule. This is now a well known effect in the high resolution spectroscopy of these complexes (Lovejoy and Nesbitt 1989, 1991) and it arises due to the very low anisotropy exhibited by many of these species. The advantage of this approach is that the chemistry can be carried out, not only from a range of geometries, but also for a selected value of the overall angular momentum, determined by the rotational state selection of the infrared step. What the present discussion adds to this type of experiment is the promise of not only being able to control the relative orientation of the reagents, but also to control the overall orientation of the system in the laboratory frame, using the pendular state methods. The combination of these methods could provide us with an approach for determining the differential state-to-state reaction cross sections as a function of the overall angular momentum and reagent approach geometry.

ACKNOWLEDGMENTS: The author is grateful for several generations of students with whom he has had the pleasure of working, and who also have made this work possible. Support for this research is gratefully acknowledged from the National Science Foundation (Grant No. CHE-89-00307) and the Donors of the Petroleum Research Fund (administered by the ACS).

REFERENCES

Alexander, M. H.; DePristo, A. E. 1976 J. Chem. Phys. 65:5009

Alfano, J. C.; Martinez, S. J.; Levy, D. H. 1992 J. Chem. Phys. 96:2522

Aoyama, T.; Yamankawa, H. 1979 Chem. Phys. Lett. 60:326

Barton, A. E.; Howard, B. J. 1982 Faraday Discuss. Chem. Soc. 73:45

Bemish, R. J.; Block, P. A.; Pedersen, L. G.; Yang, W.; Miller, R. E. 1993 J. Chem. Phys. 99:8585

Bemish, R. J.; Bohac, E. J.; Wu, M.; Miller, R. E. 1944a J. Chem. Phys. 101:9451

Bemish, R. J.; Wu, M.; Miller, R. E. 1994b Faraday Discuss. Chem. Soc. 96:97:57

Bernstein, E. R. 1994 see chapter 5 in present volume

Bernstein, E. R.; Law, K.; Schauer, M. 1984 J. Chem. Phys. 80:207

Block, P.; Bohac, E. J.; Miller, R. E. 1992 Phys. Rev. Lett. 68:1303

Bohac, E. J.; Marshall, M. D.; Miller, R. E. 1992a J. Chem. Phys. 97:4890

Bohac, E. J.; Marshall, M. D.; Miller, R. E. 1992b J. Chem. Phys. 97:4901

Bohac, E. J.; Marshall, M. D.; Miller, R. E. 1992c J. Chem. Phys. 96:6681

Bohac, E. J.; Miller, R. E. 1992 J. Chem. Phys. 99:1537

Bohac, E. J.; Miller, R. E. 1993a J. Chem. Phys. 98:2604

Bohac, E. J.; Miller, R. E. 1993b Phys. Rev. Lett. 71:54

Brouard, M.; Simons, J. P.; Wang, J. X. 1991 Faraday Discuss. Chem. Soc. 91:63

Bunker, P. R.; Carrington, T.; Gomez, P. C.; Marshall, M. D.; Kofranek, M.; Lischka, H.; Karpfen, A 1989 J. Chem. Phys. 91:5154

Bunker, P. R.; Jensen, P.; Karpfen, A.; Kofranek, M.; Lischka, H. 1990 J. Chem. Phys. 92:7432

Bunker, P. R.; Kofranek, M.; Lischka, H.; Karpfen, A. 1988 J. Chem. Phys. 89:3002

Burak, I.; Hepburn, J. W.; Sivakumar, N.; Hall, G. E.; Chawla, G.; Houston, P. J. L. 1987 J. Chem. Phys. 86:1258

Butz, K. W.; Catlett, D. L., Jr.; Ewing, G. E.; Krajnovich, D.; Parameter, C. S. 1986 J. Chem. Phys. 90:3533

Casassa, M. P.; Stephenson, J. C.; King, D. S. 1988 NATO ASI Ser., Ser B 171:367

Clary, D. C. 1992 J. Chem. Phys. 96:90

Clary, D. C.; Knowles, P. J. 1990 J. Chem. Phys. 93:6334

Cline, J. I.; Reid, B. P.; Evard, D. D.; Sivakumar, N.; Halberstadt, N.; Janda, K. C. 1988 J. Chem. Phys. 89:3535

Cohen, R. C.; Saykally, R. J. 1991a J. Chem. Phys. 95:7891

Cohen, R. C.; Saykally, R. J. 1991b Annu. Rev. Phys. Chem. 42:369

Cohen, R. C.; Saykally, R. J. 1992 J. Phys. Chem. 96:1024

Copeland, R. A.; Pearson, D. J.; Robinson, J. M.; Crim, F. F. 1982 J. Chem. Phys. 77:3974

Dayton, D. C.; Block, P. A.; Miller, R. E. 1991 J. Chem. Phys. 95:2881

Dayton, D. C.; Jucks, K. W.; Miller, R. E. 1989 J. Chem. Phys. 90:2631

Dayton, D. C.; Miller, R. E. 1988 Chem. Phys. Lett. 143:181

Dayton, D. C.; Miller, R. E. 1989 Chem. Phys. Lett. 156:578

DeLeon, R. L.; Muenter, J. S. 1984 J. Chem. Phys. 80:6092

Dvorak, M. A.; Reeve, S. W.; Burns, W. A.; Grushow, A.; Leopold, K. R. 1991 Chem. Phys. Lett. 185:399

Dyke, T. R. 1984 Top. Curr. Chem. 120:85

Dyke, T. R.; Howard, B. J.; Klemperer, W. 1972 J. Chem. Phys. 56:2442

Evard, D. D.; Bieler, C. R.; Cline, J. I.; Sivakumar, N.; Janda, K. C. 1988 J. Chem. Phys. 89:2829

Ewing, G. E. 1980 J. Chem. Phys. 72:2096

Ewing, G. E. 1987, J. Phys. Chem. 91:4662

Farrel, J. T.; Sneh, O.; Knight, A. E. W.; Nesbitt, D. J. n.d. unpublished

Felker, P. M.; Zewail, A. H. 1985 J. Chem. Phys. 82:2961

Fraser, G. T.; Pine, A. S. 1989a J. Chem. Phys. 91:633

Fraser, G. T.; Pine, A. S. 1989b J. Chem. Phys. 91:3319

Fraser, G. T.; Pine, A. S. Suenram, R. D. 1988 J. Chem. Phys. 88:6157

Friedrich, B.; Herschbach, D. R. 1991 Nature 353:412

Gericke, K. H.; Klee, S; Comes, F. J.; Dixon, R. N. 1986 J. Chem. Phys. 85:4463

Go, J.; Bethardy, G. A., Perry, D. S. 1990 J. Phys. Chem. 94:6153

Gough, T. E.; Miller, R. E.; Scoles, G. 1977 Appl. Phys. Lett. 30:338

Greene, C. H.; Zare, R. N. 1982 Annu Rev. Phys. Chem. 33:119

Gutowsky, H. S.; Chuang, C.; Keen, J. D.; Klots, T. D.; Emilsson, T. 1985 J. Chem. Phys. 83:2070

Halberstadt, N.; Beswick, J. A.; Janda, K. C. 1987 J. Chem. Phys. 87:3966

Halberstadt, N.; Brechignac, P.; Beswick, J. A.; Shapiro, M. 1986 J. Chem. Phys. 84:170

Hall, G. E.; Sivakumar, N.; Chawla, D.; Houston, P. L.; Burak, I. 1988 J. Chem. Phys. 88:3682

Hall, G. E.; Sivakumar, N.; Ogorzalek, R.; Chawla, G.; Haerri, H. P.; Houston, P. L.; Burak, I.;

Hepburn, J. W. 1986 Faraday Discuss. Chem. Soc. 82:13

Hancock, G. C.; Truhler, D. G.; Dykstra, C. E. 1988 J. Chem. Phys. 88:1786

Haughen, H. K.; Pence, W. H.; Leone, S. R. 1984 J. Chem. Phys. 80:1839

Hetzer, J. R.; Casassa, M. P.; King, D. S. 1991 J. Phys. Chem. 95:8086

Himes, J. L.; Wiggins, T. A. 1971 J. Mol. Spectrosc. 40:418

Houghen, J. T.; Ohashi, N. 1985 J. Mol. Spectrosc. 109:134

Hough, A. M.; Howard, B. J. 1987 J. Chem. Soc. Faraday Trans 2 83:173

Houston, P. L. 1987 J. Phys. Chem. 91:5388

Howard, B. J.; Dyke, T. R.; Klemperer, W. 1984 J. Chem. Phys. 81:5417

Huang, Z. S.; Jucks, K. W.; Miller, R. E. 1986a J. Chem. Phys. 85:6905

Huang, Z. S.; Jucks, K. W.; Miller, R. E. 1986b J. Chem. Phys. 85:3338

Huang, Z. S.; Miller, R. E. 1989 J. Chem. Phys. 90:1478

Hutson, J. M. 1988 J. Chem. Phys. 89:4550

Hutson, J. M. 1990 Annu. Rev. Phys. Chem. 41:123

Hutson, J. M. 1992 J. Chem. Phys. 96:6752

Insepov, Z. A.; Karayaev, E. M. 1991 Pisma. Zh. Tekh. Fiz. 17:36

Jensen, P.; Bunker, P. R.; Karpfen, A.; Kofranek, M.; Lischka, H. 1990 J. Chem. Phys. 93:6266

Johnson, K. E.; Sharfin, W.; Levy, D. H. 1981 J. Chem. Phys. 74:163

Jucks, K. W.; Huang, Z. S.; Miller, R. E.; Fraser, G. T.; Pine, A. S.; Lafferty, W. J. 1988 J. Chem. Phys. 88:2185

Jucks, K. W.; Miller, R. E. 1987a J. Chem. Phys. 87:5629

Jucks, K. W.; Miller, R. E. 1987b J. Chem. Phys. 86:6637

Jucks, K. W.; Miller, R. E. 1988a J. Chem. Phys. 88:6059

Jucks, K. W.; Miller, R. E. 1988b J. Chem. Phys. 88:2196

Kleiner, I.; Fraser, G. T.; Hougen, J. T.; Pine, A. S. 1991 J. Mol. Spectrosc. 147:155

Klemperer, W. 1978 Springer Ser. Chem. Phys. 3:398

Klemperer, W. 1987 NATO ASI Ser. Ser. C 212: 455

Kofranek, M.; Lischka, H.; Karpfen, A. 1988 Chem Phys. 121:137

Lafferty, W. J.; Suenram, R. D.; Lovas, F. J. 1987 J. Mol. Spectrosc. 123:434

Levy, D. H. 1983 In *Energy Storage and Redistribution in Molecules*, Hinze, J. ed. Plenum Press, New York, pp. 73–95

Lischka, H. 1979 Chem. Phys. Lett. 66:108

Lisy, J. M.; Tramer, A.; Vernon, M. F.; Lee, Y. T. 1981 J. Chem. Phys. 75:4733

Loesch, H. J.; Remscheid, A. 1990 J. Chem. Phys. 93:4779

Lovejoy, C. M.; Nelson, D. D. Jr.; Nesbitt, D. J. 1987 J. Chem. Phys. 87:5621

Lovejoy, C. M.; Nelson, D. D. Jr.; Nesbitt, D. J. 1988 J. Chem. Phys. 89:7180

Lovejoy, C. M.; Nesbitt, D. J. 1987 J. Chem. Phys. 87:1450

Lovejoy, C. M.; Nesbitt, D. J. 1989 J. Chem. Phys. 91:2790

Lovejoy, C. M.; Nesbitt, D. J. 1990 J. Chem. Phys. 93:5387

Lovejoy, C. M.; Nesbitt, D. J. 1991 J. Chem. Phys. 94:208

Marshall, M. D.; Bohac, E. J.; Miller, R. E. 1992 J. Chem. Phys. 97:3307

McIlroy, A.; Lascola, R.; Lovejoy, C. M.; Nesbitt, D. J. 1991 J. Phys. Chem. 95:2636

Miller, R. E. 1986 J. Phys. Chem. 90:3301

Miller, R. E. 1988 Science 240:447

Miller, R. E. 1989 Acc. Chem. Res. 23:10

Nesbitt, D. J.; Child, M. S.; Clary, D. C. 1989a J. Chem. Phys. 90:4855

Nesbitt, D. J.; Lovejoy, C. M. 1988 Faraday Discuss. Chem. Soc. 86:13

Nesbitt, D. J.; Lovejoy, C. M. 1990 J. Chem. Phys. 93:7716

Nesbitt, D. J., Lovejoy, C. M.; Lindeman, T. G.; O'Neil, S. V.; Clary, D. C. 1989b J. Chem. Phys. 91:722

Osborn, D. L.; Alfano, J. C.; Vandantzig, N.; Levy, D. H. 192 J. Chem. Phys. 97:2276

Outhouse, E. A.; Demmer, D. R.; Leach, G. W.; Wallace, S. C. 1993 J. Chem. Phys. 99:80

Parameter, C. S. 1982 J. Phys. Chem. 86:1735

Parameter, C. S. 1983 Faraday Discuss. Chem. Soc. pp. 7-22

Pine, A. S.; Fraser, G. T. 1988 J. Chem. Phys. 89:6636

Pine, A. S.; Howard, B. J. 1986 J. Chem. Phys. 84:590

Pine, A. S.; Lafferty, W. J. 1983 J. Chem. Phys. 78:2154

Pine, A. S.; Lafferty, W. J.; Howard, B. J. 1984 J. Chem. Phys. 81:2939

Pique, J.; Chen, Y.; Jonas, D. M.; Lundberg, J. K.; Hamilton, C. E.; Adamson, G. W.; Silbey, R. J.; Field, R. W. 1989 AIP Conf. Proc. 1988 191, 673

Puttkamer, K. V.; Quack, M. 1986 Faraday Discuss. Chem. Soc. 82:377

Puttkamer, K. V.; Quack, M. 1987a Proceedings of the 10th International Conference on Molecular Energy Transfer COMET 10:195

Puttkamer, K. V.; Quack, M. 1987b Mol. Phys. 62:1047

Puttkamer, K. V.; Quack, M. 1989 Chem. Phys. 139:31

Puttkamer, K. V.; Quack, M.; Suhm, M. A. 1989 Infrared Phys. 29:535

Puttkamer, K. V.; Quack, M.; Suhm, M. A. 1991 Mol. Phys. 198:1025

Quack, M.; Suhm, M. A. 1991 J. Chem. Phys. 95:28

Randall, R. W.; Walsh, M. A.; Howard, B. J. 1988 Faraday Discuss. Chem. Soc. 85:13

Rost, J. M.; Griffin, J. C.; Friedrich, B.; Herschbach, D. R. 1992 Phys. Rev. Lett. 68:1299

Rybak, S.; Jeziorski, B.; Szalewicz, K. 1991 J. Chem. Phys. 95:6576

Sharpe, S. W.; Reifschneider, D.; Wittig, C.; Beaudet, R. A. 1991 J. Chem. Phys. 94:233

Sharpe, S. W.; Sheeks, R.; Wittig, C.; Beaudet, R. A. 1988 Chem. Phys. Lett. 151:267

Shin, S. K.; Chen, Y.; Nickolaisen, S.; Sharpe, S. W.; Beaudet, R. A.; Wittig, C. 1991a In *Advances in Photochemistry*, Vol. 16, D. Volman, G. Hammond, and D. Neckers, eds. Wiley, New York, p. 249

Shin, S. K.; Wittig, C.; Goddard, W. A. 1991b J. Phys. Chem. 95:8048

Shorter, J. H.; Casassa, M. P.; King, D. S. 1992 J. Chem. Phys. 97:1824

Smalley, R. E. 1982 J. Phys. Chem. 86:3504

Steed, J. M.; Dixon, T. A.; Klemperer, W. 1979 J. Chem. Phys. 70:4095

Stuchebrukhov, A. A.; Marcus, R. A. 1993 J. Chem. Phys. 98:6044

Sun, H.; Watts, R. O. 1990 J. Chem. Phys. 92:603

Uzer, T. 1991 Phys. Rep. 2:73

Vasudev, R.; Zare, R. N.; Dixon, R. N. 1984 J. Chem. Phys. 80:4863

Vernon, M. F.; Lisy, J. M.; Krajnovich, D. J.; Tramer, A.; Kwok, H. S.; Shen, Y. R.; Lee, Y. T. 1982 Faraday Discuss. Chem. Soc. 73:387

Villarreal, P.; Miretartes, S.; Roncero, O.; Delgadobarrio, G.; Beswick, J. A.; Halberstadt, N.; Coalson, R. D. 1991 J. Chem. Phys. 94:4230

Vohralik, P. F., Miller, R. E. 1985 J. Chem. Phys. 83:1609

Wardlaw, D. M.; Marcus, R. A. 1987 Adv. Chem. Phys. 70:231

Waterland, R. L.; Skene, J. M.; Lester, M. I. 1988 J. Chem. Phys. 89:7277

Wu, M.; Bemish, R. J.; Miller, R. E. 1994 J. Chem. Phys. 101:9447

Yang, S.; Bersohn, R. 1974 J. Chem. Phys. 61:4400

Yarkony, D. R.; O'Neil, S. V.; Schaefer, H. F.; Baskin, C. P.; Bender, C. F. 1974 J. Chem. Phys. 60:855

Zare, R. N. 1982 Ber. Bunsenges. Phys. Chem. 86:422

Zhang, D. H.; Zhang, J. Z. H. 1993 J. Chem. Phys. 99:6624

Zhang, D. H.; Zhang, J. Z. H.; Bacic, Z. 1992a J. Chem. Phys. 97:3149

Zhang, D. H.; Zhang, J. Z. H.; Bacic, Z. 1992b J. Chem. Phys. 97:927

Zhang, D. H.; Zhang, J. Z. H.; Bacic, Z. 1992c Chem. Phys. Lett. 194:3113

3

Dynamics of Ground State Bimolecular Reactions

CURT WITTIG AND AHMED H. ZEWAIL

3.1. INTRODUCTION

During the past decade, the study of photoinitiated reactive and inelastic processes within weakly bound gaseous complexes has evolved into an active area of research in the field of chemical physics. Such specialized microscopic environments offer a number of unique opportunities which enable scientists to examine regiospecific interactions at a level of detail and precision that invites rigorous comparisons between experiment and theory. Specifically, many issues that lie at the heart of physical chemistry, such as reaction probabilities, chemical branching ratios, rates and dynamics of elementary chemical processes, curve crossings, caging, recombination, vibrational redistribution and predissociation, etc., can be studied at the state-to-state level and in real time.

Inevitably, understanding the photophysics and photochemistry of weakly bound complexes lends insight into corresponding processes in less rarefied surroundings, for example, molecules physisorbed on crystalline insulator and metal surfaces, molecules residing on the surfaces of various ices, and molecules weakly solvated in liquids. However, such ties to the real world are not the main driving force behind studies of photoinitiated reactions in complexed gaseous media. Rather, it is the lure of going a step beyond the more common molecular environments. Theoretical modeling, which in many areas purports to challenge experiment, must rise to the occasion here if it is to offer predictive capability for even the simplest of such microcosms. Subtleties abound.

Roughly speaking, two disparate regimes can be identified which are accessible experimentally and which correspond to qualitatively different kinds of chemical transformations. These are distinguished by their reactants: electronically excited versus ground state. For example, it is possible to study the chemical selectivity that derives from the alignment and orientation of excited electronic orbitals, albeit at restricted sets of nuclear coordinates. This is achieved by electronically exciting a complexed moiety, such as a metal atom, which then undergoes chemical transformations that depend on the geometric properties of the electronic orbitals such as their alignments and orientations relative to the other moiety (or moieties) in the complex. This approach was pioneered by Soep and coworkers (Boivineau et al. 1986a,b; Breckenridge 1989; Breckenridge et al. 1985, 1986 1987; Duval et al. 1985, 1986, 1991; Jouvet et al. 1987, 1989; Jouvet and Soep 1983, 1984, 1985; Loison et al. 1994; Soep 1994), who examined several important prototypical systems in detail by using complementary experimental methods: measurements of excited state spectral properties, product state distributions, and, more recently, time domain studies (Soep 1994).

Alternatively, a number of efforts have focused on systems that involve ground electronic state reactants, or, more precisely, systems in which the photoexcited moiety is known to evolve to ground electronic state fragments in the absence of complexation. For example, HX photodissociation has been used to prepare both hydrogen and halogen atom reactants on time scales which are short, relative to those of subsequent dynamical processes. In many of these cases, it has been possible to infer properties of the corresponding gas phase reactions. This review deals exclusively with this latter class of photoexcited reactions in neutral complexes, namely, those which parallel reactions that occur via the ground potential energy surface (PES), which we take to be the adiabatic one. Also, *binary* complexes will be stressed, though higher-than-binary complexes inevitably enter the picture. In many cases, the geometric properties of the complexes under consideration, such as equilibrium geometries and zero point excursions, are known from spectroscopic studies or can be obtained from theory with good reliability.

To focus the review, it is necessary to pass over some topics which are timely, relevant, and could be included were it not for space limitations. One of these is the area of electron photodetachment as applied to the study of the Franck–Condon region of the neutral PES accessed from the corresponding anion. This approach has been shown to provide information about the transition state region of several bimolecular reactions. This work was pioneered by Neumark and coworkers and an excellent review is already available (Manolopoulos et al. 1993; Neumark 1992). Another noteworthy area is state-to-state studies of vibrational predissociation in weakly bound complexes. Miller and coworkers have made impressive advances in which fully state and angle-resolved product distributions have been obtained (Bemish et al. 1994; Block et al. 1992; Bohac et al. 1986, 1992a,b; Bohac and Miller 1993a,b; Dayton et al. 1989), and these results have been used to bring theory and experiment into accord. The present review is limited to cases in which ultraviolet photodissociation of a complexed moiety initiates reaction.

In writing this review, it is our intention to provide an assessment of where the field stands, as well as where we see it going on the fronts most familiar to the authors. The document is not a compendium, and, as stated previously, some work had to be omitted because of the demand for brevity. Like most areas of science, progress has not been distributed homogeneously in time, nor will it be in the future. Following several early "demonstration of principle" achievements, results have flowed at a steady but modest pace. However, we believe that this is about to change, since a number of milestones have been passed in the last few years. Consequently, we envision an explosion of results that will propel this field to the same level of sophistication that presently exists for studies of stable molecules and radicals.

3.2. PHOTOINITIATED REACTIONS IN CO_2–HX COMPLEXES

3.2.1. Counterparts under Gas Phase, Single-Collision Conditions

The gas phase reaction considered to be a close counterpart to the photoinitiated reaction that occurs in CO_2–HX complexes is:

$$H + CO_2 \rightarrow HOCO^\dagger \rightarrow OH + CO \qquad \Delta H = 8960 \text{ cm}^{-1}, \qquad (1)$$

where the vibrationally excited $HOCO^\dagger$ intermediate decomposes via a unimolecular mechanism (Smith 1980). Note that reaction (1) is written for the *endoergic direction*, i.e., paralleling the reaction that is photoinitiated within weakly bound complexes. On the other hand, the *exoergic* direction is of considerable technical importance, since it is responsible for the oxidation of CO during the combustion of hydrocarbon fuels (Baulch et al. 1984; Davis et al. 1974; Gardiner 1977; Jonah et al. 1984; Mozurkewich et al. 1984a,b; Ravishankara and Thompson 1983; Warnatz 1984). This technical significance accounts in part for the large number of detailed experimental and theoretical studies that have been carried out during the last few decades with this prototypical elementary reaction.

Smith and coworkers have contributed greatly to our understanding of this system through a number of experimental and theoretical contributions, most notably comparing the results of detailed measurements of k_{-1}, that is, the rate constant for the reverse of reaction (1), to predictions made by using them oretical models based on transition state theories (Brunning et al. 1988; Smith 1977, 1980; Smith and Zellner 1973; Zellner and Steinert 1976). This work commenced in the early 1970s and stands as a landmark contribution. They were able to identify details of the potential energy surface through analyses of the various measurements of $k_{-1}(T)$, which span the temperature range $80 \leq T \leq 2000$ K (Frost et al. 1991a, 1993). Specifically, it was shown that the markedly non-Arrhenius temperature dependence shown in Figure 3-1 could be reconciled by using RRKM theory with barriers that have comparable energies for the two chemically distinct $HOCO^\dagger$ decomposition channels. Also, by measuring rates at temperatures down to 80 K (Frost et al. 1991a, 1993), they were able to demonstrate unambiguously that there is, at most, a minuscule barrier to the process whereby OH and CO combine to form the $HOCO^\dagger$ intermediate. Furthermore, they were able to discern qualitative features of the transition state leading to $H + CO_2$ by measuring the CO_2 vibrational distribution and invoking the reasonable assumption of minimal vibrational energy transfer once the reaction has progressed past the transition state (Frost et al. 1991b).

A full six-dimensional PES for the HOCO system has been developed by Schatz and coworkers, particularly L. B. Harding (Kudla et al. 1992; Schatz et al. 1987). This proved to be challenging because of numerous local minima and transition states, and consequently the development of the PES has taken several years. Many points were calculated by using large scale *ab initio* techniques and the surface was adjusted to reconcile a broad array of experimental data such as nascent product excitations, $HOCO^\dagger$ decomposition rates, overall bimolecular reaction rates, barrier heights, enthalpy changes, HOCO structural properties, and inelastic scattering data. This PES has been used in several computational studies of the reaction dynamics that employ classical (Kudla and Schatz 1991; Kudla

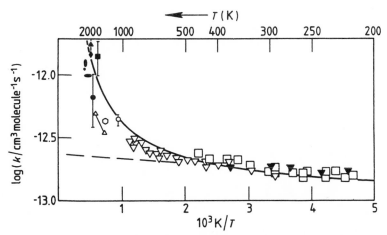

Figure 3-1. Rate constants $k(T)$ for the range $200 \leq T \leq 2000$ K, taken from Smith (1980). Note the decidedly non-Arrhenius behavior. The data below 1000 K are those of Smith and Zellner (1973) (open squares), Davis et al. (1974) (filled triangles), and Zellner and Steinert (1976) (open triangles); points above 1000 K are from the review by Warmatz. The full line shows the result of TST calculations (Smith 1980), while the dashed line is an extrapolation of the Arrhenius expression chosen by Davis et al. (1974) to represent their data in the range 220–373 K. Below 200 K, $k(T)$ remains constant until ~ 140 K; thereafter it drops to $\sim 1 \times 10^{-13}$ at 80 K (Frost et al. 1991a, 1993).

et al. 1991, 1992; Schatz 1989) and/or quantum mechanical (Clary and Schatz 1993; Hernández and Clary 1994; Schatz and Dyck 1992) methods. Despite its inherent complexities, this surface is one of the best available for a four-atom system. It can reproduce most of the experimental data on reactive and inelastic processes, at least qualitatively. The diagram given in Figure 3-2 shows the low energy reaction path.

The OH + CO → H + CO_2 reaction has also been examined experimentally by using the crossed molecular beams technique. Casavecchia and coworkers developed an intense molecular beam source of OH radicals which was crossed with a beam of CO in a standard crossed beams arrangement (Alagia et al. 1993). These measurements provided information on product translational and internal excitations for several center-of-mass (c.m.) collision energies. Moreover, it was possible to estimate reaction rates by measuring the angular distributions of products in the c.m. system and comparing these experimental observations with estimates made by using simple models to calculate the rotational periods of the HOCO† intermediate. An example of the scattering anisotropy in the c.m. system is shown in Figure 3-3 (Alagia et al. 1993). Although this is an indirect measure of the lifetime of the intermediate complex, the rates thus obtained are consistent with the measurements discussed below that employ the subpicosecond resolution pump–probe technique, albeit with weakly bound complexes as precursors.

Under 300 K ambient conditions, reactions of photolytically produced fast hydrogen atoms with CO_2 are known to yield OH + CO products via vibrationally excited HOCO† intermediates. Since the initial report by Oldershaw and Porter (1969), many groups have studied reaction (1) by using fast hydrogen atoms and

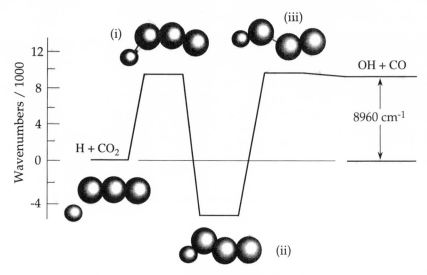

Figure 3-2. Potential energy relative to $H + CO_2$ for selected points on the HOCO PES; from left to right: (i) the transition state for $HOCO \leftrightarrow H + CO_2$, (ii) *trans*-HOCO (*cis*-HOCO and the small *cis–trans* conversion barrier are not shown), and (iii) the transition state for $HOCO \leftrightarrow OH + CO$; from the low temperature rate measurements, this transition state is known to be very loose (Frost et al. 1991a, 1993).

single-collision, arrested-relaxation conditions. Measured quantities include nascent OH and CO internal excitations, product c.m. translational excitations, and reaction cross sections. Moreover, these have been measured as a function of the c.m. collision energy. Wolfrum and coworkers reported absolute cross sections for several collision energies, thereby calibrating the relative cross sections (Jacobs et al. 1989). Additionally, they found little, if any, spatial anisotropy of the OH

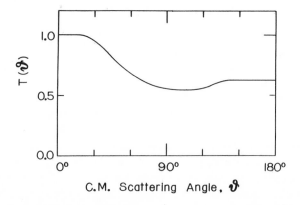

Figure 3-3. Forward/backward asymmetry in the c.m. system for the $H + CO_2$ channel, from the crossed molecular beams experiments of Casavecchia and coworkers (Alagia et al. 1993). This asymmetry can be related to the lifetime of the HOCO† intermediate.

product, in accord with the participation of the HOCO[†] intermediate (Jacobs et al. 1994). Most of the work involving hot atom $H + CO_2$ reactions has been reviewed recently (Chen et al. 1992; Shin et al. 1991a) and little has accured in the meantime. Thus, it will be summarized briefly in subsection 3.2.3.

With CO_2-HX complexes, the photoinitiated reactions are believed to parallel their uncomplexed counterparts since the halogen does not play a dominant role. Specifically, in a minority of cases Br–C attraction can lead to a short-lived HO(Br)CO[†] intermediate, as discussed below. Thus, although the photochemistry of CO_2-HX complexes involves five nuclei, it is still meaningful to make comparisons with the gas phase process depicted in reaction (1).

3.2.2. Geometric Properties of CO_2-HX Complexes

For a given precursor complex, it is most desirable to have a secure knowledge of its spectroscopic and geometric properties before embarking on studies of its photochemistry. Unfortunately, such information is not always available. In fact, some of the more attractive systems from the perspective of photoinitiated reactions in complexes have been viewed as relatively uninteresting by high resolution spectroscopists who specialize in determining and understanding structure and bonding of weakly bound complexes. As a group, they have preferred to work with light elements and simpler systems, since these are amenable to a fairly high level of theory.

On the other hand, a natural inclination of experimentalists trying to use photodissociation as a means of preparing reactants is to exploit convenient ultraviolet wavelengths. This prejudices one toward molecules that contain heavy elements. Presently, the most feasible experiments in which hydrogen atoms are required involve photoinitiated reactions in weakly bound complexes that contain molecules like HI and HBr. For example, one of our favorite sources of atomic hydrogen is HI, since direct, one-photon photodissociation occurs via a broad continuum that extends from the vacuum ultraviolet to ~ 280 nm. Hydrogen bromide runs second, while HCl is barely accessible at 193 nm and HF absorbs only in the vacuum ultraviolet. This order of preference will change in the future as double resonance photodissociation of HX moieties within complexes gains acceptance, but for the time being the ultral violet absorption spectra of the constituents of weakly bound complexes is an important, practical consideration.

The best means of determining the geometric properties of weakly bound complexes is high resolution spectroscopy in combination with theoretical models that treat explicitly the large amplitude motions. However, in some cases structural properties can also be determined with acceptable accuracy by using theory alone. Specifically, semiempirical methods based on the concepts of classical electrostatics are known to give reliable results, as has been shown, for example, by Dykstra (1990) and coworkers. In most cases, these semiempirical approaches are at least as accurate as the brute force application of large scale *ab initio* methods to calculate the weak interaction directly. However, even with these semiempirical approaches there is a reluctance to go to the heavier elements because accurate electronic structure calculations are usually needed to provide properties such as polarizabilities. Nonetheless, we believe that it is possible to make reasonable estimates of structural properties in systems for which there are no (or limited) experimental data.

The geometric properties of weakly bound CO_2–HF and CO_2–HCl complexes have been established from high resolution spectroscopic studies (Altman et al. 1982; Baiocchi et al. 1981; Lovejoy et al. 1987; Nesbitt 1988; Nesbitt and Lovejoy 1990, 1992; Shea et al. 1983). They are both quasilinear, exhibiting hinge-like, large amplitude zero point excursions for the lowest frequency bending mode. For example, Shea et al. (1983) examined the rotational Zeeman effect in CO_2–HF and CO_2–HCl by using a pulsed Fourier transform microwave spectrometer. Zero field rotational spectra of $O^{13}CO$–HF(HCl) and $^{18}OC^{18}O$–HF(HCl) were recorded, and it was found that the expectation values of the angles between the a-axis and the CO_2 and HF axes were $\langle \chi \rangle \sim 12°$ and $24°$, respectively. Note that these values are half the *full* angular widths. Thus, it is clear that large amplitude hinge-like motion is present, with a factor of approximately two difference in the $\langle \chi \rangle$ values for CO_2 and HF. Furthermore, since the dominant intermolecular attractive force for these complexes is hydrogen bonding, CO_2–HCl is floppier than CO_2–HF.

On the other hand, CO_2–HBr was found to be different. The bromine faces the carbon with a Br–C–O angle of approximately $90°$, in what we refer to as an inertially T-shaped structure (Sharpe et al. 1990; Zeng et al. 1992). Unfortunately, the accuracy of our measurements was insufficient to locate the hydrogen. It is reasonable for it to be localized near the oxygen, even though hydrogen bonding is not the *dominant* interaction.

This qualitative change in going from CO_2–HCl to CO_2–HBr was attributed to the presence of two attractive interactions that compete to determine the equilibrium geometry: hydrogen bonding versus polarization of the halogen electron density by the substantial CO_2 quadrupole moment. The latter interaction dominates in CO_2–HBr, though from electronic structure calculations we concluded that the competition is close (Zeng et al. 1992). Unfortunately, no spectroscopic evidence has ever been reported which supports the existence of a quasilinear CO_2–HBr isomer.

This competition between weak hydrogen bonding and polarization of the bromine depends sensitively on molecular properties. For example, consider the case of SCO–HX complexes, for which both SCO–HF and SCO–HCl have been found to be linear as shown; that is, the HX is hydrogen-bonded to the oxygen (Altman et al. 1982; Baiocchi et al. 1981; Fraser et al. 1989; Legon and Willoughby 1985; Shea et al. 1983). These are analogous to CO_2–HF and CO_2–HCl. However, it has been discovered recently that SCO–HBr is *also* linear as shown (Hu and Sharpe 1994; Walker et al. submitted), in contrast to inertially T-shaped CO_2–HBr. This qualitative structural difference underscores both the subtle competitions that occur in such systems and the need for experimental structural determinations.

For CO_2–HBr, rotation about the a-axis, which lies close to a line connecting the carbon and bromine nuclei, is analogous to the rotation of free CO_2. However, it was found that the oxygen atoms do not occupy equivalent sites in the complex; that is, the hydrogen stays near one oxygen for the time scale of the spectroscopic observation. Specifically, if the wavefunction of the complex possesses C_{2v}

symmetry, Bose–Einstein statistics require that all odd K_a values are missing in the lower state, as has been observed with CO_2/rare gas complexes (Randall et al. 1988; Sharpe et al. 1988). This is definitely not the case. Both even and odd K_a states are found in the lower and upper states, indicating that the H atom probability density is not symmetric about the C_{2v} axis on the time scale of the measurement. We interpret this as an indication that the H atom is probably localized near one of the oxygen atoms.

By exploiting the higher resolution available with microwave spectroscopy, it was possible to locate the HBr axis relative to the a-axis and it was found that this angle has an average value of approximately 100° (Rice et al. 1995). However, by itself, the measurement is *also* consistent with an average H–Br–C angle of approximately 80°. In principle, if the zero point amplitude is very large, both equilibrium angles could prove to be consistent with the photoinitiated reaction studies. By obtaining slices on the intermolecular potential surface, it was possible to calculate approximate wavefunctions for some of the intermolecular degrees of freedom (Zeng et al. 1992), and this confirmed the large amplitude hydrogen motion.

From the above considerations, it follows that CO_2–HI will most probably be inertially T-shaped, with at least as large a hydrogen zero point amplitude as for the case of CO_2–HBr. There is no fundamental reason why the structure of this complex has not yet been measured. We attempted this once by using high resolution infrared tunable diode laser spectroscopy and confirmed the inertially T-shaped character (Lin et al. n.d.). However, before accurate rotational constants could be obtained, the HI ruined the vacuum pump, as it is prone to do. This lessened our appetite for a more precise structural determination.

3.2.3. Frequency Domain Studies

Since photoinitiated reactions in CO_2–HX complexes have been studied more than those in any other weakly bound complex, they will be one of our main points of concentration. The first studies of this system examined the disposal of the available energy into product degrees of freedom, as well as a steric effect; specifically, how the reaction probability changes for end-on versus broadside hydrogen approaches (Shin et al. 1990). This was followed by time domain studies in which the close proximity of the complexed molecules provided a well defined $t = 0$ marker (Scherer et al. 1987, 1990). Although this concept is universal, it is especially appropriate when photodissociation yields atomic hydrogen, with its high characteristic speed. Here, we review the status of frequency domain measurements made between 1985 and the present, before moving to the more recent time domain studies.

As stated above, high resolution spectroscopic studies indicate that a qualitative change occurs in going from the quasilinear complexes containing HF and HCl to inertially T-shaped CO_2–HBr, and it is safe to assume that CO_2–HI is also inertially T-shaped. Because HF forms very directional hydrogen bonds with Lewis bases, complexes such as CO_2–HF are especially attractive. However, there is a serious experimental complication: HF absorption lies deep in the vacuum ultraviolet, peaking near 125 nm (Lee 1985). Thus, its use in studies of weakly bound complexes will likely be restricted to double resonance photodissociation, as described below. Nonetheless, we predict that it will play a prominent role in future research.

a. Nascent $OH(X^2\Pi)$ excitations

For the CO_2–HX systems under consideration, every measurement to date of nascent OH excitations obtained under complexed conditions has shown that there is less OH internal energy than for the case of the corresponding gas phase reaction. This is to be expected, on the basis of the naive argument that complexed systems have larger heat capacities than their uncomplexed gas phase counterparts. Consequently, if statistics prevails, OH deriving from complexes will have less internal energy than OH deriving from the gas phase reaction (1) at the same total energy.

On the basis of simple dynamical and kinematic considerations, as well as a minimal entropy analysis of the product state distributions, we concluded that $HOCO^\dagger$ intermediates formed via complexes had less internal energy than those formed via the corresponding gas phase reaction (Wittig et al. 1988a,b). This was easy to rationalize. One obvious factor is the squeezed atom effect (Wittig et al. 1988a). Here, the photoinitiated transfer of the hydrogen atom from the halogen to CO_2 results in recoil between the HOCO and halogen atom products. Specifically, in the process of being transferred from the halogen to CO_2, the light hydrogen pushes against its heavier surroundings, and the resulting increase in X–HOCO recoil comes at the expense of HOCO internal excitation. An analogous effect has been observed in clusters of HBr with Ar, where the recoil energy was observed to be $\sim 10\%$ of the total available energy (Segall et al. 1993). Furthermore, this is consistent with theoretical predictions (Alimi and Gerber 1990; Garcia-Vela et al. 1991, 1992).

Additionally, we speculated that the lowest OH internal excitations (i.e., $v = 0$, low N) might derive preferentially from higher-than-binary complexes and/or a mechanism that involves a five-atom $HO(Br)CO$ intermediate produced by a multicenter process following photoexcitation (Hoffmann et al. 1990).

Because of the entrance channel regiospecificity, the question has often been asked: is it possible that the different OH level distributions relative to the gas phase are simply due to the restricted set of hydrogen approach angles and impact parameters? We think not, because this would require the $HOCO^\dagger$ intermediate to behave quite nonstatistically, and under the present experimental conditions this seems unlikely. This will be discussed further in subsection 3.2.4.

b. Higher-than-binary complexes

Over the years, several experiments were carried out to address the issue of higher complexes. Originally, mass spectrometer signals were recorded simultaneously with the OH LIF signals, while the experimental conditions (i.e., signal strengths) were varied (Radhakrishnan et al. 1986). It was found that the CO_2HBr^+ mass spectrometer signal varied linearly with the OH LIF signal. On the other hand, signals from $(CO_2)_2HBr^+$ and $CO_2(HBr)_2^+$ varied faster than linearly with the OH LIF signal. This was taken as support for the contention that the OH yield derived mainly from binary complexes. However, this experiment did not provide *proof* because of the unclear relationship between the concentration of a given neutral cluster and the mass spectrometer signal that appears at that mass. For example, does CO_2HBr^+ derive from CO_2-HBr or from higher neutral clusters, which are known to crack into smaller units following electron impact ionization? Because of this ambiguity, it is possible that the CO_2HBr^+ signal is not a monitor of CO_2HBr, but is some unknown function of the cluster size distribution. Later, it was shown that the CO_2HBr^+ signal varied linearly with the CO_2-HBr concentration (Shin et al. 1990). This was determined by monitoring CO_2HBr^+ signals while measuring relative CO_2-HBr concentrations using high resolution tunable diode laser spectroscopy with a pulsed slit expansion. Again, this evidence is strongly suggestive but not conclusive. For example, if neutral trimers yield a large CO_2HBr^+ signal and if their concentration tracks that of CO_2-HBr, then their contribution to the OH LIF signal cannot be discerned by measuring the variation of the CO_2HBr^+ signal versus the OH LIF signal. Another caveat is the possibility that the OH yield for one or more of the higher clusters is significantly higher than that for binary clusters. Though there is no evidence suggesting that this is the case, it is best to leave open the possibility.

From the above considerations, we conclude that isolating contributions from clusters of different sizes is a task of serious proportions. A foolproof diagnostic is called for, and this leads inevitably to methods that enable weakly bound complexes to be size selected. Without reliable size selection, every case will be "special" and progress will accrue slowly, since many checks will be required to build a case based on circumstantial evidence.

Ultimately, we would like to *choose* different complexes (dimers, trimers, tetramers, etc.), but given the present state of affairs, isolating contributions from binary complexes will suffice, at least for the time being. Because of its paramount importance, this issue will be discussed further and proposed solutions will be outlined in section 3.4, which is devoted to prospects for future research. Of the two approaches that are most viable, that is, molecular beam deflection (Buck 1994) and laser-based double-resonance methods, the former is superior for larger clusters while the latter is superior for binary complexes, as well as offering state selection. Consequently, we prefer the latter for the near future. Specifically, an infrared laser can be used to excite HX moieties within binary complexes, thereby tagging the complexes via the spectral resolution that enables them to be distinguished from monomers, trimers, etc. These vibrationally excited HX moieties are then photodissociated at wavelengths that leave the untagged complexes unaffected. In this way, weakly bound complexes are both size selected, and, in many cases, state selected. This method is depicted schematically in Figure 3-4. It is our judgment that the double resonance method is well suited to a broad range of systems and will yield qualitative advances. Furthermore, the approach is applicable to ultrafast as well as nanosecond techniques. Such experiments are within reach using technology that is essentially off-the-shelf, and examples will be given of recent experimental results that establish the viability of this approach.

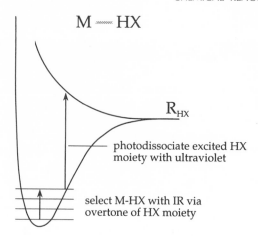

Figure 3-4. Schematic of size and state selection of weakly bound complexes. Weak M–HX intermolecular coupling and the corresponding long predissociation lifetimes enable ro-vibrational levels of the complex to be isolated. Ultraviolet photons at sufficiently long wavelengths act on the excited (but not the unexcited) HX moieties.

3.2.4.   Time Domain Studies

In 1987, the picosecond-resolution pump–probe technique was applied to the CO_2-HI system, in which reaction is photoinitiated by ultraviolet photodissociation of the HI moiety (Scherer et al. 1987). This was the first case in which the $t = 0$ clocking method, which had been developed earlier for the study of photofragmentation dynamics (Zewail 1988, 1993), was applied to the study of a *bimolecular* reaction, albeit in the unique environment of a complex. A schematic drawing of the concept is given in Figure 3-5. This experiment demonstrated unambiguously that reaction was not direct, that is, it occurred via an intermediate having lifetimes in the picosecond regime. Furthermore, these lifetimes depend on the energy of the photon used to initiate the reaction. The pump and probe pulses were both of several picoseconds duration, and the probe was of sufficiently narrow linewidth that individual OH rotational levels could be resolved. Specifically, data were collected for $N = 1$ and 6. Despite the low photon count rates (typically less than one per laser firing) and the difficulty of the measurements, it was demonstrated

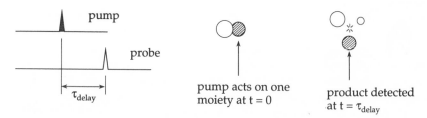

Figure 3-5. Schematic of ultrafast pump–probe method applied to photoinitiated reactions in binary complexes.

that such experiments were indeed possible. Consequently, this result spawned further efforts along similar lines.

Subsequent experiments by Ionov et al. (1992, 1993b) were carried out with better time resolution, enabling OH buildup to be determined with subpicosecond accuracy. These rates were in accord with predictions made by using RRKM theory, where it was assumed that the nearby halogen atom plays no role other than to lower the $HOCO^\dagger$ energy by a small amount as per the squeezed atom effect (Wittig et al. 1988a). The parameters used in the RRKM calculations were those obtained by Smith and coworkers from extensive modeling of $k_{-1}(T)$ for the overall $OH + CO \rightarrow H + CO_2$ reaction (Brunning et al. 1988; Smith 1977; Smith and Zellner 1973; Zellner and Steinert 1976). Because of the uncertainty of the $HOCO^\dagger$ energy, as well as the likelihood of a distribution of $HOCO^\dagger$ energies for a single photolysis wavelength, it was not possible to make an exact comparison between experiment and theory. However, despite the latitude in assigning $E^\dagger$ values, agreement was claimed. At the very least, there was no serious *disagreement*.

The time resolved OH buildup measurements all confirmed a unimolecular decomposition mechanism for the photoinitiated reaction in weakly bound CO_2–HI complexes. Though the Ionov et al. (1992, 1993b) rates are larger than those reported by Scherer et al. (1990) (see Figure 3-6), the experiments differ in important respects. For example, in the Ionov et al. experiments, a broadband probe monitors essentially all of the OH rotational levels simultaneously. Therefore, these experiments are sensitive to the average behavior of all processes that yield OH. On the other hand, the Scherer et al. experiments were carried out with OH rotational state resolution, which enabled them to show that the buildup rates for the $N = 1$ rotational level were slower than those for $N = 6$ by as much as 70%. This is noteworthy because unimolecular rate theories predict that all products are produced with the same rate for a microcanonical system undergoing unimolecular decomposition. Thus, one might ask whether or not: (1) the system is behaving nonstatistically, (2) $HOCO^\dagger$ intermediates are produced having significantly different $E^\dagger$ values, that is, via different mechanisms, and (3) other sources of OH are involved, for example, $HOBr^\dagger$. Here, we discuss these possibilities.

It has been known for many years that in low dimensional systems undergoing unimolecular decomposition, reaction rates fluctuate about a mean value which is generally taken to be the statistical one, such as the RRKM rate (Green et al. 1991, 1992; Hernandez and Miller 1993; Manthe and Miller 1993; Polik et al. 1988, 1990a,b). Likewise, product state distributions display irregularities that can be attributed to both the interferences that occur between the different decay paths (i.e., through the different transition state levels) as well as the rapid changes in the product state distributions that are manifestations of so-called chaotic behavior of the parent in the small molecule limit. For example, in the case of NO_2, it has been shown that product state distributions appear to be statistical when sufficient averaging is done over the decaying parent resonances, while with good parent energy resolution, the product level distributions fluctuate as per the mappings and interferences of transition state levels projected onto the product degrees of freedom (Hunter et al. 1993; Peskin et al. 1994; Reid et al. 1993a,b; Reid

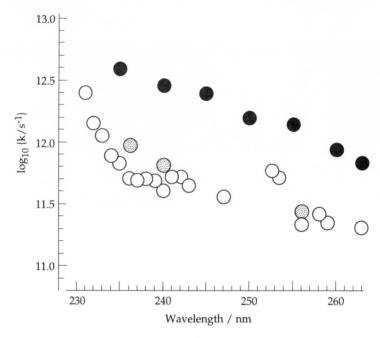

Figure 3-6. Rate coefficients $k(E)$ for HOCO† decomposition. The filled circles are from Ionov et al. (1992, 1993b), who monitored all OH $(v = 0)$ rotational levels. The open and shaded circles are from Scherer et al. (1990) (i.e., values of τ_2^{-1}) for $N = 1$ and $N = 6$, respectively.

and Reisler 1994; Reisler et al. 1994). Likewise, Moore and coworkers have observed fluctuations in $k(E)$ for the case of H_2CO unimolecular decomposition, again related to the sparseness of the manifold of parent energy levels (Hernandez et al. 1993; Miller et al. 1990).

It must be emphasized that such phenomena are to be expected for a statistical system only in the regime of low level densities. Theories like RRKM and phase space theory (PST) (Pechukas and Light 1965) are applicable when such quantum fluctuations are absent: for example, due to a large density of states and/or averaging over experimental parameter such as parent rotational levels in the case of incomplete expansion-cooling and/or the laser linewidth in ultrafast experiments. However, in the present case, it is unlikely that such phenomena can be invoked to explain why different rates are obtained when using ultrafast pump–probe methods that differ only in experimental detail.

The observed differences in the OH buildup rates at a given photolysis wavelength can be attributed to OH rotational levels having different parentages. For example, if higher-than-binary complexes yield mainly low N, this could be reflected in the product-state-selective measurements, even if the number of higher complexes is smaller than the number of binary complexes. In this case, there would be differences between results obtained with product state resolution versus those obtained by using a broadband probe. The latter would only be sensitive to the low N contribution in direct proportion to its fraction of the total product yield, whereas spectral resolution enables the low N component to be isolated. A hypothetical situation is depicted in Figure 3-7 for two production mechanisms that yield cold and hot rotational distributions. Note that although the filled-bars contribution accounts for only 25% of the total population, it accounts for 80% of the $N = 1$ signal. Clearly, product spectral resolution offers a valuable addition to time domain studies when both can be used simultaneously.

To test for the possible participation of higher-than-binary complexes, Ionov et al. (1992, 1993b) recorded OH buildup times for different expansion conditions, that is, covering the range from modest to considerable clustering. The observed buildup times for a given photolysis wavelength did not change, even when clustering was so severe that the OH LIF signal level dropped. Our interpretation is that higher-than-binary complexes inhibit reaction and/or yield the same rates as do binary complexes. Inhibition is deemed most likely, since the reaction is quite endoergic and the gas phase cross section is known to be small near the thermodynamic threshold, increasing sharply only at significantly higher energies, as shown in Figure 3-8 (Chen et al. 1989). With its high entrance barrier, the $H + CO_2$ system is somewhat unique in this regard. In general, it is expected that the product buildup rate will depend on the size of the cluster. Additionally, as was the case in the nanosecond resolution experiments, variations of mass spectrometer signals at the mass of the mixed binary cluster versus OH LIF signal intensities were seen to be linear—a result that is pleasing but not conclusive. As will be discussed in section 3.4, double resonance selection of binary complexes will enable contributions from higher complexes to be eliminated. This will provide benchmark cases against which theoretical calculations can be tested. Also, by exploiting OH detection via LIF *as well as* H atom detection via the high-n Rydberg time-of-flight (HRTOF) method (Ashfold et al. 1992; Schneider et al.

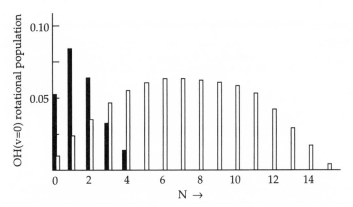

Figure 3-7. Hypothetical situation in which two distinct mechanisms yield hot and cold OH rotational distributions (open and filled bars, respectively). Though the open and filled bars represent 75% and 25%, respectively, of the total OH, the filled bar dominates at $N = 1$. Thus, state specific detection of $N = 1$ senses the minor channel in preference to the major channel. This might occur when two pathways yield hot and cold OH, for example, $Br + HOCO^\dagger$ versus $HO(Br)CO^\dagger$, respectively.

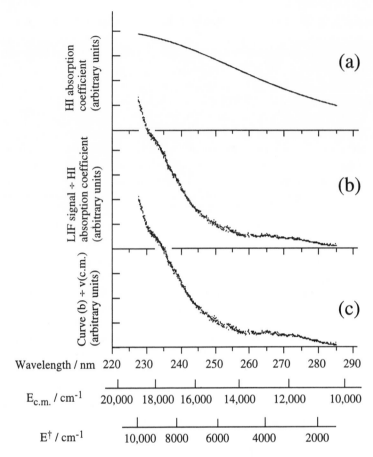

Figure 3-8. Production of OH versus photolysis wavelength for room temperature CO_2/HI samples (from Shin et al. 1991a): (a) HI 300 K absorption coefficient (arbitrary units) versus wavelength; the I* contribution has been removed since there is no contribution from slow hydrogen associated with I*, (b) normalized OH LIF signals divided by curve (a), yielding relative OH ($v = 0$, low N) production rates, (c) curve (b) divided by $v_{c.m.}$, yielding relative OH ($v = 0$, low-N) reaction probabilities. Also shown are the c.m. collision energies and $HOCO^\dagger$ energies in excess of the reaction endoergicity. Note the small reaction probabilities for $E^\dagger < 2000$ cm^{-1}. These data can be converted to absolute cross sections by using the values obtained by Wolfrum and coworkers (e.g., 0.4 Å^2 at a collision energy of 15,000 cm^{-1}) (Jacobs et al. 1994).

1990; Wen et al. 1994), it will be possible to measure the reaction probability. Such a feat has not yet been accomplished.

    Quite recently, time resolved measurements were carried out with complexes formed by expanding $N_2O/HI/He$ mixtures (Ionov et al. 1994), where the reaction of interest is (Böhmer et al. 1992; Marshall et al. 1987, 1989):

$$H + N_2O \rightarrow OH + N_2 \qquad \Delta H = -63 \text{ kcal mol}^{-1}. \tag{2}$$

This system had been studied previously under gas phase, single-collision conditions, where it was shown that the dominant reactive pathway yields $OH + N_2$ via a 1,3-hydrogen shift mechanism (Böhmer et al. 1992). Specifically, by recording OH Doppler profiles, the N_2 product was found to contain a very large amount of internal excitation, which could be reconciled by assuming that hydrogen attaches initially to the terminal nitrogen and the $HNNO^\dagger$ intermediate thus formed then transfers the hydrogen to the oxygen via a transition state having a long N–N bond. Figure 3-9 shows the N_2 vibrational distribution calculated using a simple Franck–Condon model and a mean N–N separation of 1.23 Å, that is, the 1,3-hydrogen shift transition state value (Walch 1993). The fit to the OH sub-Doppler resolution lineshape is very good. We are aware of no other mechanism that can account for this high degree of N_2 internal excitation.

With complexes, reaction (2) was shown to be overwhelmingly dominant over the endoergic NO + NH channel (Hoffman et al. 1989a,b; Shin et al. 1992). Specifically, at the 255 nm photolysis wavelength of the time resolved measurements, the contribution from the NH + NO channel lies between the detection limit. It was our belief that participation of the $HNNO^\dagger$ intermediate *might* yield measurable OH production times (i.e., > 100 fs).

As in the case of CO_2–HI, OH buildup rates were monitored as the degree of complexation was varied. However, this time a clear variation was observed in which the OH buildup time increased monotonically with the degree of complexation, as shown in Figure 3-10. This observation can be interpreted straightforwardly: whereas higher-than-binary complexes strongly inhibited reaction in the case of $H + CO_2$, reaction is less inhibited in the case of $H + N_2O \rightarrow OH + N_2$, with its smaller entrance barrier. In the latter case, slowing of the H atoms does not inhibit reaction, while the increased heat capacity of the larger complexes results in smaller rates. Such conclusions are reasonable, but the evidence is circumstantial. Unequivocal proof is needed.

Participation of an $HO(X)CO^\dagger$ intermediate can also account for the observation that different OH rotational levels display different production rates, since this intermediate might yield rotationally cold OH via stepwise decomposition processes:

$$CO_2\text{–}HX + h\nu \rightarrow HO(X)CO^\dagger \rightarrow X + HOCO^\dagger \tag{3a}$$

$$\rightarrow XCO + OH \tag{3b}$$

$$\rightarrow HOX^\dagger + CO, \tag{3c}$$

followed by OH formation via decomposition of vibrationally excited $HOCO^\dagger$ and $HOX^\dagger$ deriving from reactions (3a) and (3c):

$$HOCO^\dagger \rightarrow OH + CO \tag{3d}$$

$$HOX^\dagger \rightarrow OH + X. \tag{3e}$$

Figure 3-11 provides a schematic description of these processes for the case of CO_2–HBr.

It was decided that the best way to examine this possibility was by carrying out electronic structure calculations for CO_2–HBr complexes in the ground and

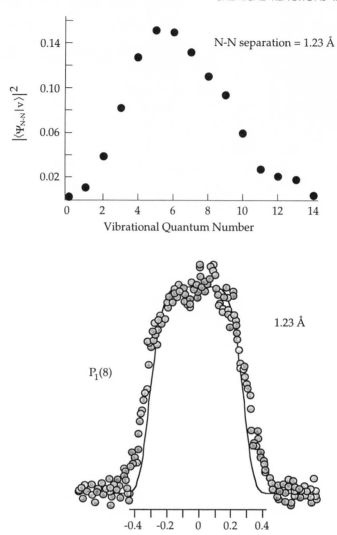

Figure 3-9. Sub-Doppler resolution OH LIF spectra indicate a modest amount of c.m. translational energy for all of the OH levels monitored in the reaction $H + N_2O \rightarrow OH + N_2$; see Böhmer et al. (1992) for details. A Franck–Condon projection of the 1,3-hydrogen shift transition state N–N separation of 1.23 Å (solid line, lower part) predicts a high degree of N_2 vibrational excitation (upper part) and yields good agreement with the data (shaded circles, lower part).

photoexcited states (Shin et al. 1991b). The motivations were: (1) to examine the possible role of the HO(Br)CO† intermediate formed by simultaneous O–H and C–Br bond formation during HBr dissociation, (2) to obtain approximate wavefunctions for the intermolecular degrees of freedom and to use these to estimate the distribution of hydrogen approaches, and (3) to estimate the extent of the squeezed atom effect. Computational details are given elsewhere (Shin et al.

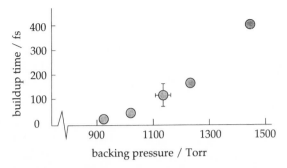

Figure 3-10. Ultrafast pump–probe measurements with supersonically expanded $HI/N_2O/$ He mixtures show that OH buildup times vary with the degree of complexation. At the lowest backing pressures, the buildup times are ≤ 100 fs. For buildup times below 100 fs, the uncertainties associated with fitting the data to assumed lifetimes are too large to allow other than an upper limit to be obtained.

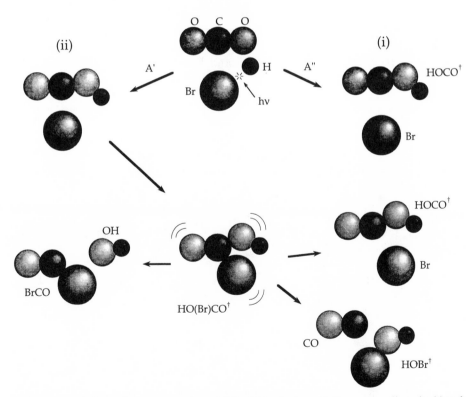

Figure 3-11. Schematic representation of CO_2–HBr photoexcitation proceeding via A′ and A″ PESs, neglecting spin–orbit interaction. There is no C–Br bonding on the A″ surface and the system evolves to $HOCO^\dagger$ and Br moving away from each other as per the squeezed atom effect, as depicted in (i). The A′ surface supports weak C–Br bonding that can evolve to a vibrationally excited $HO(Br)CO^\dagger$ intermediate, as depicted in (ii), which can decompose via several channels; see text for details.

1991b); the methodology employed was that developed over many years by Goddard and coworkers.

In the case of (1), it was found that both attractive and repulsive C–Br interactions were possible, depending on the orientations of the bromine orbitals (i.e., A′ versus A″, respectively). Our original intuition was that hydrogen capture might be facilitated by the presence of the C–Br interaction, due to rehybridization of the orbitals on the carbonyl. This turned out to be a small effect because of the initial geometry afforded by the weakly bound precursor. The attractive C–Br interaction was found to be only ~ 170 cm^{-1} at the 3.6 Å C–Br distance of the weakly bound precursor, as shown in Figure 3-12. Nonetheless, C–Br attraction can facilitate the formation of a highly excited HO(Br)CO† intermediate whose decomposition is expected to be dominated by the Br + HOCO† and/or CO + HOBr† channels. The HOCO† and/or HOBr† thus produced may account for OH product being formed with low N and with longer buildup times than OH deriving from HOCO† formation that occurs without the participation of the HO(Br)CO† intermediate.

In the case of (2), it was found that several percent of the hydrogen atoms make approaches that have high reaction probabilities. The angular potential is very shallow (see Figure 3-13), thus accommodating large amplitude zero point fluctuations, and Schatz has shown that many of these are likely to have appreciable reaction probability (Schatz and Fitzchartes 1988). This confirmed our intuition, but did little to improve our understanding. In the case of (3), it

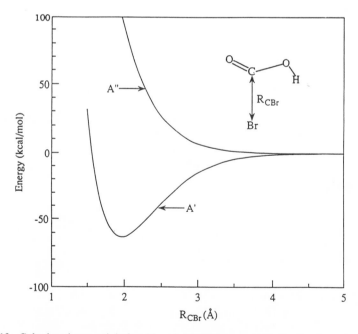

Figure 3-12. Calculated potentials for A′ and A″ PESs versus C–Br distance, showing the repulsive and attractive natures of these surfaces; from Shin et al. (1991b). Note that the C–Br distance in CO₂–HBr is 3.6 Å (Sharpe et al. 1990; Zeng et al. 1992).

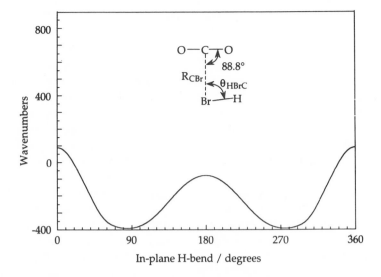

Figure 3-13. Optimized reaction path for in-plane hydrogen motion, showing the shallowness of the potential which gives rise to large amplitude zeropoint fluctuations.

was shown that the potential energy along the direction of hydrogen departure was consistent with what one would expect for the degree of Br–HOCO recoil inferred from the experiments.

The above calculations, in which the possible participation of a $HO(Br)CO^\dagger$ intermediate was confirmed, provide a means of reconciling the different rate measurements. For example, with equal A' and A'' photoexcitation yields, $HO(Br)CO^\dagger$ can be formed as much as half the time. However, this is an upper limit, and it is more likely that $HO(Br)CO^\dagger$ is formed less than half the time. If these intermediates contribute rotationally cold OH, for example, via reactions (3d) and (3e), the spectrally resolved observation of OH ($N = 1$) can yield longer lifetimes than the measurements in which all rotational levels are probed simultaneously. This explanation is consistent with all of the experimental observations. However, though it *can* reconcile the Scherer et al. and Ionov et al. rate measurements, we emphasize that proof is still needed.

3.2.5. Theoretical Modeling of the Reaction Dynamics

For single-collision conditions (i.e., no complexes), this system has been examined at levels of dynamical theories ranging from classical trajectories (Kudla and Schatz 1991; Kudla et al. 1991, 1992; Schatz 1989) to quantum scattering (Clary and Schatz 1993; Schatz and Dyck 1992), and resonance stabilization (Hernández and Clary 1994). As long as the PES is good, the trajectories should give reasonable values for the reaction rates (i.e., within a factor of 2 or 3) at the experimental energies of the hot atom reactions, which lie well above the threshold region. This seems to be the case, though when using classical mechanics with an imperfect

PES it is hard to assign fault uniquely when the calculations fail to reproduce accurately the known facts.

Since there are only four atoms, this case can act as a useful bridge between classical and quantum dynamics calculations. Since nonadiabatic effects are expected to be minor, the quantum calculations are limited by the PES and the computational precision. Consequently, they will ultimately offer the most meaningful comparison between experiment and theory. However, the quantum dynamics calculations carried out to date have been limited to two dimensions. Therefore, they have little to do with the HOCO system per se, except in the general sense of exploring the physics of vibrational resonances coupled to continua.

Our belief is that $k(E)$ can be predicted for HOCO† decomposition reasonably well by using RRKM theory. Though not intellectually appealing, this is expected to yield good numbers. Since $k(T)$ has been fitted over the temperature range $80 \leq T \leq 2000$ K, the use of RRKM theory amounts to essentially an extrapolation of known rates to the higher energies that are accessed in the hot atom experiments. This is the reason that the RRKM approach is presently the most accurate; that is, it is a representation of reliable experimental rate data over a range of temperatures. Caveats about IVR time scales, etc., limit the accuracy and useful range of the extrapolation, but do not change this conclusion.

With complexes, the only dynamical calculations to date have been classical trajectories in which it was assumed that there is no significant C–Br chemical interaction following photoexcitation (Schatz and Fitzcharles 1988). Consequently, the role of the complex has been limited to H + CO$_2$ interactions sampled over the probability density for the intermolecular degrees of freedom, as well as the squeezed atom effect. These calculations have yielded reaction probabilities versus attack angle and nascent V, R, T excitations which are in reasonable agreement with the experimental results. Much work is still needed and the challenges are daunting.

3.3. THE REACTION Br + I$_2$ → IBr + I VIA THE WEAKLY BOUND HBr–I$_2$ PRECURSOR

In section 3.2, experiments were described in which photodissociation of the HX moiety within a weakly bound complex liberated hydrogen atoms which then went on to react with the other moiety in the complex. Alternatively, photodissociation of HX moieties within weakly bound complexes can be used to prepare halogen atom reactants. This provides advantages which derive mainly from the rapid removal of the hydrogen.

In the first demonstration of this approach, Sims et al. photodissociated HBr within HBr–I$_2$ complexes in order to initiate the reaction (Gruebele et al. 1991; Sims et al. 1992):

$$Br + I_2 \rightarrow IBr + I \qquad \Delta H = -6.7 \, \text{kcal mol}^{-1} \qquad (4)$$

where reaction is believed to transpire via a long-lived Br–I–I† intermediate. This and similar reactions involving three halogen atoms have served as valuable prototypes since the early days of reaction dynamics (Beck et al. 1968; Blais and Cross 1970; Carter et al. 1973; Cross and Blais 1969, 1971; Firth and Grice 1987a,b; Firth et al. 1987a,b; Fisk et al. 1967; Girard et al. 1987, 1991; Hoffmann et al. 1983; Lee et al. 1968, 1969, 1977; Loesch and Beck 1971; Trickl and Wanner 1983). Specifically, a number of groups have demonstrated the important role played by the trihalogen intermediate, mainly by using the crossed molecular beams technique. In these studies, it was possible to infer the lifetime of the intermediate only indirectly, that is, by comparing the observed angular distribution in the c.m. system to calculations that require, as input, an estimated average rotational period (Lee et al. 1968). However, by exploiting the $t = 0$ clocking method, Sims et al. were able to obtain the IBr buildup time directly. In the absence of processes additional to reaction (4), this is the same as the Br–I–I† decomposition lifetime.

Although the structure of the HBr–I$_2$ complex has not yet been determined experimentally, it seems probable that the hydrogen faces outward (Sims et al. 1992). For example, the analogous cases of HF–ClF and HF–Cl$_2$ have been examined by Klemperer and coworkers who used the molecular beam electric resonance technique to determine the average structures (Baiocchi et al. 1982; Novick et al. 1977). In both cases, the three halogen atoms were found to lie along a straight line. For HF–ClF, they deduced an average H–F–Cl angle of 125° for both the deuterated and undeuterated cases, and from this they concluded that the H–F–Cl equilibrium angle is $\sim 125°$, albeit with a large hydrogen zero point amplitude. Recently, Blake and coworkers examined HF–Cl$_2$ and found it to be quasilinear (Stockman and Blake 1993).

With the hydrogen facing outward, its role differs qualitatively from the case of CO$_2$–HX complexes. Namely, HBr photodissociation ejects the hydrogen *away from* the I$_2$, while giving the Br atom a modest push *toward* the I$_2$, as shown in Figure 3-14. This rapid ejection of ejection of hydrogen places the system on the triatom PES near the linear geometry, whereas the equilibrium angle is expected to be $\sim 150°$ with a small barrier to linearity (Sannigrahi and Peyerimhoff 1986; Viste and Pyykkö 1984). Reaction follows, unaffected by the hydrogen which has long since departed. This system is clean experimentally in the sense that the hydrogen atom leaves on a time scale that is nearly two orders of magnitude shorter than that of the subsequent dynamics. Note that this is true regardless of the dynamical details of how the hydrogen escapes.

The photodissociation of uncomplexed HBr leaves the Br atom in one or both of the spin–orbit states (i.e., the $^2P_{3/2}$ ground state and the $^2P_{1/2}$ excited state, hereafter referred to as Br*) with relative populations that depend on the photolysis wavelength (Magnotta et al. 1981; Xu et al. 1987, 1988). With HBr–I$_2$ complexation, rapid hydrogen departure places the trihalogen system at a geometry in which the Br–I$_2$ distance is large and Br–I$_2$ attraction is weak relative to the bromine spin–orbit splitting. Thus, the system begins evolving on potentials which, to a first approximation, preserve the Br and Br* atomic spin–orbit electronic configurations in the entrance channel. For Br* it is unlikely that direct reactive channels are open, since it has been shown that a propensity exists in which spin–orbit excitation is preserved in such trihalogen exchange reactions (Gordon et al. 1982; Haugen et al. 1985; Hofmann and Leone 1978; Wiesenfeld and Wolk 1978a,b). Indeed, the reaction I* + Br$_2$ → IBr + Br* even leads to a

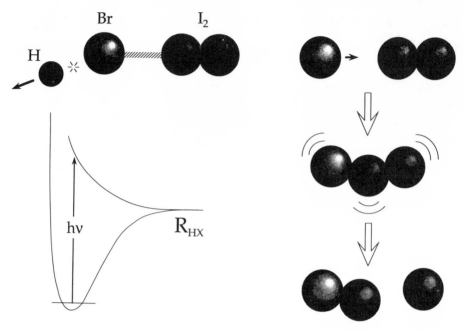

Figure 3-14. Schematic drawing of the photoinitiated reaction of Br with I_2. Hydrogen removal is rapid, relative to the characteristic time scales of the motions of the heavier nuclei. Photoexcitation at 220 nm yields primarily $Br(^2P_{3/2})$, though some $Br(^2P_{1/2})$ is also produced. As shown on the right, reaction transpires via the vibrationally excited trihalogen intermediate.

population inversion and consequent laser oscillation on the Br ← Br* transition at $3685 \, cm^{-1}$ (Spencer and Wittig 1979). The main participation of Br* in producing IBr is expected to be due to quenching of Br* followed by reaction (4).

The distribution of orbital alignments and orientations relative to the I_2 moiety are dictated by the zero point distribution of HBr axes in the weakly bound precursor. It is reasonable to assume that this distribution of HBr axes results in an effective Br–I_2 long range potential which is less attractive than for the most favorable orbital arrangement. This brings up a valuable aspect of this approach, namely, that the trihalogen system begins its life on a long-range part of the PES, which is one of the hardest regions to study, both experimentally and theoretically. Since observables such as rates can depend sensitively on this part of the PES, it is possible that the elusive long-range interaction problem can be addressed. Consequently, while important in its own right because of the scientific significance of the trihalogens, this experiment also opens the door to complementary studies.

In the experiments of Sims et al., it was found that IBr was formed with a risetime of 53 ± 4 ps when photolyzing HBr–I_2 complexes at 218 nm. A typical experimental trace is shown in Figure 3-15 (lower trace). In this case, the Br + I_2 "collision energy" in the c.m. system is $145 \, cm^{-1}$, neglecting the van der Waals interaction in the entrance channel. A similar result (44 ± 4 ps) was obtained with DBr–I_2 (upper trace), for which the collision energy is $286 \, cm^{-1}$. Assuming that most of the observed IBr derives from Br rather than Br* photoproducts, these

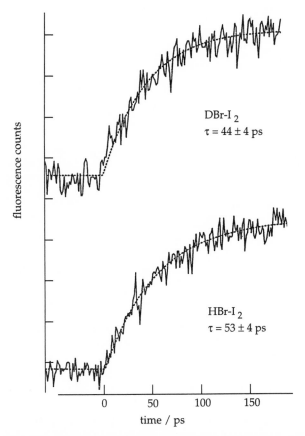

Figure 3-15. Typical experimental traces showing IBr buildup following 218 nm photo-excitation of HBr–I_2 and DBr–I_2 complexes.

data confirm the long IBr buildup times that result from the presence of a significant well along the reaction coordinate. Moreover, they support the contention that the hydrogen faces outward; otherwise, the opposite trend would be expected, that is, a longer lifetime for DBr–I_2 than HBr–I_2.

The van der Waals attraction between Br and I_2 is estimated to be $\sim 400\ \mathrm{cm}^{-1}$ by analogy with halogen/rare gas complexes (Bieler and Janda 1990; Bieler et al. 1991). This ensures that photodissociation of the HBr moiety cannot produce Br + I_2 except via quenching of Br* or the unlikely instance in which the hydrogen is trapped efficiently between the heavy particles. With the Br atom unable to escape from the I_2 because of the Br–I_2 van der Waals attraction, the system is ensured of an essentially unity quantum yield.

It was possible to construct a PES which was consistent with the earlier work on trihalogens and which was able to reconcile the main findings of the time resolved studies by using the method of classical trajectories (Sims et al. 1992). A range of parameters was explored, and the experimental results were reproduced best by using wells lying 13–17 kcal mol^{-1} below the entrance channel with exit barriers of 3–4 kcal mol^{-1}. A schematic energy diagram is shown in

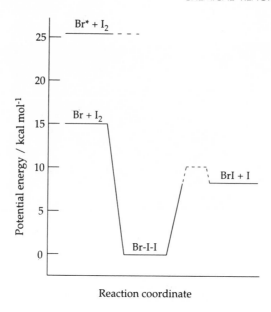

Figure 3-16. Partial energy diagram for the Br + I$_2$ system. The Br(^{2}P$_{1/2}$) is not expected to react directly, though reaction can occur following quenching to Br(^{2}P$_{3/2}$).

Figure 3-16. Minor changes in the parameters that define the PES did not result in large differences in calculated lifetimes, for example, approach angles of 0° (collinear) and 55° gave results that differed by only 20%, with 0° yielding the smaller lifetimes. The conclusion is that a qualitatively intuitive PES with a well lying ~ 15 kcal mol^{-1} below the entrance channel and ~8 kcal mol^{-1} below the exit channel can reconcile the main experimental finding of a lifetime of ~50 ps for the Br–I–I$^+$ intermediate. The exit barrier of 3–4 kcal mol^{-1} provided the best fit, but since nuances such as the van der Waals interaction in the entrance channel were suppressed, this should be viewed as providing grist for further theoretical work. For example, recall that the measured lifetimes for the 145 and 286 cm^{-1} "collision energies" are 53 ± 4 ps and 44 ± 4 ps, respectively. Likewise, computational results for the 145 cm^{-1} collision energy yielded consistently longer lifetimes (30% on average) than those obtained with the 286 cm^{-1} collision energy. Consequently, the inclusion of the Br–I$_2$ van der Waals attraction into the trajectory calculations is expected to lengthen the lifetime by starting the reaction at a lower energy, thereby lessening the size of the exit barrier needed to fit the data.

Recently, Wright et al. (1994) used ultrafast photoionization techniques to detect vibrationally excited I$_2$ following 220 nm photoexcitation of HBr–I$_2$ complexes. They attributed this to quenching of Br* by I$_2$, pointing out that their measured lifetime of 51 ± 5 ps is close to those reported by Sims et al. (1992). This raises exciting possibilities. It seems inevitable that Br*–I$_2$ complexes will find their way eventually to the ground state PES since there is nowhere else to go. However, although the Br* yield at 220 nm is unknown, it is expected to be modest. Specifically it is only 15% at 193 nm and is expected to diminish at

longer wavelengths (Magnotta et al. 1981; Xu et al. 1987, 1988). Thus, if the Br*
photoproduct proves to be the main source of IBr, this implies that there is a
barrier in the entrance channel for the $Br + I_2$ reaction. Given the delicate
interplay between the atomic halogen spin–orbit interaction and the forces present
in the region of the triatom well, it will be important to examine this system in
more detail.

Finally, we wish to point out that the modest well depth of the $Br–I–I^\dagger$
intermediate places this system in the category of a unimolecular decomposition
reaction in the limit of a low density of states. This is not unlike the case of NO_2,
which has received considerable attention (Hunter et al. 1993; Peskin et al. 1994;
Reid et al. 1993a,b; Reid and Reisler 1994; Reisler et al. 1994). In such cases, the
rates are expected to fluctuate as per the overlap of the wavefunctions of the
$Br–I–I^\dagger$ resonances with those of the $IBr + I$ exit channel continuum, with chaotic
mixtures of the continuum contributions. The manner in which the experiments
are carried out leads to some averaging over $Br–I–I^\dagger$ rotational levels and total
energies, but this may not wash out the fluctuations in rates that occur as the
energy is tuned. Therefore, it will be very useful to extend the rate measurements
to cover a continuous range of photon energies. Additionally, small characteristic
rates like those observed (2×10^{10} s^{-1}) raise the spectre of observing step-like
structure in $k(E)$, corresponding to the successive openings of reactive channels.
These have been observed in photoinitiated reactions of H_2CO (Lovejoy et al.
1992) and NO_2 (Brucker et al. 1992; Ionov et al. 1993a; Miyawaki et al. 1990,
1991, 1993; Wittig and Ionov 1994) and can be assigned to quantized features of
the transition state. With the trihalogens, such steps would reveal dynamical
features that have yet to be observed under scattering conditions, despite numerous
predictions (Chatfield et al. 1991a,b, 1992a,b, 1993; Friedman and Truhlar 1991;
Lynch et al. 1991; Truhlar et al. 1990).

3.4. PROSPECTS FOR FUTURE RESEARCH

Many factors and considerations are germane to future research in this area. On
the technical side, achieving cluster size selection stands as one of the most
important and sought-after goals. It would be most desirable to achieve this while
maintaining sufficiently high densities for studies of photoinitiated reactions to be
carried out with product state resolution and/or ultrafast time resolution. The two
methods that, in our opinion, are most viable are molecular beam deflection, as
pioneered by Buck (1994) and coworkers, and laser-based double-resonance
methods. Less direct approaches are deemed inferior.

The molecular beam deflection method is shown schematically in Figure 3-17
(Buck et al. 1985). It is based on momentum transfer between clusters entrained
in a molecular beam and rare gas atoms which are the constituents of a second
molecular beam at 90° to the cluster beam. Collisions between the rare gas atoms
and the clusters under single-collision conditions deflect a small percentage of the
clusters from their original path. The maximum deflection angle depends on the
mass of the cluster. For example, binary clusters may be deflected into a broad
range of angles with a well defined upper limit set by the momentum conservation

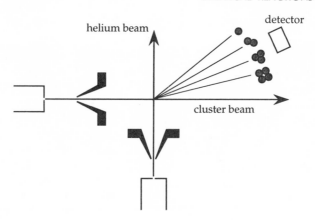

Figure 3-17. Schematic of the molecular beam deflection method. Under single-collision conditions, collisions between clusters and He atoms deflect the clusters to angles whose maxima are determined by momentum conservation.

condition, while higher clusters are scattered into narrower ranges of angles. As the number of subunits in the cluster increases, the upper limits to the deflection angle decrease successively.

Since the present experiments that use this approach employ mass spectrometric detection, it is essential that the neutral cluster of interest yields a signal at the parent mass. It is not problematic if ion fragmentation in the mass spectrometer is severe, as long as the neutral clusters yield enough ions at the parent masses to provide acceptable count rates. In this case, the cluster ion signal at the maximum-allowed deflection angle is a faithful signature of the neutral cluster. Higher clusters may fragment severely in the ionizer of the mass spectrometer, yielding large ion concentrations at the mass of interest; however, these higher clusters will be deflected to smaller angles than the neutral cluster having the same mass as that monitored mass spectrometrically.

This method of separating clusters has become an important experimental technique which has been applied to the study of several important classes of clusters: van der Waals (e.g., ethylene) (Alrichs et al. 1990; Buck et al. 1987, 1988c), dipolar (e.g., acetonitrile) (Buck 1992; Buck and Ettischer 1994; Buck et al. 1990a), and different hydrogen bonds (e.g., ethylene–acetone) (Buck et al. 1993) and isomers have also been identified (e.g., methanol versus hydrazine) (Beu et al. 1994; Buck et al. 1988a,b, 1990b,c; Huisken et al. 1991; Huisken and Stemmler 1988, 1992). It can be used with mixed clusters (e.g., methanol hexamer) (Buck et al. 1990b; Buck and Hobein 1993). Moreover, it is very general. It can be applied to almost any species and is generous in the range of cluster sizes that can be accommodated. An example from the work of Buck and coworkers is given in Figure 3-18 showing the time dependence used to identify different clusters (Buck et al. 1985).

One of the disadvantages is that the clusters thus separated have internal excitations. This can be detrimental in cases where one would like to use the clusters after they have been separated, particularly if it is desirable to maintain, to the extent possible, a well defined geometry. Even at 0 K, large zero point

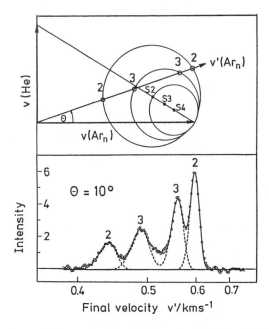

Figure 3-18. Experimental demonstration of the molecular beam deflection method, from Buck et al. (1985). The Newton circles show the peak positions associated with the different clusters; S_i denotes the velocity of the center-of-mass for different clusters. The TOF data were recorded at a lab angle of 10°; note the correspondence between peaks 2 and 3 and the corresponding intersection points on the Newton circles.

amplitudes preclude the possibility of assigning the "geometry" to the equilibrium value, and excitation of intermolecular vibrations exacerbates this problem severely. For example, a modest amount of vibrational excitation in the internal degrees of freedom can cause the cluster to have liquid-like behavior (Buck 1994). Moreover, this problem is most severe for small clusters when using the molecular beam deflection method.

Another disadvantage is that, though certain spectroscopic studies can be carried out with the selected clusters, it is difficult to envision the present embodiment of this method as providing high enough concentrations of selected clusters to apply product-state-selective and/or ultrafast techniques to photo-initiated reactions within the selected complexes. This is because the process whereby the clusters are differentiated employs strong, momentum-changing collisions, that is, such collisions tag the clusters' translational degree of freedom. It is possible that a precision Doppler method could catch the clusters thus tagged before they disperse from the interaction region, but this has yet to be demonstrated. Once they have separated in space, the low concentrations that result make it difficult to use such selected clusters in further photochemical studies.

The Buck method is most useful for medium sized clusters, and in experiments that do not attempt detailed photochemical studies of the size selected clusters. High resolution spectroscopy of the selected clusters is impractical because the number per quantum state is small, due to the many occupied ro-vibrational

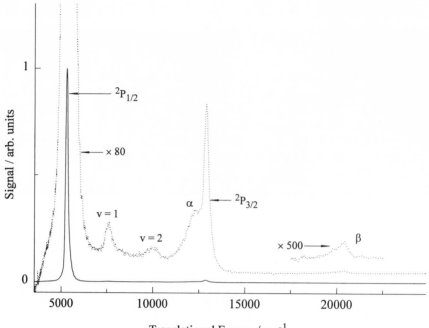

(a)

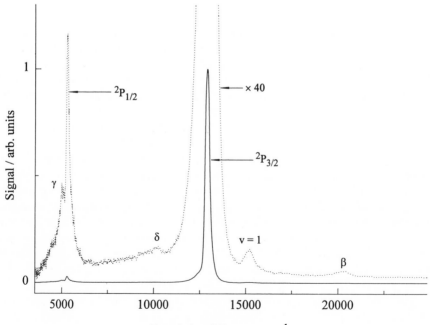

(b)

levels. However, low resolution spectroscopic studies can be revealing (Buck 1994), though interpretation requires care.

At least for the near future, we believe that the method of choice for overcoming the size selection problem for binary complexes is double resonance photoinitiation using pulsed lasers—one operating in the infrared, the other in the ultraviolet. This is analogous the methods developed by Crim and coworkers (Brouwer et al. 1987; Butler et al. 1986; Crim 1984; Dübal and Crim 1985; Rizzo et al. 1983; Sinha et al. 1990; Tichich et al. 1986, 1987) and Rizzo and coworkers (Fleming et al. 1991a,b,c; Fleming and Rizzo 1991; Luo et al. 1990; Luo and Rizzo 1990, 1991). The infrared pulse tags complexes by exciting (primarily) a moiety whose ultraviolet absorption is red-shifted by virtue of the implanted vibrational excitation. This step is also innately state selective and, by using nanosecond lasers, rotational resolution of the complexes will usually be possible. In the presence of ultraviolet radiation, complexes thus tagged can experience two fates: (1) vibrational predissociation, yielding free molecules which, in general, will possess vibration and rotation excitation, and (2) photodissociation of one moiety within the complex, namely, the one containing the implanted vibrational excitation. With ultrafast lasers, there is a trade-off between vibrational predissociation and spectral resolution. To ensure the latter, one might choose transform-limited infrared pulses of ~ 100 ps duration, which is short enough to overcome most vibrational predissociation rates. This is compatible with the subsequent application of ultrafast pump–probe techniques.

With nanosecond lasers, predissociation rates are usually slow enough to permit individual rotational lines to be resolved, though they are often fast compared with typical pulse durations. Therefore, one must be wary of the possibility that vibrational predissociation removes most of the tagged complexes. For example, a homogeneous linewidth of only 30 MHz is sufficiently narrow that it may be hard to observe in spectroscopic studies, but could prove devastating were there a delay of tens of nanoseconds between the tagging and photodissociation pulses. Since it will not always be possible to determine predissociation lifetimes a priori, the recommended approach is to use temporally overlapped pulses and high ultraviolet intensities. In this case, the fate of the tagged complexes is determined by the relative rates of ultraviolet photoexcitation versus predissociation. Under optimal conditions, most of the tagged complexes will be further excited with the UV radiation with only a slight loss of spectral resolution due to the additional lifetime broadening caused by efficient up-pumping. Numerical estimates suggest that it will not be difficult to overcome subnanosecond predissociation lifetimes. For example, consider 266 nm HI photodissociation, for

---

Figure 3-19. Photodissociation of HI monomers and clusters. The solid traces indicate the substantial discrimination available when using polarized photolysis radiation; note the high S/N. Under conditions of such minimal clustering, it is reasonable to assume that most of the clusters are binary. Peaks labeled $v = 1$ and $v = 2$ are due to inelastic H + HI collisions within the cluster. The superelastic peak β is assigned tentatively to secondary photolysis of I*–HI complexes, in which the escaping hydrogen deactivates the nearby I*. (a) Vertical and (b) horizontal polarization of the photolysis radiation relative to the molecular beam. The plenum pressure is 1900 torr.

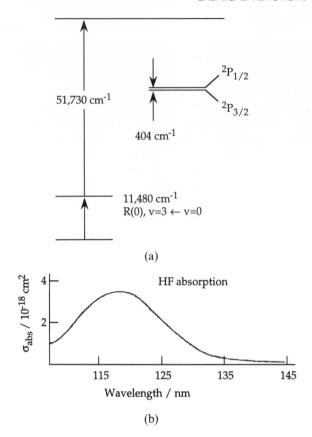

Figure 3-20. (a) Schematic of sequential two-photon HF excitation scheme and (b) HF absorption spectrum. Note that HF absorption above 150 nm (66, 700 cm^{-1}) is very weak; from Lee (1985).

which $\sigma_{abs} = 3 \times 10^{-19}$ cm^2. A 20 mJ pulse of 5 ns duration focused into a 0.1 mm^2 spot results in optical excitation rates $> 10^9$ s^{-1}. Such experimental conditions are easily met, ensuring that a multitude of experiments are possible in the nanosecond regime.

Figure 3-19 shows an experimental result obtained by expanding HI/He mixtures under conditions that discriminate against higher-than-binary complexes (Segall et al. 1994). Of course, higher-than-binary clusters are present to some extent, but every effort was made to minimize their presence, namely, the very high S/N enabled the experiments to be carried out under conditions of minimal clustering.

In these experiments, there was no double-resonance tagging, so all of the HI moieties were available for photodissociation. What is observed experimentally is the resulting atomic hydrogen time-of-flight distribution which is obtained by using the HRTOF technique (Ashfold et al. 1992; Schneider et al. 1990; Wen et al. 1994). This method provides excellent resolution and S/N compared to other TOF methods. The dominant features (solid curves) are due to HI monomer. However, upon magnification, features are seen that are due to

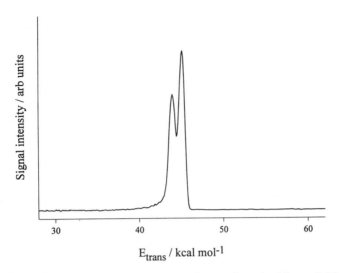

Figure 3-21. Experimental realization of the scheme given in Figure 3-20. The peaks correspond to the fluorine spin–orbit states. The low resolution relative to the data shown in Figure 3-19 is due to the large linewidth of the 193.3 nm excitation laser.

complexes. The photodissociation of an HI moiety in $(HI)_2$ produces HI in $v = 1$ and 2 via intracluster collisions of photolyfically produced hydrogen atoms with the nearby HI moiety. This inelastic process results from "missed reactions." In this example, the peaks labeled $v = 1$ and 2 derive from the photodissociation of the vibrationally excited HI which is produced by the first photochemical event in the $(HI)_2$ complex. It was verified by H → D substitution that the peaks shown in Figure 3-19 are due to HI vibrations, and the sequential two-photon nature of the $v = 1$ and 2 peaks was confirmed by fluence dependence measurements. Furthermore, there is a small feature which is offset from the monomer by approximately the iodine spin–orbit energy. This is assigned to the scattering of hydrogen from nearby I* that is produced by the first ultraviolet photon, deactivating the I* in the process.

It is noteworthy that the photolytic removal of the second hydrogen atom from the complex is quite efficient. Thus, even under "ordinary" experimental conditions, it will be possible to ensure efficient ultraviolet photoexcitation in the face of subnanosecond predissociation lifetimes.

The initial step in the double-resonance scheme is the excitation of a local mode hydrogen stretch vibration localized in a hydrogen halide moiety. In principle, this can be done either at the fundamental or one of the overtones. With presently available Ti:sapphire lasers and parametric oscillators (OPOs), it is possible to saturate fundamentals and first overtones, thus ensuring maximum population transfer. Second overtones cannot be pumped as efficiently, but offer enormous discrimination against background and can be used to shift frequencies out of the vacuum ultraviolet and into a more user-friendly part of the ultraviolet. Thus, first and second overtones are very attractive.

As with the HI example given above, a "demonstration of principle" experiment is needed, and Figure 3-20 shows just this (Zhang et al. n.d). Hydrogen fluoride gas was cooled by supersonic expansion and the $\Delta v = 3$ R(0) transition was excited by using a narrow bandwidth pulsed Ti:sapphire laser. These molecules were photodissociated by using just a few millijoules of 193 nm radiation. The resulting hydrogen translational energy distribution is shown in Figure 3-21. The two peaks are due to the atomic fluorine $^2P_{3/2}$ and $^2P_{1/2}$ states and the low resolution is due to the broad excimer laser linewidth. The important point is the high S/N, which exceeds 10^2 for the data presented in the figure and can be improved to $\geq 10^3$. This ensures that experiments with tagged complexes will yield adequate S/N for a number of systems, e.g., Rg–HX, CO_2–HX, N_2O–HX, HX–HY, etc.

It is our conclusion, based on the considerations and preliminary results given above, that the way is clear for size and state selection of weakly bound complexes using infrared photoexcitation. Ultrashort pulse excitation can then provide dynamics on the ground state and reach the PES at different points.

ACKNOWLEDGMENT: This article is dedicated to the memory of Richard B. Bernstein, who would have been a coauthor were he with us. The consummate perfectionist, he saw achievement as vehicular to greater progress. His tireless enthusiasm was infectious. Dick Bernstein was an optimist.

REFERENCES

Alagia, M.; Balucani, N.; Casavecchia, P.; Stranges, D.; Volpi, G. G. 1993 J. Chem. Phys. 98:8341

Alimi, R.; Gerber, R. B. 1990 Phys. Rev. Lett. 64:1453

Alrichs, R.; Brode, S.; Buck, U.; DeKieviet, M.; Lauenstein, C.; Rudolph, A.; Schmidt, B. 1990 Z. Phys. D 15:341

Altman, R. S.; Marshall, M. S.; Klemperer, W. 1982 J. Chem. Phys. 77:4344

Ashfold, M. N. R.; Lambert, I. R.; Mordaunt, D. H.; Morley, G. P.; Western, C. 1992 J. Phys. Chem. 96:2938

Baiocchi, F. A.; Dixon, T. A.; Joyner, C. H.; Klemperer, W. 1981 J. Chem. Phys. 74:6544

Baiocchi, F. A.; Dixon, T. A.; Klemperer, W. 1982 J. Chem. Phys. 77:1632

Baulch, D. L.; Cox, R. A.; Hampson, R. A., Jr.; Kerr, J. A.; Troe, J.; Watson, R. T. 1984 J. Phys. Chem. Ref. Data 13:1359

Beck, D.; Engelke, F.; Loesch, H. J. 1968 Ber. Bunsenges. Phys. Chem. 72:1105

Bemish, R. J.; Wu, M.; Miller, R. E. 1994 Faraday Discuss. Chem. Soc. 97:57

Beu; Buck, U.; Hobein, M. 1994 J. Chem. Phys. in press

Bieler, C. R.; Janda, K. C. 1990 In Atomic and Molecular Clusters, E. R. Bernstein, ed., Elsevier, p. 455

Bieler, C. R.; Spence, K. E.; Janda, K. C. 1991 J. Phys. Chem. 95:5058

Blais, N. B.; Cross, J. B. 1970 J. Chem. Phys. 52:3580

Block, P. A.; Bohac, E. J.; Miller, R. E. 1992 Phys. Rev. Lett. 68:1303

Bohac, E. J.; Marshall, M. D.; Miller, R. E. 1986 J. Chem. Phys. 85:4890

Bohac, E. J.; Marshall, M. D.; Miller, R. E. 1992a J. Chem. Phys. 96:6681

Bohac, E. J.; Marshall, M. D.; Miller, R. E. 1992b J. Chem. Phys. 97:4901

Bohac, E. J.; Miller, R. E. 1993a J. Chem. Phys. 98:2604

Bohac, E. J.; Miller, R. E. 1993b Phys. Rev. Lett. 71:54

Böhmer, E.; Shin, S. K.; Chen, Y.; Wittig, C. 1992 J. Chem. Phys. 97:2536

Boivineau, M.; Calve, J. L.; Castex, M. C.; Jouvet, C. 1986a Chem. Phys. Lett. 128:528

Boivineau, M.; Calve, J. L.; Castex, M. C.; Jouvet, C. 1986b Chem. Phys. Lett. 130:208

Breckenridge, W. H. 1989 Acc. Chem. Res. 22:21

Breckenridge, W. H.; Duval, M. C.; Jouvet, C.; Soep, B. 1985 Chem. Phys. Lett. 122:181

Breckenridge, W. H.; Duval, M. C.; Jouvet, C.; Soep, B. 1987 In Structure and Dynamics of Weakly Bound Molecular Complexes, A. Weber, ed., D. Reidel, Amsterdam, p. 213

Breckenridge, W. H.; Jouvet, C.; Soep, B. 1986 J. Chem. Phys. 84:1443

Brouwer, L.; Cobos, C. J.; Troe, J.; DiAbal, H. R.; Crim, F. F. 1987 J. Chem. Phys. 86:6171

Brucker, G. A.; Ionov, S. I.; Chen, Y.; Wittig, C. 1992 Chem. Phys. Lett. 194:301

Brunning, J.; Derbyshire, D. W.; Smith, I. W. M.; Williams, M. D. 1988 J. Chem. Soc. Faraday Trans. 2 484:105

Buck, U. 1992 Ber. Bunsenges. Phys. Chem. 96:1275

Buck, U. 1994 J. Phys. Chem. 98:5190

Buck, U.; Ettischer, I. 1994 Faraday Discuss. Chem. Soc. 97:215

Buck, U.; Gu, X. J.; Hobein, M.; Lauenstein, C. 1988a Chem. Phys. Lett. 163:453

Buck, U.; Gu, X. J.; Hobein, M.; Lauenstein, C. 1990a Chem. Phys. Lett. 174:247

Buck, U.; Gu, X. J.; Hobein, M.; Lauenstein, C.; Rudolph, A. 1990b J. Chem. Soc. Faraday Trans. 2 286:1923

Buck, U.; Gu, X. J.; Lauenstein, C.; Rudolph, A. 1988b J. Phys. Chem. 92:5561

Buck, U.; Gu, X. J.; Lauenstein, C.; Rudolph, A. 1990c J. Chem. Phys. 92:6017

Buck, U.; Hobein, M. 1993 Z. Phys. D 28:331

Buck, U.; Hobein, M.; Schmidt, B. 1993 J. Chem. Phys. 98:9425

Buck, U.; Huisken, F.; Lauenstein, C.; Meyer, R.; Sroka, R. 1987 J. Chem. Phys. 87:6276

Buck, U.; Lauenstein, C.; Rudolph, A.; Heijmen, B.; Stolte, S.; Reuss, J. 1988c Chem. Phys. Lett. 144:396

Buck, U.; Meyer, H.; Pauly, H. 1985 In *Unsteady Fluid Motions*, Springer Lecture Notes in Physics, F. Obermeier and G. E. A. Meier, eds., Springer

Butler, L. J.; Tichich, T. M.; Likar, M. D.; Crim, F. F. 1986 J. Chem. Phys. 85:2331

Carter, C. F.; Levy, M. R.; Woodall, K. B.; Grice, R. 1973 Faraday Discuss. Chem. Soc. 55:381

Chatfield, D. C.; Friedman, R. S.; Lynch, G. C.; Truhlar, D. G. 1992a J. Phys. Chem. 96:57

Chatfield, D. C.; Friedman, R. S.; Lynch, G. C.; Truhlar, D. G. 1993 J. Chem. Phys. 98:342

Chatfield, D. C.; Friedman, R. S.; Schwenke, D. W.; Truhlar, D. G. 1992b J. Phys. Chem. 96:2414

Chatfield, D. C.; Friedman, R. S.; Truhlar, D. G. 1991a Faraday Discuss. Chem. Soc. 91:289

Chatfield, D. C.; Friedman, R. S.; Truhlar, D. G.; Garrett, B. C.; Schwenke, D. W. 1991b J. Amer. Chem. Soc. 113:486

Chen, Y.; Hoffmann, G.; Oh, D.; Wittig, C. 1989 Chem. Phys. Lett. 159:426

Chen, Y.; Hoffmann, G.; Shin, S. K.; Oh, D.; Sharpe, S.; Zeng, Y. P.; Beaudet, R. A.; Wittig, C. 1992 In *Advances in Molecular Vibrations and Collision Dynamics*, Vol. 1, J. Bowman, ed., JAI Press, Greenwich, p. 187

Clary, D. C.; Schatz, G. C. 1993 J. Chem. Phys. 99:4578

Crim, F. F. 1984 Annu. Rev. Phys. Chem. 35:657

Cross, J. B.; Blais, N. B. 1969 J. Chem. Phys. 50:4108

Cross, J. B.; Blais, N. B. 1971 J. Chem. Phys. 55:3970

Davis, D. D.; Fischer, S.; Schiff, R. 1974 J. Chem. Phys. 61:2213

Dayton, D. C.; Jucks, K. W.; Miller, R. E. 1989 J. Chem. Phys. 90:2631

Dübal, H. R.; Crim, F. F. 1985 J. Chem. Phys. 83:3863

Duval, M. C.; Benoist D'Azy, O.; Breckenridge, W. H.; Jouvet, C. 1986 J. Chem. Phys. 85:6324

Duval, M. C.; Jouvet, C.; Soep, B. 1985 Chem. Phys. Lett. 119:317

Duval, M. C.; Soep, B.; Breckenridge, W. H. 1991 J. Phys. Chem. 95:7145

Dykstra, C. E. 1990 J. Phys. Chem. 94:6948, and references therein

Firth, N. C.; Grice, R. 1987a Mol. Phys. 60:1261

Firth, N. C.; Grice, R. 1987b Mol. Phys. 60:1273

Firth, N. C.; Keane, N. W.; Smith, D. J.; Grice, R. 1987a Faraday Discuss. Chem. Soc. 84:53

Firth, N. C.; Smith, D. J.; Grice, R. 1987b Mol. Phys. 61:859

Fisk, G. A.; McDonald, J. D.; Herschbach, D. R. 1967 Faraday Discuss. Chem. Soc. 44:228

Fleming, P. R.; Li, M.; Rizzo, T. R. 1991a J. Chem. Phys. 94:2425

Fleming, P. R.; Li, M.; Rizzo, T. R. 1991b J. Chem. Phys. 95:865

Fleming, P. R.; Luo, X.; Rizzo, T. R. 1991c In *Mode Selective Chemistry*, B. Pullman and J. Jortner, eds., Kluwer, Dordrecht

Fleming, P. R.; Rizzo, T. R. 1991 J. Chem. Phys. 95:1461

Fraser, G. T.; Pine, A. S.; Suenram, R. D.; Dayton, D. C.; Miller, R. E. 1989 J. Chem. Phys. 90:1330

Friedman, R. S.; Truhlar, D. G. 1991 Chem. Phys. Lett. 183:539

Frost, M. J.; Salh, J. S.; Smith, I. W. M. 1991a J. Chem. Soc. Faraday Trans. 2 87:1037

Frost, M. J.; Sharkey, P. ; Smith, I. W. M. 1991b Faraday Discuss. Chem. Soc. 91:305

Frost, M. J.; Sharkey, P.; Smith, I. W. M. 1993 J. Phys. Chem. 97:12254

Garcia-Vela, A.; Gerber, R. B.; Valentini, J. J. 1991 Chem. Phys. Lett. 186:223

Garcia-Vela, A.; Gerber, R. B.; Valentini, J. J. 1992 J. Chem. Phys. 97:3297

Gardiner, W. C., Jr. 1977 Acc. Chem. Res. 10:326

Girard, B.; Billy, N.; Gouédard, G.; Vigué J. 1987 Faraday Discuss. Chem. Soc. 84:65

Girard, B.; Billy, N.; Gouédard, G.; Vigué, J. 1991 Europhys. Lett. 14:13

Gordon, E. B.; Nadkhin, A. I.; Sotnichenko, S. A.; Boriev, I. A. 1982 Chem. Phys. Lett. 86:209

Green, W. H., Jr.; Moore, C. B.; Polik, W. F. 1992 Annu. Rev. Phys. Chem. 43:307

Green, W. H.; Mahoney, A. J.; Zheng, Q.-K.; Moore, C. B. 1991 J. Chem. Phys. 94:1961

Gruebele, M.; Sims, I. R.; Potter, E. D.; Zewail, A. H. 1991 J. Chem. Phys. 95:7763

Haugen, H. K.; Weitz, E.; Leone, S. R. 1985 Chem. Phys. Lett. 119:75

Hernández, M. I.; Clary, D. C. 1994 J. Chem. Phys. 101:2779

Hernandez, R.; Miller, W. H. 1993 Chem. Phys. Lett. 214:129

Hernandez, R.; Miller, W. H.; Moore, C. B.; Polik, W. F. 1993 J. Chem. Phys. 99:950

Hoffmann, G.; Ohr D.; Chen, Y.; Engel, Y. M.; Wittig, C. 1990 Israel J. Chem. 30:115

Hoffmann, G.; Oh, D.; Iams, H.; Wittig, C. 1989a Chem. Phys. Lett. 155:356

Hoffmann, G.; Oh, D.; Wittig, C. 1989b J. Chem. Soc. Faraday Trans. 2 85:1141

Hoffmann, S. M. A.; Smith, D. J.; Grice, R. 1983 Mol. Phys. 49:621

Hofmann, H.; Leone, S. R. 1978 Chem. Phys. Lett. 54:314

Hu, A; Sharpe, S. W. 1994 unpublished

Huisken, F.; Kulcke, A.; Laush, C.; Lisy, J. M. 1991 J. Chem. Phys. 95:3924

Huisken, F.; Stemmler, M. 1988 Chem. Phys. Lett. 144:391

Huisken, F.; Stemmler, M. 1992 Z. Phys. D 24:277

Hunter, M.; Reid, S. A.; Robie, D. C.; Reisler, H. 1993 J. Chem. Phys. 99:1093

Ionov, S. I.; Brucker, G. A.; Jaques, C.; Chen, Y.; Wittig, C. 1993a J. Chem. Phys. 99:3420

Ionov, S. I.; Brucker, G. A.; Jaques, C.; Valachovic, L.; Wittig, C. 1992 J. Chem. Phys. 97:9486

Ionov, S. I.; Brucker, G. A.; Jaques, C.; Valachovic, L.; Wittig, C. 1993b J. Chem. Phys. 99:6553

Ionov, S. I.; Ionov, P. I.; Wittig, C. 1994 Faraday Discuss. Chem. Soc. 97:391

Jacobs, A.; Volpp, H. R.; Wolfrum, J. 1994 Chem. Phys. Lett. 218:51

Jacobs, A.; Wahl, M.; Weller, R.; Wolfrum, J. 1989 Chem. Phys. Lett. 158:161

Jonah, C. D.; Mulac, W. A.; Zeglinski, P. 1984 J. Phys. Chem. 88:4100

Jouvet, C.; Boivineau, M.; Duval, M. C.; Soep, B. 1987 J. Phys. Chem. 91: 5416

Jouvet, C.; Duval, M. C.; Soep, B.; Breckenridae, W. H.; Whitham, C.; Visticot, J. P. 1989 J. Chem. Soc. Faraday Trans. 2 85:1133

Jouvet, C.; Soep, B. 1983 Chem. Phys. Lett. 96:426

Jouvet, C.; Soep, B. 1984 J. Chem. Phys. 80:2229

Jouvet, C.; Soep, B. 1985 Laser Chem. 5:157

Kudla, K.; Koures, A. G.; Harding, L. B.; Schatz, G. C. 1992 J. Chem. Phys. 96:7465

Kudla, K.; Schatz, G. C. 1991 J. Phys. Chem. 95:8267

Kudla, K.; Schatz, G. C.; Wagner, A. F. 1991 J. Chem. Phys. 95:1635

Lee, L. C. 1985 J. Phys. B 18:L293

Lee,Y. T.; LeBreton, P. R.; McDonald, J. D.; Herschbach, D. R. 1969 J. Chem.Phys. 51:455

Lee,Y. T.; McDonald, J. D.; LeBreton, P. R.; Herschbach, D. R. 1968 J. Chem. Phys. 49:2447

Lee, Y. T.; Valentini, J. J.; Auerbach, D. J. 1977 J. Chem. Phys. 67:4866

Legon, A. C.; Willoughby, L. C. 1985 J. Mol. Struct. 131:159

Lin, Y.; Wittig, C.; Beaudet, R. A. n.d. unpublished

Loesch, H. J.; Beck, D. 1971 Ber. Bunsenges. Phys. Chem. 75:736

Loison, J. C.; Dedonder-Lardeux, C.; Jouvet, C.; Solgadi, S. 1994 Faraday Discuss. Chem. Soc. 97:379

Lovejoy, C. M.; Schuder, M. D.; Nesbitt, D. J. 1987 J. Chem. Phys. 86:5337

Lovejoy, E. R.; Kim, S. K.; Moore, C. B. 1992 Science 256:1541

Luo, X.; Fleming, P. R.; Seckel, T. A. ; Rizzo, T. R. 1990 J. Chem. Phys. 93:9194

Luo, X.; Rizzo, T. R. 1990 J. Chem. Phys. 93:8620

Luo, X.; Rizzo, T. R. 1991 J. Chem. Phys. 94:889

Lynch, G. C.; Halvick, P.; Zhao, M.; Truhlar, D. G.; Yu, C. H.; Kouri, D. J.; Schwenke, D. W. 1991 J. Chem. Phys. 94:7150

Magnotta, F.; Nesbitt, D. J.; Leone, S. R. 1981 Chem. Phys. Lett. 83:21

Manolopoulos, D. E.; Stark, K.; Werner, H. J.; Arnold, D. W.; Bradforth, S. E.; Neumark, D. M. 1993 Science 262:1852

Manthe, U.; Miller, W. H. 1993 J. Chem. Phys. 99:3411

Marshall, P.; Fontijn, A.; Melius, C. F. 1987 J. Chem. Phys. 86:5540

Marshall, P.; Ko, T.; Fontijn, A. 1989 J. Phys. Chem. 93:1922

Miller, W. H.; Hernandez, R.; Moore, C. B.; Polik, W. F. 1990 J. Chem. Phys. 93:5657

Miyawaki, J.; Tsuchizawa, T.; Yamanouchi, K.; Tsuchiya, S. 1990 Chem. Phys. Lett. 165:168

Miyawaki, J.; Yamanouchi, K.; Tsuchiya, S. 1991 Chem. Phys. Lett. 180:287

Miyawaki, J.; Yamanouchi, K.; Tsuchiya, S. 1993 J. Chem. Phys. 99:254

Mozurkewich, M.; Lamb, J. J.; Benson, S. W. 1984a J. Phys. Chem. 88:6429

Mozurkewich, M.; Lamb, J. J.; Benson, S. W. 1984b J. Phys. Chem. 88:6435

Nesbitt, D. J. 1988 Chem. Rev. 88:843, and references therein

Nesbitt, D. J.; Lovejoy, C. M. 1990 J. Chem. Phys. 93:7716

Nesbitt, D. J.; Lovejoy, C. M. 1992 J. Chem. Phys. 96:5712

Neumark, D. M. 1992 Acc. Chem. Res. 26:33

Novick, S. E.; Janda, K. C.; Klemperer, W. 1977 J. Chem. Phys. 65:5115

Oldershaw, G. A.; Porter, D. A. 1969 Nature 223:490

Pechukas, P.; Light, J. C. 1965 J. Chem. Phys. 42:3281

Peskin, U.; Reisler, H.; Miller, W. H. 1994 J. Chem. Phys. 101:8874

Polik, W. F.; Guyer, D. R.; Miller, W. H.; Moore, C. B. 1990a J. Chem. Phys. 92:3471

Polik, W. F.; Guyer, D. R.; Moore, C. B. 1990b J. Chem. Phys. 92:3453

Polik, W. F.; Moore, C. B.; Miller, W. H. 1988 J. Chem. Phys. 89:3584

Radhakrishnan, G.; Buelow, S.; Wittig, C. 1986 J. Chem. Phys. 84:727

Randall, R. W.; Walsh, M. A.; Howard, B. J. 1988 Faraday Discuss. Chem. Soc. 85:1

Ravishankara, R. R.; Thompson, R. L. 1983 Chem. Phys. Lett. 99:377

Reid, S. A.; Brandon, J. T.; Hunter, M.; Reisler, H. 1993a J. Chem. Phys. 99:4860

Reid, S. A.; Reisler, R. 1994 J. Chem. Phys. 101:5683

Reid, S. A.; Robie, D. C.; Reisler, H. 1994 J. Chem. Phys. 100:4256

Reisler, H.; Keller, H. M.; Schinke, R. 1994 Comments At. Mol. Phys. 30:191

Rice, J. K.; Lovas, F. J.; Fraser, G. T.; Suenram, R. D. 1995 J. Chem. Phys. 103:3877

Rizzo, T. R.; Hayden, C. C.; Crim, F. F. 1983 Faraday Discuss. Chem. Soc. 75:276

Sannigrahi, A. B.; Peyerimhoff, S. D. 1986 Int. J. Quantum Chem. 30:413

Schatz, G. C. 1989 Rev. Mod. Phys. 61:669

Schatz, G. C.; Dyck, J. 1992 Chem. Phys. Lett. 188:11

Schatz, G. C.; Fitzcharles, M. S. 1988 In *Selectivity in Chemical Reactions*, J. C. Whitehead, ed., Kluwer, Dordrecht, p. 353

Schatz, G. C.; Fitzcharles, M. S.; Harding, L. B. 1987 Faraday Discuss. Chem. Soc. 84:359

Scherer, N. F.; Khundkar, L. R.; Bernstein, R. B.; Zewail, A. H. 1987 J. Chem. Phys. 87:1451

Scherer, N. F.; Sipes, C.; Bernstein, R. B.; Zewail, A. H. 1990 J. Chem. Phys. 92:5239

Schneider, L.; Meier, W.; Welge, K. H.; Ashfold, M. N. R.; Western, C. M. 1990 J. Chem. Phys. 92:7027

Segall, J.; Wen, Y.; Singer, R.; Wittig, C.; Garcia-Vela, A.; Gerber, R. B. 1993 Chem. Phys. Lett. 207:504

Segall, J.; Zhang, J.; Dulligan, M.; Beaudet, R. A.; Wittig, C. 1994 Faraday Discuss. Chem. Soc. 97:195

Sharpe, S. W.; Sheeks, R.; Wittig, C.; Beaudet, R. A. 1988 Chem. Phys. Lett. 151:267

Sharpe, S. W.; Zeng, Y. P.; Wittig, C.; Beaudet, R. A. 1990 J. Chem. Phys. 92:943

Shea, J. A.; Read, W. G.; Campbell, E. J. 1983 J. Chem. Phys. 79:614

Shin, S. K.; Chen, Y.; Böhmer, E.; Wittig, C. 1992 In *The Dye Laser: 20 Years*, M. Stuke, ed., Springer-Verlag, Berlin, p. 57

Shin, S. K.; Chen, Y.; Nickolaisen, S.; Sharpe, S. W.; Beaudet, R. A.; Wittig, C. 1991a In *Advances in Photochemistry*, Vol. 16, D. Volman, G. Hammond, and D. Neckers, eds., Wiley, New York, p. 249

Shin, S. K.; Chen, Y.; Oh, D.; Wittig, C. 1990 Philos. Trans. R. Soc. Lond. Ser. A 332:361

Shin, S. K.; Wittig, C.; Goddard, W. A., III. 1991b J. Phys. Chem. 95:8048

Sims, I. R.; Gruebele, M.; Potter, E. D.; Zewail, A. H. 1992 J. Chem. Phys. 97:4127

Sinha, A.; Vander Wal, R. L.; Crim, F. F. 1990 J. Chem. Phys. 92:401

Smith, I. W. M. 1977 Chem. Phys. Lett. 49:112

Smith, I. W. M. 1980 *Kinetics and Dynamics of Elementary Gas Reactions*, Butterworths, London, p. 202

Smith, I. W. M.; Zellner, R . 1973 J. Chem. Soc. Faraday Trans. 2 69:1617

Soep, B. 1994 unpublished

Spencer, D. J.; Wittig, C. 1979 Optics Lett. 4:1

Stockman, P. A.; Blake, G. A. 1993 Chem. Phys. Lett. 212:298

Tichich, T. M.; Likar, M. D.; Dtbal, H. R.; Butler, L. J.; Crim, F. F. 1987 J. Chem. Phys. 87:5820

Tichich, T. M.; Rizzo, T. R.; DUbal, H. R.; Crim, F. F. 1986 J. Chem. Phys. 84:1508

Trickl, T.; Wanner, J. 1983 J. Chem. Phys. 78:6091

Truhlar, D. G.; Schwenke, D. W.; Kouri, D. J. 1990 J. Phys. Chem. 94:7346

Viste, A.; Pyykkö, P. 1984 Int. J. Quantum Chem. 25:223

Walch, S. P. 1993 J. Chem. Phys. 98:1170

Walker, A. R. H.; Chen, W.; Novick, S. E.; Bean, B. D.; Marshall, N. D. submitted

Warnatz, J. 1984 In *Combustion Chemistry*, W. C. Gardiner, ed., Springer-Verlag, New York, p. 197

Wen, Y.; Segall, J.; Dulligan, M.; Wittig, C. 1994 J. Chem. Phys. 101:5665

Wiesenfeld, J. R.; Wolk, G. L. 1978a J. Chem. Phys. 69:1797

Wiesenfeld, J. R.; Wolk, G. L. 1978b J. Chem. Phys. 69:1805

Wittig, C.; Engel, Y. M.; Levine, R. D. 1988a Chem. Phys. Lett. 153:411

Wittig, C.; Ionov, S. I. 1994 J. Chem. Phys. 100:4714

Wittig, C.; Sharpe, S.; Beaudet, R. A. 1988b Acc. Chem. Res. 21:341

Wright, S. A.; Tuchler, M. F.; McDonald, J. D. 1994 Chem. Phys. Lett. 226:570

Xu, Z.; Koplitz, B.; Wittig, C. 1987 J. Chem. Phys. 87:1062

Xu, Z.; Koplitz, B.; Wittig, C. 1988 J. Phys. Chem. 92:5518

Zellner, R.; Steinert, W. 1976 Int. J. Chem. Kin. 8:397

Zeng, Y. P.; Sharpe, S. W.; Shin, S. K.; Wittig, C.; Beaudet, R. A. 1992 J. Chem. Phys. 97:5392

Zewail, A. H. 1988 Science 242:1645, and references therein

Zewail, A. H. 1993 J. Phys. Chem. 97:12427, and references therein

Zhang, J.; Dulligan, M.; Wittig, C. n.d. unpublished

4

Photochemistry of van der Waals Complexes and Small Clusters

C. JOUVET AND D. SOLGADI

4A. REACTION OF EXCITED VAN DER WAALS COMPLEXES

4A.1. INTRODUCTION

In a chemical reaction, the shape of the potential energy surface (PES) dictates the reaction rate and energy disposal in the products. Not only does the dynamics depend crucially upon the features of the surface, but, ultimately one seeks to influence the course of the reaction by preparing selectively certain regions of the surface. For harpooning reactions, the propensity rules for energy disposal in the products (influence of the entrance kinetic energy, effect of the early or late barrier) have been established by Polanyi (1972) and have been used later as guidelines. Here, the surface may easily be modeled in simple terms using long-range electrostatic interaction in the entrance valley. There was, then, need of an experimental method which allows the possibility of observing directly the characteristic regions of this potential energy surface, but also to investigate precisely the surface in other types of reaction. The study of the reactivity of van der Waals complexes is intended to fulfil this purpose.

In classical experiments, the surface is obtained by inversion of the experimental data which are differential cross sections and internal energy distribution of the products. This procedure is difficult and not unambiguous. The first step is to determine the correlation between the entrance channel's parameters (kinetic energy, internal energy, angular momentum) and the final states of the products (kinetic energy, internal energy, angular distribution). This requires a precise control of the entrance channel. Therefore, the goal of many experiments is to reduce the initial states to a small subset, and to measure the energy disposal in the products with the greatest accuracy. This was first achieved by controlling the kinetic energy of the reactants in crossed beam experiments. Later, a certain control of the collision geometry was obtained by orienting the molecules or the atomic orbitals in crossed beam experiments or by using prealigned systems in a van der Waals complex: this subject is discussed in Buelow et al. (1986).

Access to the final states is achieved by the use of standard techniques such as angular, time-of-flight mass spectrometry for kinetic energy and angular distribution, and laser-induced fluorescence or multiphoton ionization for the internal degrees of freedom.

In the last ten years, the idea of getting direct spectroscopic information on the reactive surface by optically probing the collision complex during its formation, and its decay into products, has emerged. Some experiments, in which a laser photon absorbed by the collision pair induces the chemical reaction, have been successful; the frequency of the exciting photon being directly related to the shape of the surface. For example, Brooks and coworkers have performed such an experiment of $K + HgBr_2 + hv \rightarrow KBr + HgBr$ (Hering et al. 1980) and $K + NaCl + hv \rightarrow Na + KCl$ (Maguire et al 1983) in a crossed beam apparatus. The same kind of experiments performed in gas phase are easier as the number of collisions is greater, but they are still difficult. They have been performed on such systems as $Xe + Cl_2$ (Dubbov et al. 1981; Inoue et al. 1984) or $Mg + H_2$ (Kleiber et al. 1985, 1986). In each case, the qualitative theoretical interpretation is difficult, sometimes because of the low signal-to-noise ratio, but mainly because the spectroscopic information is averaged over all geometries and energies of the collision pair; hence, the deconvolution of the data is neither obvious nor unambigous.

On the other hand, photodissociation of the molecule is often considered as half collision. Optical excitation of the stable molecule prepares the collision partners in a state with a well defined symmetry, geometry, and angular momentum. Photodissociation experiments are in constant development, and can now give very precise information on the dissociative potential energy surface and on the dynamics. The distribution of the internal energy in the products is then easily obtained (REMPI or LIF experiments) and the development of Doppler profile spectroscopy allows measurements of the kinetic energy. The method of vector correlation gives very detailed information on the dissociation: as an example, in the photodissociation of H_2O_2, it has been determined that the rotation J_{OH} vector is aligned with the recoil velocity v_{OH} vector. This indicates that the fragment rotation is predominantly generated by the torsion about the O–O axis (Docker et al. 1986; Klee et al. 1986) and not by the repulsion forces along the O–O fragmentation axis, as could be expected.

Moreover, the absorption spectra or the emission during the dissociation [O_3, (Imre et al. 1982), NaI (Foth et al. 1982)] give direct information on the intermediate states of reaction, and on the dynamics, not only at the turning point as for the laser-assisted collisions.

The study of reactive excited states of van der Waals complexes is the link between the laser-assisted collision and the photodissociation approach; it brings the collisional problem into a much simpler photodissociation problem. Here, a cold complex which has a defined geometry is formed between the collision partners and optically excited to trigger the reactive process. This creates the photodissociation of a molecule with very weakly bound ground state. The van der Waals spectroscopy has already allowed the accurate determination of the interatomic potential [Na–Ar (Smalley et al. 1977; Tellinghuisen et al. 1979), HgAr (Breckenridge et al. 1985, 1994; Fuke et al. 1984)]. More complex collisional problems, such as collision-induced vibrational (Brumbaugh et al. 1983; Halberstadt and Soep 1984; Stephenson and Rice 1984) or electronic (Goto et al. 1986; Jouvet and Soep 1981; Rice 1986) relaxation have been already studied by dissociation of the relevant van der Waals complex.

If one considers the van der Waals complexes as a way to study binary collisions, the possibility of the formation of clusters of given size is a way to probe the role of the environment of other molecules on the reactivity. It is well known that solvent effects play an important role, not only in the kinetics but also in the results of chemical reaction. The study of molecular clusters in supersonic jet experiments allows step-by-step solvation of reactants: as will be shown in this chapter, most of the reactions which have been studied occur when a finite number of molecules is reached—this number being often small (less than ten molecules).

4A.2. PREPARATION OF THE COMPLEX

The cold complexes are obtained in the supersonic expansion. It is obvious that in order to obtain the van der Waals complex, the chemical reaction should not occur in the gas mixture before expansion. The technique is therefore limited to systems in which the ground state partners do no react, or react very slowly in the gas mixture before expansion. However, even if a small percentage of the reactants has disappeared in the mixture, it is still possible to perform the experiment by discriminating the cold fragment formed directly before the expansion from the hot fragment issued from the reaction in the van der Waals complex. The stability of the ground state van der Waals complex on a reactive surface with an early barrier has been demonstrated in the case of $Hg + Cl_2$ (Jouvet and Soep 1983). At the temperature (400–500 K) necessary to get a sufficiently high vapor pressure of mercury, the reaction $Hg + Cl_2 \rightarrow HgCl_2$ proceeds (Leighton and Leighton 1935; Ogg et al. 1936) through a barrier estimated at 5 kcal. It was nevertheless possible to obtain the cold $Hg–Cl_2$ complex by mixing the reactants just before the supersonic expansion. In the jet, both $HgCl_2$ molecules and $Hg–Cl_2$ complexes were present, but only the excited complexes yielded the observed fluorescence of the $HgCl(B^2\Sigma^+)$ product.

The first point to be considered is how this kind of half collision is connected to other types of collisional experiments. As already mentioned, in order to describe a collision the parameters of the entrance channel have to be known and those of the output channels have to be precisely measured. In the van der Waals method they can, in principle, all be known, but only a few can be varied.

4A.2.1. The Entrance Channel

The relative kinetic energy of the reagent is very small; it cannot be reduced to less than half a quantum of the vibration of the internuclear van der Waals mode. On the other hand, this energy can be increased by exciting the optically active van der Waals mode in the upper surface. This is usually the case in metal–molecule complexes (Breckenridge et al. 1985; Fuke et al. 1984) where a long progression of this mode is observed.

The total angular momentum J can be well defined if the upper rotational levels are resolved. This is possible when the upper surface is not totally repulsive. On the other hand, if the upper state is dissociative, and therefore unstructured, the angular momentum is related to the thermal distribution in the jet. The maximum value of J is given by the square root of KT_{rot}/B, where T_{rot} is the rotational temperature in the jet and B the rotational constant of the ground state complex. Typically, T_{rot} is 5–10 K and $B = 0.1$ cm^{-1} leading to $J < 10$.

The initial geometry is the geometry of the ground state complex which can, in principle, be determined.

In the case of the atom–atom problem, the initial geometry coresponds to the initial distance from which the partners are going to evolve on the reactive surface, but depends also on the symmetry of the prepared electronic state. The distance is defined as the ground state equilibrium distance and the uncertainty on this value is the width of the vibrational wavefunction of the vibrationless gound state level.

The symmetry selection is an essential parameter in orienting the chemical reaction, as it defines the orientation of the orbitals of the incoming reactants, and therefore the shape of the potential energy surface. This alignment effect was demonstrated, for example, in the reactivity of excited Ca with HCl by Rettner and Zare (1981). This symmetry of the electronic state selection is easily obtained in the excitation of the van der Waals complex. For example, when one of the atoms is excited to a P state, the two electronic states of the complex (Σ and Π) can be easily assigned as they have neither the same binding energy nor the same rotational selection rules. For example, in the Hg/rare gas systems, the two states issued from the excitation of the $^3P_1$ state of Hg can be unambiguously assigned through the rotational study of Hg–He (Duval et al. 1985): the more strongly bound state being the $\Omega = 0$ state and the shallower one the $\Omega = 1$ state (Hund's case C notation).

For the atom–diatom case, considerations for the distance between the two reactants are the same as for the atom–atom case. The geometry of the complex is kept fixed by the anisotropy of the atom–diatom potential. This geometry, often linear- or T-shaped, can be obtained by spectroscopic measurements (infrared or microwave). Nevertheless, for light diatoms (H_2 and D_2) the anisotropy of the ground state potential is much smaller than the rotational constant, and then the diatom is considered as freely rotating: the diatom does not necessarily rotate but the geometry is totally undefined. In this case, it is going to be convenient to label the electronic state in C ∞ v notation. Nonetheless, even when the anisotropy is stronger than the rotational constant of the diatom, the van der Waals molecule is fairly floppy. For example, in the Hg–HCl complex (Shea and Campbell 1984), the microwave spectrum shows that the molecule is linear—the H atom being between Hg and Cl—but that a wide bending amplitude (35°) is still present. In the few examples studied by this technique the geometry was unknown, but it was necessary to guess one to understand the results.

For more complex systems (Xe–CClBr$_3$ (Richmann et al. 1993), the geometry of the ground state complex can induce a strong selectivity of the chemical reaction (on the products). The cold complex can be locked in one of the possible conformers (corresponding to a preferential bond with one of the halogen atoms) which will favor the reaction with that closest atom.

4A.2.2. The Final States

All the techniques already developed for photodissociation dynamics can be used here (laser-induced fluorescence, time resolved spectroscopy, multi photon spectroscopy, etc).

4A.2.3. The Intermediate State of Reaction

In the photodissociation of a stable molecule, the absorption spectrum gives some information on the main features (direct dissociation, predissociation) of the energy potential surface. In the same manner, the absorption spectrum of the reactive van der Waals complexes will display the characteristic features of the potential energy surface of the collision pair. The portion of the surface which will be characterized by the optical excitation is the region which is Franck–Condon accessible from the ground state complex. This corresponds to a distance between the two reactants of typically 3.5–5 Å. In a harpoon type reaction, this distance is in the vicinity of the crossing radius between the covalent and the ionic potential curve. In that case, the excitation of the van der Waals complex will give directly spectroscopic information on the mixed state resulting from the avoided crossing. In a covalent type of reaction, occurring at a shorter distance, the intermediate state precursor of the reaction is prepared and observed, that is, the entrance channel collision complex is studied. As will be seen later, this is a very sensitive way to demonstrate the presence of a barrier on the surface.

It can be pointed out that the van der Waals technique is the equivalent of the excitation of the collision pair or the laser-assisted collision, and we will discuss the differences and similarities between the excitation of the complex and the collisional process.

In the van der Waals complex technique, a smaller region of the reactive surface is explored but with no, or very small, kinetic energy. For example, in the complex case, the potential energy surface which lies below the energy of the free reactants can be studied with variable but small kinetic energy, depending on the van der Waals stretching quantum excited. In a collision experiment, the same region will be studied with larger kinetic energy. On the contrary, when the complex is excited above the dissociation limit, the process is very similar to the collision one. The direct dissociation of the complex into the initial reactants is the equivalent of the collisional elastic process.

An important aspect of this kind of experiment is the time evolution of the reactive excited complex. It can be expected that for a reaction near the threshold, fine tuning of the optical excitation should yield to drastic changes in the reactive decay time as the spectroscopy already shows for the Ca–HCl system (Keller 1991, Soep et al. 1991, 1992). Real time evolution of binary reactive collisions can be studied through van der Waals complexes, since time $t = 0$ is defined by the excitation laser, as well as the starting internuclear distance between reactants which is fixed by the ground state geometry. This approach has been used for ground state reactions (Buelow et al. 1986; Shin et al. 1991; Takayanagi and Hanazaki 1991; Zewail 1991) and on "nonreactive" species such as HgAr (Krim 1994, Krim et al. 1994).

We shall discuss some examples of reactions of excited van der Waals complexes. Up to now, only a few examples of a atom–diatom reactions have been studied—the atom being Hg, Ca, or Xe, and the diatom being H_2 and halogen-containing molecules. These examples show clearly the new features in the reaction dynamics, such as orbital specificity, selectivity in the products, products state distribution, and observation of the intermediate states.

4A.3. ORBITAL SELECTIVITY IN THE Hg–H_2 VAN DER WAALS COMPLEX REACTION

It was known that, in the gas phase, the excitation of the resonant $Hg(6^1S_0-6^3P_1)$ transition in the presence of H_2 leads to the formation of $HgH(^2\Sigma^+)$ through a direct mechanism (Breckenridge 1983; Callear and McGurk 1972; Callear and Wood 1972; Vikis and le Roy 1973). The quenching cross section of $Hg(^3P_1)$ by H_2 (30 Å^2) suggested that the reaction in the complex should be slow enough for a structured intermediate state to be observed in the region accessible from the ground state complex.

The reactivity of $^3P_1$ mercury within the Hg–H_2 complex depends crucially upon the mercury 6p orbital orientation with the complex axis, as has been demonstrated (Breckenridge et al. 1986, 1987; Jouvet and Soep 1985b). The pπ preparation of the reaction provokes a rapid direct insertion of Hg in the H–H bond, while in the pσ preparation the reaction is much slower and indirect. It is important to mention how straightforwardly the orbital orientation may be achieved in a complex of a p-excited atom. The combination of the attractive and repulsive forces between the metal and the molecule result in the formation of two different electronic states of different energies depending upon the orbital orientation of the atomic orbital with respect to the complex axis. In the very simple case of the diatomic Hg/rare gas complexes, those states on both sides of the mercury $^3P_1$ line have been observed—the pπ state being the deepest state. In Hund's case c for mercury complexes, the p orbital is not as well aligned as it would be in case a—the $\Omega = 0$ state corresponds to a $^3\Pi$ state, and the $\Omega = 1$ state is a linear combination of $^3\Sigma + ^3\Pi^*$. Also, in Hund's case c, as in case a, the rotation of the complex may destroy the orbital orientation: this effect of the Coriolis coupling will vary with the ratio of the rotational spacing ($2BJ$) to the Ω states separation. At the very low rotational temperatures achieved in our experiments (≈ 3 K), this condition will always be met.

On the other hand, in collisions involving an important rotational momentum, as at room temperature, the distinction between the pπ and pσ excitations may be reduced. Hence, reactions within complexes provide a good and simple means of studying orbital stereoselectivity. The reaction of $Hg(^3P_1)$ within the Hg–$H_2(^3P_1)$ complex is a good example, as shown in Figure 4-1. The action spectrum displayed in the figure measures the efficiency to produce the HgH molecule as a function of the exciting frequency: the rapid reaction in the $^3\Pi$ ($\Omega = 0$) potential region is

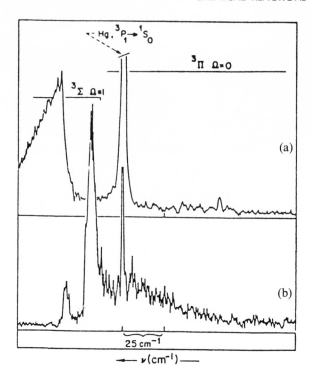

Figure 4-1. (a) Fluorescence excitation spectrum of the Hg–H$_2$ complex. (b) Action spectrum of the Hg–H$_2$ complex: the pump laser is varied while the probe laser sits on the band head of the 0–0 transition in Hg–H($^2\Pi_{1/2} \leftarrow ^2\Sigma^+$). In the $\Omega = 1$ domain extending into the continuum seen in (a), one sees vibrational bands—the reaction is then slow; for the $\Omega = 0$ excitation (in the red), no structure appears—indicating a fast reaction.

shown by a continuous spectrum, while the structure in the $^3\Sigma$ ($\Omega = 1$) region indicates a process slower than 3 ps (Jouvet and Soep 1985b).

Moreover, a small barrier to the reaction is observed in the entrance valley of the $^3\Sigma$ approach, amounting at least to 100 cm^{-1} (Breckenridge et al. 1987). This barrier is evidenced by a direct dissociation of the complex into Hg($^3P_1$) + H$_2$ when the exciting photon provides greater energy than the Hg–H$_2$ $^3\Sigma$ well. These results have been confirmed by the *ab initio* calculations of Bernier and Millié (1987). It is found that the the Π surface in the Hg($^3P_1$) + H$_2$ is attractive and that the reaction proceeds through the lengthening of the H–H bond followed by mercury insertion at a distance close to 1.5 Å (Jouvet and Soep 1985b). Furthermore, a barrier is also observed in the other configuration $^3\Sigma$ ($\Omega = 1$).

This example shows that use of the van der Waals technique brings out important features on the potential surface, in the differences in reactivity, and in energy distribution upon the symmetry of the entrance channel. It would be very interesting to see if the orbital specificity observed in the complex persists in the collisional regime.

4A.4. THE VAN DER WAALS TECHNIQUE AS A PROBE FOR A HARPOON-TYPE CHEMICAL REACTION

The harpoon mechanism was suggested by Polanyi (1932) for reactions between alkalis and halogens and has been often considered for many systems in which the reactive cross section is very large (100, 200 Å^2). In this mechanism, reaction results from the crossing between covalent and ionic curves in the entrance channel within the collision complex: an electron jumps from one of the reagents to the other. Large electron affinity and low ionization potential yield large crossing radii and then large reactive cross sections.

Harpoon-type reactions which have been studied by the van der Waals technique are:

$$Hg^* + Cl_2 \rightarrow HgCl(B\ ^2\Sigma^+) + Cl$$

(Jouvet and Soep 1983; Jouvet 1985; Jouvet and Soep 1985a)

$$Xe^* + X_2 \quad \text{or} \quad Xe + X_2^* \rightarrow XeX(B, C) + X \qquad (X_2 = Cl_2, Br_2, I_2)$$

(Boivineau 1987; Boivineau et al. 1986a,b,c)

$$Xe^* + BrCCl_3 \rightarrow XeBr + CCl_3 \text{ (Richmann et al. 1993)}$$

$$Ca^* + HX \rightarrow CaX^* + H \text{ (Keller 1991; Soep et al. 1991, 1992)}$$

4A.4.1. The Hg–Cl₂ Case

The $Hg(6^3P_{0,1,2}) + Cl_2$ leading to the HgCl(B) state has already been extensively studied in gas (Dreiling and Setser 1984; Husain et al. 1980; Wadt 1980) and in crossed beam experiments (Krause et al. 1975), and the cross section was measured to be in the order of 100 Å^2, which suggests a harpoon-type mechanism. This system was chosen since the crossing between the covalent $Hg(6^3P_1) + Cl_2$ and the ionic $Hg^+ + Cl_2^-$ curve occurs at an internuclear distance (about 4 Å) similar to the expected ground state equilibrium distance of the $Hg(^1S_0)$–Cl_2 van der Waals complex (Figure 4-2). Thus, the intermediate state resulting from the

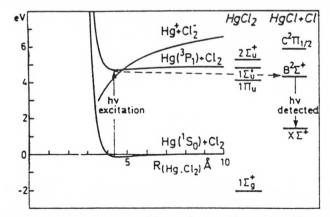

Figure 4-2. Potential energy diagram of the $Hg(^3P_1) + Cl_2$ reaction. The entrance channel (on the left) and the exit channel ($HgClB^2\Sigma^+ \rightarrow X\Sigma^+$ fluorescence) are indicated.

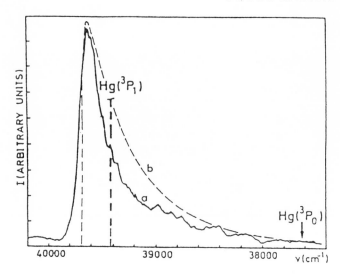

Figure 4-3. (a) Action spectrum of the Hg–Cl$_2$ complex: excitation spectrum of the Hg–Cl$_2$ complex while monitoring the B → X emission of the HgCl product. (b) Simulation of the action spectrum. The upper reactive state is obtained by mixing the ionic (Hg$_2^+$, Cl$^-$) potential and covalent (van der Waals) curve.

crossing should be easily observed. Indeed, the intermediate state is revealed through the action spectrum displayed in Figure 4-3. This spectrum, in which the total emission HgCl(B$^2\Sigma^+$ + –X$^2\Sigma^+$) of the product HgCl(^{2}B) is observed as a function of the exciting laser frequency, corresponds to the excitation of the Hg–Cl$_2$ van der Waals complex onto the reactive surface. This spectrum is broad, more than 1000 cm^{-1} wide, stopping abruptly 300 cm^{-1} above the Hg resonance transition, and no fluorescence, even in the UV, is observed above this cutoff. As for the Hg–H$_2$ ($^3\Pi$) excitation, the spectrum is totally continuous and structureless indicating a fast reaction without any activation barrier.

These results can be understood using a simple model where the accessed state is a mixed state resulting from the avoided crossing of the covalent and ionic states, the covalent one bearing the oscillator strength. The simulation shown in Figure 4-3 results from the interaction of one weakly bound van der Waals potential and the ionic Hg$^+$–Cl$_2$ curve.

The important parameter is the coupling coefficient between the ionic and the covalent curves. Since the observed spectrum lies in the vicinity of the covalent curve, this implies that the coupling term is small ($\cong 0.1$ eV), which has been interpreted as the effect of the ground state—a small coupling being expected for the T-shaped complex (Jouvet et al. 1987).

This experiment shows that the harpoon mechanism can be observed through the spectroscopic observation of the charge transfer state Hg$^+$Cl$_2^-$. Moreover, it shows the interest of starting from a fixed geometry to understand the spectroscopy of the reactive collision complex.

4A.4.2. The Rare Gas (Rg) Halogen (X₂) Reactions

During the last ten years, reactive collisions $Rg^* + X_2$ or $Rg + X_2^*$ leading to the RgX* excimer through the harpoon mechanism have been widely studied (Ishiwata et al. 1984; Johnson et al. 1986a,b, 1987; Le Calvé et al. 1985; O'Grady and Donovan 1985; Tamagake et al. 1979; Wilkinson et al. 1986; Yu et al. 1983). These systems and mainly the $Xe + Cl_2$ or Br_2 have been the first ones where the collision complex was excited in the gas phase by one photon (Dubbov et al. 1981; Grieneisen et al. 1983; Wilcomb and Burnham 1981) or two photons (Inoue et al. 1984; Ku et al. 1983; Setser and Ku 1985). The same two-photon excitation technique has been used for the excitation of the $Xe–X_2$ van der Waals complexes $(X_2 = Cl_2, Br_2, I_2)$. These results can be then compared with the collisional ones.

The reactions observed in the $Xe–X_2$ complexes bear some striking features due to the excitation within the complex, but also to the fact that either Xe or X_2 can be chromophores. This is shown in the action spectra, the energy disposal in the products, and in the reactions with the various halogens.

a. *The action spectra*

The formation of XeX electronically excited molecules has a high energetical threshold and involves either the excitation of metastable Xe atoms (^3P) or high lying states of the halogen. These states can be reached by two or three photons' excitation, mediated by low lying states. As in the preceeding reactions, the accessed states are correlated with the electronic states of the components. Surprisingly, no chemiluminescence of XeCl is observed while exciting the $XeCl_2$ complex in the 440 nm region (Figure 4-4), where the three-photon atomic transition $Xe(^1S_0–^3P_1)$ (Faisal et al. 1977) is expected. This situation is distinct from the preceding $Hg–Cl_2$ case where the absorption arising from the $Hg(^1S_0–^3P_1)$ transition could shine upon the crossing with the $Hg^+ + Cl_2^-$ potential (Figure 4-2). On the contrary, this crossing radius between the $Xe^+ + Cl_2^-$ and $Xe(^3P_1) + Cl_2$ potential curves lies at 5 Å, outside the Franck–Condon accessible region located close to 3.5 Å. Thus, the optical excitation views the inside region of this well, where other crossings must occur that do not lead to the chemiluninescent reaction.

The influence of the relay states to form the XeCl B or C luminescent products is observed in Figure 4-5. While monitoring the B state formation, the action spectrum is structured with a vibrational spacing close to the one observed in the X–B transition in XeCl. On the other hand, the observation of the C state yields a broad action spectrum. These results are striking and specific to the complex excitation: as in the Setser and Kulaser assisted collision experiment, no difference was noticed in the action spectra observing the B or C states of XeCl (Boivineau et al. 1986b). This interesting specificity is explained by the fixed geometry in the ground state complex and the two-photon excitation. The two-photon transition is enhanced by the Cl_2 $(X^1\Sigma_g–^1\Pi_u)$ transition leading to an unstructured intermediate state. The other two-photon excitation must involve the XeCl bond, the same vibrational structure as in the XeCl(X–B) absorption being observed. Thus, in the latter case the absorption of the second photon should occur while the new XeCl bond is being formed; this closely resembles the Imre and Kinsey experiment (Imre et al. 1982).

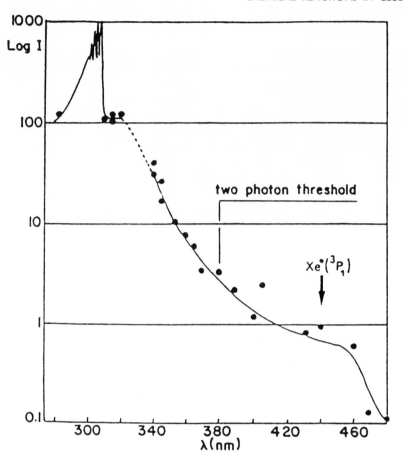

Figure 4-4. Action spectrum obtained under multiphoton excitation of the Xe–Cl$_2$ complex. The observable peak is XeCl(B–X) emission at 308 nm. The arrow corresponds to the three-photon Xe(^{3}P$_1$) excitation, and the two-photon energetical threshold is given by the bar.

b. *The vibrational distribution in the* XeCl *product*
In the collisional process, the XeCl excimer is found to be vibrationally excited; vibrational levels as high as $v = 100$ are populated (Ishiwata et al. 1984; Johnson et al. 1986a,b, 1987; Le Calvé et al. 1985; O'Grady and Donovan 1985; Tamagake et al. 1979; Wilkinson et al. 1986; Yu et al. 1983). In the complex case, the product is vibrationally cold and the highest v' level populated reaches only $v' = 10$, through the excess of energy is similar in both cases (≈ 2 eV). The difference was explained (Boivineau et al. 1986a) in terms of initial geometry difference in the entrance channel. In the collisional experiment involving a harpoon mechanism, the system reaches the triatomic ion-pair surface in the crossing region at ≈ 5 Å where the Xe$^+$ Cl$^-$ ion pair is formed correlating to the B and C states of XeCl. This distance, 5 Å, is far from the equilibrium distance of the B and C states, 2.9 Å (Boivineau et al. 1986c), and will lead to a highly vibrationally excited XeCl

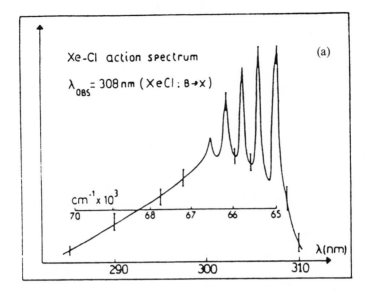

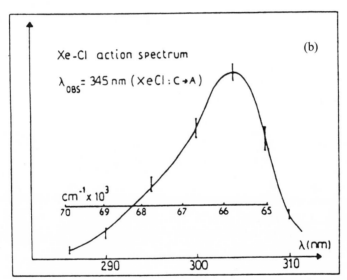

Figure 4-5. (a) XeCl (b) action spectrum under two-photon excitation of the Xe–Cl$_2$ complex ($\lambda_{pbs.}$ = 308 nm). (b) XeCl (c) action spectrum (λ = 345 nm). One can see that the structured spectrum appears only when the XeCl (B–X) emission is monitored.

product. On the other hand, in the complex, the system is promoted on an ion-pair surface with an Xe–Cl distance of the ground state complex (≈ 3.5 Å),[1] which is not too far from the XeCl(B, C) equilibrium distance, and yields a vibrationally cold product.

[1] Assuming that the ground state equilibrium distance of the Xe–Cl$_2$ distance is similar to the Xe–Cl(X) distance (3.3 Å), as in this state, the interaction between Cl and Xe is typically van der Waals (Sur et al. 1979).

4A.4.3.   The Xe BrCCl₃ Reaction

Comparison between collisions with the metastable Xe atom, laser-assisted collision and van der Waals excitation has been performed by Richman et al. (1993) on the Xe + BrCCl₃ system. Excitation of xenon with BeCCl₃ leads to two possible channels: formation of XeBr or XeCl. The relative proportion of each product is highly sensitive to the entrance channel and to the experimental conditions, as listed below.

(1) The full collision of metastable $Xe(6s, {}^3P_2)$ atoms with $BrCCl_3$ favors the formation of XeBr(B, C) rather than XeCl by a ratio of 2:1 (Setser and Qim 1991). This behavior is expected since in the ground state of $BrCCl_3^-$ the electron is more likely to be located on the C–Br antibonding orbital. In this process, the reaction starts on the covalent Xe^*–RX potential and evolves to the Xe^+ RX^- potential, thus leading to XeX(B, C) products.

(2) Laser-assisted excitation of xenon with halogen-containing molecules has been studied with thermal collision pairs. With $BrCCl_3$, the XeCl(B, C) product seems to be strongly favored. However, controlling the initial conditions can affect the chemical product branching and the product energy distribution. Even for thermal conditions, the photoexcitation of the collision pair to the reactive ion-pair potential leads to some selectivity with respect to the X–RX bond distance (which is estimated to be two or three times the van der Waals distance) in the entrance channel.

(3) In the Xe–BrCCl₃ complex, excitation of the complex at 247 nm leads exclusively to the XeBr(B, C) formation. The fixed geometry of the complex can explain this result. If, in the complex, the xenon atom is directly bonded to the bromine atom, then the proximity of these two atoms can induce this selectivity.

This reaction also shows the importance of the entrance channel for the reaction and the resulting products, and it is the first evidence of product control of the reaction path using the geometrical selectivity achieved in van der Waals complexes.

4A.5.   OBSERVATION OF THE REACTIVE POTENTIAL ENERGY SURFACE OF THE Ca–HX* SYSTEM THROUGH VAN DER WAALS EXCITATION

Reactions within a van der Waals (vdW) complex of calcium with hydrogen halides (HCl and HBr) lead to electronically excited calcium halides. These reactions have been quite extensively studied in full collisions of excited calcium beams (Brinckmann et al. 1980; Brinckmann and Telle 1977; Rettner and Zare 1981, 1982; Telle and Brinckmann 1990). The electronic excitation of the calcium atom results in a strong chemiluminescence under collisional conditions. The efficiency of this chemiluminescence depends upon the electronic state and the fine structure component, and the final product state is influenced by the preparation conditions of the collision. In the reaction $Ca(4s4p^1P_1) + HCl$, the direction of the polarization of the P orbital with respect to the collision relative velocity (p_π or p_σ) has an effect on the branching ratio to the products CaCl, $A^2\Pi$ or $B^2\Sigma^+$ (Rettner and Zare 1981, 1982).

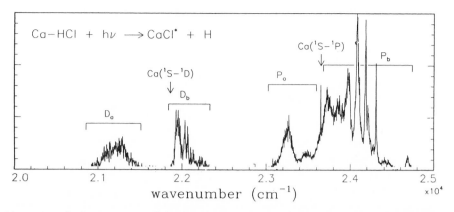

Figure 4-6. Action spectrum of the Ca–HCl van der Waals complex. The exciting laser has been varied between 2100 and 25000 cm^{-1} while recording the intensity of the CaCl(A–X) transition (Soep et al. 1991). One observes vibrational progressions associated with two van der Waals electronic states for each excited Ca electronic state. For the P_b labeled state, the width of the vibrational progression decreases as the energy increases.

In the vdW complex (Keller 1991; Soep et al. 1991, 1992), transitions to electronic states correlating at infinite distance to the $4s4p^1P_1$ and $4s3d^1D_2$ states of calcium have been observed in the spectra of the laser-induced reactions. For example, the action spectrum in the 4000–5000 Å range, obtained by monitoring the CaCl A–X (or B–X) emission, is shown in Figure 4-6. This region covers the Ca ($^1S_0-^1D_2$) and ($^1S_0-^1P_1$) transitions and does not overlap with CaCl transitions. No chemiluminescence signal is observed at longer wavelengths than 5000 Å due to the energy threshold for CaX* formation.

The action spectrum presented in Figure 4-6 reveals two sets of bands localized near each of the two calcium transitions. Both sets extend over about 1000 cm^{-1} with broad structures in their red part and narrow structures in their blue part. As seen in Figure 4-6, the Ca($^1S_0-^1P_1$) transition clearly exhibits a vibrational structure.

The large spacing between the bands in Figure 4-6 ($^1P_1$ region) must correspond to vibrational transitions in the excited Ca–HCl complex. In order to assign these bands, HCl has been replaced by DCl. This experiment has shown a strong isotopic effect resulting in an important decrease of the band spacings.

The spectra have been analyzed in terms of a local excitation of the calcium and internal van der Waals coordinates (stretching and bending). Indeed, strong resonances are observed in these spectra, which can be assigned to van der Waals modes. The excitation of these vibrations allows the exploration of an important region in the excited PES. This exploration is more extensive when the laser-excited coordinate is different from the reaction coordinate. In such a case, the longer lived the reaction intermediate is, the more of the PES will be explored. The excitation of different resonances is reflected in the different widths of the transitions—an indication of their reactivities.

Correlating with the Ca(^1P) state, a progression in the blue part of the Ca–HCl action spectrum is observed with an irregular spacing of ca. 200 cm^{-1}, a value implying a van der Waals mode progression. Moreover, the experiments exhibit a large H/D isotope effect on the band spacing. Furthermore, the relative positions of the four bands in the bluemost part of the spectrum closely match the energy separations of free HCl rotor lines for $J = 4 \ldots 7$ [$E_J = b.J(J + 1)$, where b is the HCl ground state rotational constant]. This has been confirmed by the observation in the corresponding Ca–DCl spectrum of four transitions matching the $J = 5 \ldots 8$ line positions of a free DCl rotor. This correspondence indicates that these ensembles of transitions lead to upper levels beyond the barrier to free rotation of HCl/DCl within the excited state of the complex. On the contrary, the other members of the progression relate to the bending motion associated with the anisotropy of the excited Ca–HCl potential.

It has been shown that this spectrum can be modeled using the bending potential $V(\theta)$ of the excited state to reproduce the positions and intensities of most of the transitions of the Ca–HCl ^1P region (Soep et al. 1991).

A spectacular result can be seen in the action spectrum (Figure 4-6): the vibrational bands become narrower when the energy excess in the system increases, which indicates that the reaction time is longer as the energy in the intermediate state is larger. As discussed earlier, this surprising behavior has been interpreted in terms of rotational excitation of the HCl molecule within the complex. This

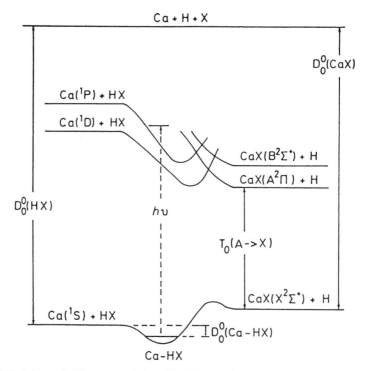

Figure 4-7. Schematic diagram of the Ca–HX reaction. The optical excitation of the Ca–HX van der Waals complex toward an excited state correlating with Ca(^1P) or (^1D) states leads to the reaction with excited CaX* production. This reaction is followed by detection of the CaX* emission.

rotational motion prevents the calcium atom from "catching" the halogen atom and slows down the reaction.

These results show that one could represent the motion of the complex by one-dimensional modes on surfaces correlated to excited atomic calcium (see Figure 4-7), and the spectra are characteristic of the entrance valley of this reaction. On the other hand, the chemiluminescent reaction of excited calcium with HCl is known to occur with very high cross sections: 25 Å^2 ($^1D_2$) (Brinckman et al. 1980) to 68 Å^2 (^1P) (Rettner and Zare 1982). These cross sections agree with a passage of the ionic covalent crossing without a barrier at 3.5 Å. Hence, there should be a smooth passage from the reagents to the products and the observation of action spectra with distinct features, not a continuum, can be interpreted as *the excitation of local modes perpendicular to the reaction coordinate.*

The selectivity on the products' energy distribution can be observed through the emission spectra. The fluorescence spectrum (Figure 4-8) is characteristic of the CaCl $A^2\Pi$–$X^2\Sigma^+$ and $B^2\Sigma^+$–$X^2\Sigma^+$ emission. It comprises two series of short vibrational progressions that are well separated.

From this spectrum, it has been possible to estimate the branching ratio of the reaction to the A or B states of CaCl. This ratio is approximately statistical and is the same as for the gas phase reaction Ca($4s4p^1P_1$) + HCl (Rettner and Zare 1982). This ratio remained constant throughout the excitation spectrum.

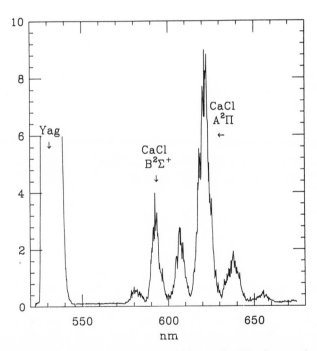

Figure 4-8. Fluorescence spectrum of the Ca–HCl van der Waals complex (Soep et al. 1991). The excitation laser has been fixed on an absorption band of the complex and the fluorescence has been dispersed. The saturated line at 532 nm corresponds to the scattered light of the evaporation laser.

The Ca–HBr excitation spectrum is very similar to that of Ca–HCl, and the observation of the A and B states of CaBr* is resolved by similar fluorescence measurements. However, in contrast with Ca–HCl, the ratio of the A to B emission depends strikingly on the excitation region ($^1P_1$, $^1D_2$) or on which set of bands is excited in each region. A strong selectively upon the final electronic state can be achieved by very small changes in excitation energy in the case of Ca–HBr. This different selectivity behavior for Ca–HCl and CaHBr is not yet well understood.

4A.6. CONCLUSIONS

In part 4A, we have shown a few examples of the work that has been done on excited van der Waals complexes. Ground state reaction induced through the photodissociation of one molecule within the complex is also a very important way to study chemical processes. This subject has not been covered in this chapter as well as real time experiment.

This excited state van der Waals complex experiment should have a promising future. The systems studied so far have given some information on the accessible Franck–Condon region on the reactive surface. This region can be extended to the vibrational excitation of the intermolecular modes of the complex. This vibrational excitation would create minima in the continuous spectrum, allowing a more accurate description of the intermediate state. Similarly, it is worthwhile to note that these kinds of system, where the reactants are frozen in a small (van der Waals) well at the bottom of the entrance channel barrier, should be quite ideal for the study of the role of specific internal degress of freedom with regard to barrier-crossing product formation.

This method should find its most challenging developments in large poly-atomic systems, as well as larger systems (cluster) where the selective approaches in organic photochemistry could be mimicked through the formation and excitation of the equivalent isomers, nucleophilic substitution, etc.

4B. CHEMICAL REACTIONS IN SMALL CLUSTERS

4B.1. INTRODUCTION

In part 4A, it has been shown that the excitation of small molecular complexes can be successfully used in order to get very detailed information on the reactive processes. The reaction in van der Waals complexes is one method, among others, of characterizing binary gas phase reactive collisions.

The understanding of chemical reaction mechanisms in solution is often based on the nature of the interactions between reactants and solvent, which are governed by the physical properties of molecules, such as polarity, or by the possibility of "bonds" formation (e.g., hydrogen-bonding) and their dynamical evolution. The goal of the majority of works on molecular clusters is to try to fill the gap between the gas phase reaction and the condensed phase reaction by a step-by-step solvation of the reactive system. This approach will give useful information on these interactions at the microscopic level; information is also needed for modeling of macroscopic systems.

The role of the solvent in a chemical event can be divided in two main aspects, described briefly, as follows.

First, there is the *thermal bath effect*. In the gas phase, the energy within the reactive partners is conserved and therefore the atoms "fly" above all the intermediate states without being caught in the deep well in the reactive surface. For example, in a simple reaction like the one described previously, $Hg^* + H_2$, the product of the reaction is HgH, whereas the same reaction performed in Ar matrices leads to the formation of the strongly bonded H–Hg–H molecule (Legay-Sommaire and Legay 1993). Such an effect can also be invoked for the recombination after dissociation, for example, for $I_2^-/(CO_2)_n$ clusters as reported by Lineberger and coworkers (Papanikolas et al. 1993).

Second, there is an *energetic effect*. This effect is particularly important in a reaction in which charge separation occurs. Then, the ability of the solvent to accommodate the charges, the electron affinity, or the proton affinity can become the key factor to initiate the reaction.

Both effects can be studied quite independently in molecular clusters: adding more and more neutral atoms (often Ar) around the reactive precursor can change the thermal bath environment, only weakly changing the energetics. On the other hand, more polar molecules will gradually change the energetics of the reactive process.

In the following, we will focus our discussion on reactions occurring in clusters in which one chromophore is surrounded by the solvent molecules; the reactions occur when the chromophore is excited in its first excited state or ionized. The interest of using a chromophore within the cluster is that multiphoton ionization can be used in connection with mass spectrometry. In this case, the ionization is a soft process and the spectroscopy of the cluster can give information on the size of the cluster, which is excited and responsible for the reactive event. This assigment can be difficult in other methods (electron bombardment) due to fragmentation processes associated with ionization.

Many interesting ionic reactions have also been studied in clusters in which no chromophore molecule is present. These are discussed in the other chapter of this volume.

4B.2. PROTON TRANSFER

Intermolecular charge- and proton-transfer processes, that is, the motion of the proton associated with the change of electronic structure in hydrogen-bonded complexes, are involved in many chemical reactions in solution. The study of such processes in small clusters leads to very detailed information on the reactive event.

It is well known that proton transfer can be induced by electronic excitation of molecules which present a strong difference in acidity between their ground and excited states. Generally, the absorption spectrum of the hydrogen-bonded complex $AH\cdots B$ is only slightly modifed compared with the free AH molecule.

On the other hand, if the proton transfer occurs, the fluorescence of the complex is close to that of the A^{*-} anion, as involved in the following process:

$$AH \cdots B \xrightarrow{h\nu} AH^* \cdots B \longrightarrow A^- * \cdots HB^+$$

$$\xrightarrow{-h\nu_{fluo.}} A^- \cdots HB^+ \longrightarrow AH \cdots B.$$

This has been shown by Weller and coworkers (Beens et al. 1965) and Mataga and coworkers (Mataga and Kaifu 1964; Mataga et al. 1964) for a number of aromatic acids and bases in liquid phase.

These reactions involve both intramolecular rearrangement of the acid and base coupled with motions of the solvent molecules due to creation of the two ions. Such studies have been initiated by Leutwyler and coworkers (Cheshnovsky and Leutwyler 1985; 1988; Knochenmuss et al. 1988; Knochenmuss and Leutwyler 1989) who investigated heterogeneous free clusters, $AH \cdots B_n$ (where AH = 1-naphthol, B = ammonia, water, methanol, or piperidine). These works have been followed by many experiments on 1-naphthol (Bernstein 1992; Kim et al. 1991b; Hineman et al. 1992, 1993; Plusquellic et al. 1992) or similar alcohols [AH = phenol (Jouvet et al. 1990; Solgadi et al. 1988), 2-naphthol (Droz et al. 1990; Plusquellic et al. 1992), allyl-, propenyl-, or propylphenols (Kim et al. 1991a). In many cases, size selective proton transfer has been evidenced and the dynamics of this proton transfer (i.e., time scale measurements) has been determined (Bernstein 1992; Breen et al. 1990; Hineman et al. 1992, 1993; Steadman and Syage 1990; Syage 1990, 1993; Syage and Steadman 1991).

4B.2.1. Principles

It is well known that the acidity of many hydroxy-aromatic compounds is strongly affected by electronic excitation and that these molecules become significantly more acidic in the excited states than in the ground state. For example, phenol and 1-naphthol

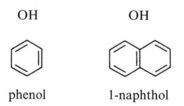

phenol   1-naphthol

are very weak acids in their ground state (phenol $pK_a = 9.94$, 1-naphthol $pK_a = 9.4$). They become strong acids in their first excited singlet state [phenol $4 \leq pK_a^* \leq 5.7$ (Bartok et al. 1962; Grabner et al. 1977; Wehry and Rogers 1965), 1-naphthol $pK_a = 0.5 \pm 0.2$ (Förster 1950a; Harris and Selinger 1980; Selinger and Weller 1977; Weller 1952, 1958)]. These molecules and the proton transfer reaction have been widely studied in solution (see, e.g., Reichardt 1988). In clusters, the proton transfer, that is, the ionic character of the hydrogen bond between AH acid and B_n basic cluster will depend on the acidity of AH, which is modified by

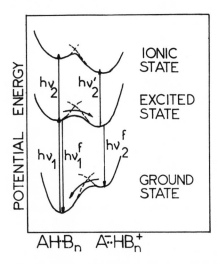

Figure 4-9. Double-minimum potential energy curves corresponding to the neutral $AH \cdots B_n$) and ion-pair $(A^- \cdots HB_n^+)$ structures in the ground, excited, and ionic states. The term hv_1 corresponds to the $S_1 \leftarrow S_0$ excitation, hv_2 and hv_2' correspond to the energy necessary to ionize the excited "neutral" (AH^*B_n) or "ionic" $(A^{-*}B_nH^+)$ clusters; hv_1^f and hv_2^f correspond to the emission AH^*B_n and $A^{-*}B_nH^+$, respectively.

electronic excitation (AH^*) or by ionization (AH^+), and the basicity of B_n, which depends on the nature of B and the cluster size.

The resulting structural changes are currently discussed in terms of a double-minimum potential surface with minima corresponding to the neutral $(AH \cdots B_n)$ and ion-pair $(A^- \cdots B_n^+)$ structures (Figure 4-9).

The processes under study may be schematically represented as:

$$
\begin{array}{c}
\text{VIII} \\
AH^+\text{–}(B)_n \xrightarrow{} A^\circ \ BH^+(B)_{n-1} \\
VI \uparrow {\scriptstyle hv_2} \qquad\qquad VII \uparrow {\scriptstyle hv_2'} \\
AH\text{–}(B)_n \xleftarrow[IV]{} AH^*\text{–}(B)_n \xrightarrow[III]{} A^{*-}BH^+(B)_{n-1} \xrightarrow[V]{} A^-BH^+(B)_{n-1} \\
II \uparrow {\scriptstyle hv_1} \\
AH\text{–}(B)_n \xrightarrow[I]{} A^-BH^+(B)_{n-1}
\end{array}
$$

(Scheme 1)

As the hydroxy-aromatic molecule is a weak acid in the ground state, one can assume that in all $AH \cdots B_n$ clusters the equilibrium is mainly displaced towards the neutral form $[AH(B)_n$—process I]. When the AH molecule in the cluster is excited by a first photon (process II), two types of behavior can be observed:

1. Relaxation of the neutral excited complex towards the ground state with emission of fluorescence (process IV). The red shift in the absorption (excitation) spectra of the AH–B_n complex with respect to the absorption of the bare AH molecule will only measure the increase of the binding energy in the excited neutral form AH*$\cdots B_n$ of the cluster. Then, the emission spectrum will be similar to the fluorescence of the free molecule.
2. Proton transfer (process III) in the excited state, which can be followed by the emission of the ionic form of a hydrogen-bonded phenolate anion $[A^- * \cdots BH^+(B)_n$—process V]. A large shift is then expected in the excitation spectrum and the dispersed fluorescence spectra typical of $A^- *$.

Another characterization procedure of the excited clusters can be obtained by ionization by a second photon and detected by mass spectrometry (processes VI and VII). Tuning this second photon, the first one being fixed on the $S_1 \leftarrow S_0$ transition of a given cluster, allows one to determine the ionization threshold of this cluster. The ionization potentials of AH and A^- being significantly different, the ionization process $A^- * \cdots HB_n^+ \rightarrow A \cdots HB_n^+ + e^-$ will occur at lower energies than the AH*$\cdots B_n \rightarrow AH^+ \cdots B_n + e^-$ process. The two-photon ionization techniques can provide a mass selective way of detecting proton transfer in clusters.

4B.2.2. Proton Transfer in the Excited State of Phenol and 1-Naphthol

Most studies of proton transfer in aromatic molecules concern phenol (Abe et al. 1982a,b,c; Fuke and Kaya 1983; Gonohe et al. 1985; Jouvet et al. 1990; Lipert and Colson 1988; Mikami et al. 1987, 1988; Oikawa et al. 1983; Solgadi et al 1988; Steadman and Syage 1990; Syage 1990; Syage and Steadman 1991) or 1-naphthol (Cheshnovsky and Leutwyler 1985, 1988; Knochenmuss et al. 1988; Knochenmuss and Leutwyler 1989). These compounds can be associated with various proton acceptors: ammonia, water, methanol, monoethylamine, piperidine, etc., differing essentially by their different gas phase proton affinities.

First evidence of an excited state proton transfer can be obtained by comparing the emission spectrum of the complex with those of the excited state acid and its conjugated base. In aqueous solution, the excited 1-naphthol (1-NpOH*) presents an emission in the range 320–380 nm while the excited naptholate anion (1-NpO*$^-$) is characterized by a broad emission in the range 400–500 nm and red-shifted by about 7000 cm^{-1} (Weller 1952; Förster 1950b) as compared with free naphthol. In solution, in the ground state, there is an equilibrium between AH and A^- (process I), largely displaced towards AH. In small clusters, such an equilibrium does not exist. The studies of Leutwyler et al. in supersonic jets are based on the observation of the large red shifts in fluorescence emissions of 1-NpOh–$(B)_n$ (processes IV and V). Correlation of this fluorescence with the size of the clusters has been performed by one-color resonant two-photon ionization.

This proton transfer can be seen in both the excitation and fluorescence spectra.

Figures 4-10 and 4-11 show the R2PI excitation spectra of the smaller clusters in the vicinity of the $S_1 \leftarrow S_0$ transition of 1-naphthol (frequency range

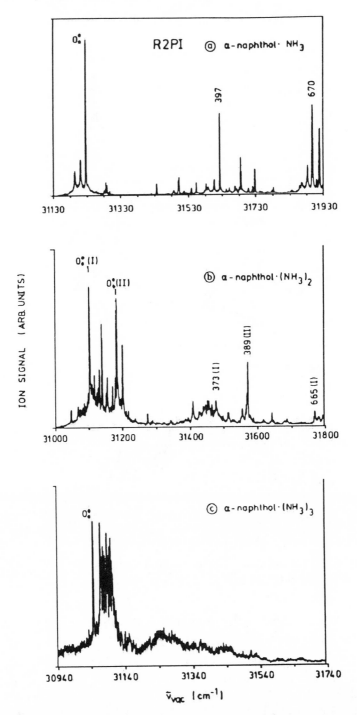

Figure 4-10. Resonance enhanced two-photon ionization spectra of 1-naphthol$(NH_3)_n$ clusters ($n = 1$–3). 1-naphthol$(NH_3)_2$ shows two 0_0^0 origins corrseponding to two different isomers (from Cheshnovsky and Leutwyler 1988).

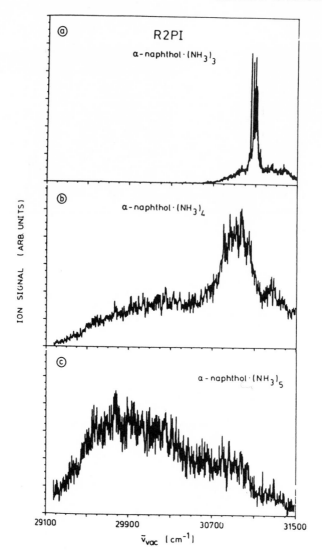

Figure 4-11. Resonance enhanced two-photon ionization spectra of 1-naphthol(NH$_3$)$_n$ clusters ($n = 3$–5) (from Cheshnovsky and Leutwyler 1988).

29,100–32,500 cm^{-1}). From these spectra, the authors pointed out the following observations and comments:

1. For the $n = 1$–3 clusters, the excitation spectra exhibit well structured spectra similar to the vibrational structure of the S$_1 \leftarrow$ S$_0$ transition of the bare 1-naphthol.
2. For $n \geq 4$ clusters, a broadening and a net red-shift of the vibrational bands is observed, indicative of a charge transfer.

In order to check the assumption of a proton transfer occurring in the excited state of the clusters, selective resolved emission spectra of these clusters have been recorded. Figure 4-12 shows the fluorescence spectra for 1-NpOH(NH$_3$)$_{n \leq 4}$ (Cheshnovsky and Leutwyler 1988), the excitation being localized on the 0_0^0 band for the clusters with $n \leq 3$ and at 30,985 cm^{-1} (322.7 nm) for $n = 4$.

For the first three clusters, the structures of these emission spectra are clearly similar to the one observed for bare 1-naphthol S$_1 \leftarrow$ S$_0$ state, with a Franck–Condon maximum situated on the 0–0 band. The small increase of complexity of the spectra is certainly due to intermolecular vibrations.

For $n = 4$, the emission spectrum shows a broad structure centered at about 430 nm, that is, largely displaced towards the red. The same kind of spectra are obtained when $n > 4$ clusters are excited. These excited clusters give rise to an emission characteristic of naphtholate anion fluorescence emission which corresponds to an emission following the intracluster proton transfer.

Phenol(NH$_3$)$_n$ or phenol(MEA)$_n$ clusters exhibit the same kind of spectroscopic behavior (Jouvet et al. 1990; Solgadi et al. 1988). Excitation spectra of the smallest clusters ($n > 3$) are well structured and show emission spectra matching that of phenol. Larger clusters' excitation spectra become broad and the emission spectra become more typical of that of the phenolate anion (see Figure 4-13).

The 2-naphthol/ammonia clusters studied by Droz et al. (1990) show a proton transfer also occurring for $n \geq 4$, but, in absorption, the broad S$_1 \leftarrow$ S$_0$ bands are not red shifted with respect to the bare molecule as in 1-naphthol(NH$_3$)$_n$ clusters.

Ionization threshold measurements are also indicative of the proton transfer reaction. This has been seen for the phenol–B$_n$ system (Jouvet et al. 1990; Solgadi et al. 1988). The ionization thresholds depend on the structure (AH*$\cdots$B$_n$ or A$^-$*$\cdots$HB$_n^+$) of the excited state (processes VI and VII in scheme 1).

Ionization cross sections versus $h(\nu_1 + \nu_2)$ energy are presented in Figure 4-14 for phenol(NH$_3$)$_n$ and phenol(MEA)$_n$ clusters. From these results, the following observations can be made.

1. The ionization threshold for each cluster size is lower than the ionization potential of free phenol (Mikami et al. 1987).
2. The clusters may be divided in two groups: (a) the smallest ones ($n = 1$–3) for ammonia and $n = 1$, 2 for MEA) with ionization threshold lowered by ca. 0.6–1 eV with respect to bare phenol, (b) the larger ones ($n \geq 4$ for ammonia and $n \geq 3$ for MEA) with a further decrease of ionization threshold by 0.8–1 eV.

Since the energy to ionize PhO$^-\cdots$B$_n^+$ is much lower than that necessary to ionize PhOH$\cdots$B$_n$, these different behaviors have been interpreted as being due to the proton transfer reaction at the excited state leading to drastic changes in the ionization energies of the different clusters. The same kind of experiments have been reported for 1-naphthol, leading to the same conclusion (Kim et al. 1991b).

The ionization thresholds of 1-naphthol(NH$_3$)$_{n=1-4}$ vary as a function of n. A significant lowering of the threshold in comparison with the one of $n = 1$ and 2 [I.P. = 7.15 eV and 7.13 eV for $n = 1$ and 2, respectively, I.P. of free

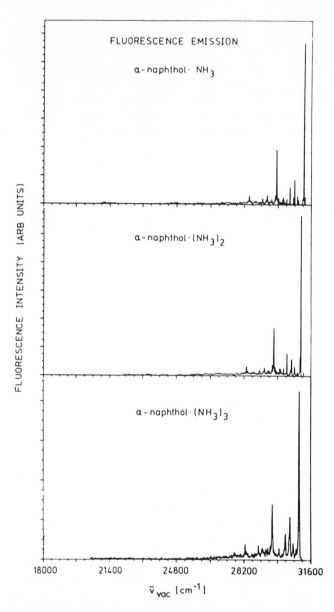

Figure 4-12. Fluorescence emission spectra of 1-naphthol(NH_3)$_n$ clusters ($n = 1$–4). The excitation laser is fixed on the 0_0^0 bands for $n = 1$–3. For $n = 4$, the excitation frequency is fixed at $30,985\ cm^{-1}$ (laser-scattered light is marked with an asterisk) (from Cheshnovsky and Leutwyler 1988). This clearly shows that the emission spectrum of $n = 4$ is strongly displaced towards the red characteristic of the naphtholate emission.

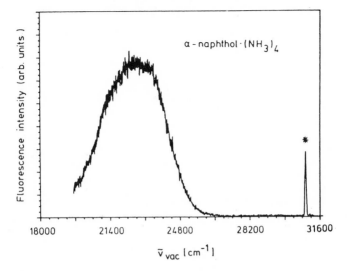

Figure 4-12. (*continued*).

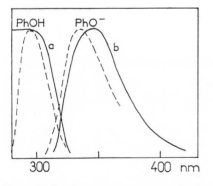
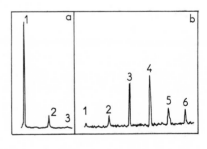

Figure 4-13. Fluorescence spectra of phenol(MEA)$_n$ clusters (MEA = monoethylamine). The spectra of the clusters (full line) were obtained by measurement of the fluorescence cut by a series of filters. (a) Fluorescence of phenol(MEA)$_n$ with $n \leq 2$ (excitation wavelength 280.9 nm). For comparison, the fluorescence spectrum in a solution of phenol in ethanol (excitation wavelength = 280 nm) is represented by a dashed line. (b) Fluorescence spectrum for a larger phenol(MEA) clusters (excitation wavelength = 281.5 nm). The fluorescence spectrum in a solution of phenolate anion in NaOH 10^{-4} mol l^{-1} in ethanol is represented by a dashed line. The maxima of these curves have been normalized to unity. The mass spectra of the clusters corresponding to these excitation conditions are given in the lower part of the figure.

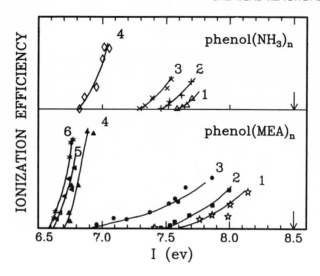

Figure 4-14. Ionization efficiency versus $h(v_1 + v_2)$ of phenol$(NH_3)_n$ $(n = 1-4)$ and phenol$(MEA)_n$ clusters $(n = 1-6)$. The arrow at 8.5 eV represents the ionization potential of free phenol. The ionization thresholds (in eV) are respectively:

n	phenol$(NH_3)_n$	phenol$(MEA)_n$
0	8.50	8.50
1	7.85 ± 0.1	7.60 ± 0.1
2	7.69 ± 0.1	7.50 ± 0.1
3	7.51 ± 0.1	6.93 ± 0.1
4	6.89 ± 0.1	6.68 ± 0.1
5	6.68 ± 0.1	$6.5-6.6 \pm 0.1$
6		$6.5-6.6 \pm 0.1$

One observes a first gap between free phenol and the clusters in the thresholds for $n = 1-3$ for ammonia and $n = 1-2$ for MEA assigned to the hydrogen bonding. The gap increases for larger clusters assigned to the proton transfer in the excited state.

1-naphthol $= 7.76$ eV (Kim et al. 1991b)] is observed for $n = 3$ and 4 (I.P. $\cong 6.93$ eV) and has been attributed to the proton transfer. However, the case of $n = 3$ is very interesting due to the probable existence of two isomers: one with a lower ionization threshold (6.93 eV) and the other with an ionization threshold very close to that of 1-naphthol$(NH_3)_n$ with $n = 1$ and 2 (I.P. $= 7.11$ eV). It seems that the $n = 3$ cluster can be really considered as the frontier case between smaller clusters where no transfer occurs and larger ones where it does: each isomer of naphthol$(NH_3)_3$ seems to belong to one of these categories.

Time dependent, excited state proton transfer for phenol/ammonia (Steadman and Syage 1990; Syage 1990; Syage and Steadman 1991) and for the

1-naphthol/ammonia (Bernstein 1992; Breen et al. 1990; Hineman et al. 1992, 1993) has been also studied. In these experiments, the following reaction

$$ROH-M_n \xrightarrow{h\nu_1} ROH^*-M_n \xrightarrow{k} RO^{*-}-H^+M_n$$
$$h\nu_2 \downarrow$$
$$ROH^+-M_n$$

is followed at a real time scale. The principle is as follows: one picosecond laser is used to pump the excited states of the clusters and a second one is used to probe them by ionizing the clusters. Time resolution is achieved by delaying, in time, the two lasers. The time evolution of cluster-ion signals is observed as a function of the cluster size.

Figure 4-15 shows the results obtained for the phenol–ammonia system (Steadman and Syage 1990; Syage 1990; Syage and Steadman 1991). Two types of behavior are observed. First, for clusters with $n < 5$, a relatively long time monoexponential decay is evidenced. This decay is characteristic of the lifetime of the excited state of the clusters. Second, the figure clearly shows that this decay is strongly perturbed for $n \geq 5$ ammonia molecules. The proton transfer reaction, which is responsible for this shortening of the excited state lifetime, occurs with a constant of 60 ± 10 ps. However, it subsists a discrepancy between these dynamical results and the previous ones: the ionization threshold lowering indicative of a proton transfer reaction in the excited state was observed for $n = 4$ (Jouvet et al. 1990; Solgadi et al. 1988); phenol$(NH_3)_4$ shows a long-lived excited state [>1 ns (Syage and Steadman 1988)]. Excited lifetimes for partially or fully deuterated clusters $C_6H_5OD-(ND_3)_n$ and $C_6D_5OD-(ND_3)_n$ are significantly longer (>1 ns for $n > 4$) than for the protonated clusters. The barrier to reaction is due to a crossing between a covalent state and a solvent-stabilized ion-pair state of phenol (Syage 1993).

For 1-naphthol, excited state proton transfer has been detected for $n = 3$ and 4 clusters only. A biexponential decay is obtained for $n = 3$. The short component ($\tau_1 \cong 60$ ps when the 0–0 transition is excited) is attributed to the proton transfer itself. This transfer time decreases when an excess of vibrational energy is given to the system (40 and 10 ps for 800 and 1400 cm^{-1}, respectively). The long lifetime component ($\tau_2 \cong 500$ ps) is more likely attributed to either a cluster reorganization subsequent to the proton transfer event or a proton transfer in a cluster of different geometry (Hineman et al. 1992, 1993). For $n = 5$, it seems that proton transfer may also occur in the ground electronic state, as suggested by the increase of $(NH_3)_5^+$ fragments and the large red shift in the $n = 5$ cluster mass resolved spectrum.

As for the phenol, a discrepancy in the behaviour of clusters is evidenced when comparing static and dynamic experiments: the limit size ($n = 4$ and 3, respectively) for which proton transfer is observed might be due in this case, to the existence of isomers for the $n = 3$ cluster (Kim et al. 1991b). It should also be mentioned that the size value indicated by these experiments may be questioned due to important fragmentation processes as shown by Hineman et al. (1993).

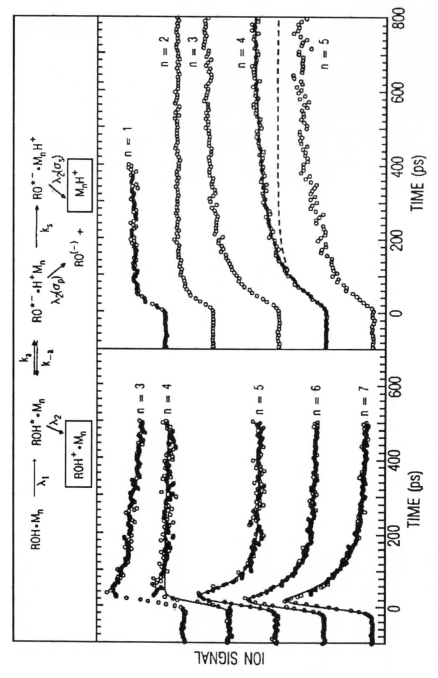

Figure 4-15. Picosecond measurements for decay of reactant clusters ROH*M_n (as detected by two photon ionization of the cluster ROH–M_n^+) and formation of product RO*–H^+M_n (as detected by the fragment M_nH^+) (from Steadman and Syage 1990).

How can the size dependent reactivity be rationalized and what are the important factors which are responsible for this cluster size reactivity? Experiments have been performed with 1-naphthol associated with other molecules (Knochenmuss et al. 1988; Knochenmuss and Leutwyler 1989). Naphthol undergoes proton transfer with two molecules of piperidine. For phenol, one needs three of four ammonia or three monoethylamine (MEA) molecules for the same process.

A correlation can be made between the gas phase proton affinities (PA) of the B_n clusters which are strongly dependent on their sizes and the propensity of the $AH-B_n$ clusters to undergo proton transfer in the excited state. These proton affinities of clusters (water, methanol, ammonia, and piperidine) which are estimated (e.g., Knochenmuss and Leutwyler 1989) or deduced from experiment (Bisling et al. 1987; Ceyer et al. 1979; Kamke et al. 1988) are reported in Figure 4-16.

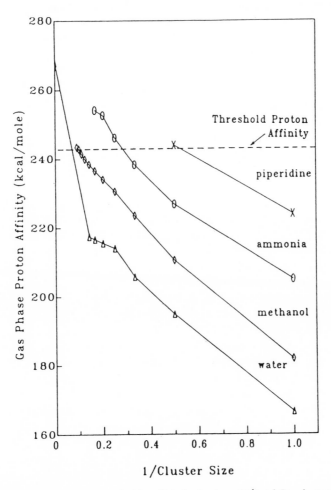

Figure 4-16. Gas phase proton affinities (PA in kcal mol^{-1}) of B_n clusters versus $1/n$ (B = piperidine, ammonia, methanol, and water) (from Knochenmuss and Leutwyler 1989). The threshold proton affinity corresponds to the energetic limit for which excited state proton transfer occurs for 1-naphthol in small clusters.

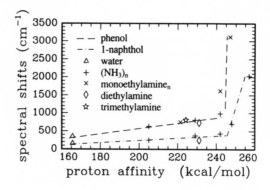

Figure 4-17. Relation between gas phase proton affinities (in kcal mol^{-1}) of base molecules B or B$_n$ clusters and spectral shifts of the S$_1 \leftarrow$ S$_0$ state of phenol(B$_n$) and naphthol(B$_n$) (Cheshnovsky and Leutwyler 1988) clusters (in cm^{-1}). ($\triangle$) water; (+) (NH$_3$)$_n$; ($\times$) monoethylamine$_n$; ($\diamond$) diethylamine; ($\star$) trimethylamine.

For the stronger proton acceptors (ammonia, monoethylamine, and piperidine) a relation between the B$_n$ proton affinities and the spectral shifts of the S$_1 \leftarrow$ S$_0$ states of phenol(B)$_n$ or naphthol(B)$_n$ shows a linear dependence for proton affinities lower than a limit value situated around 10.4 eV ($\cong$240 kcal mol^{-1}) for both phenol or naphthol molecules. Above this limit, the spectral shift is much larger and is different for phenol and 1-naphthol (see Figure 4-17). Nevertheless, this limit seems to correspond with the energetical limit of the proton transfer reaction.

Methanol and water are weaker bases than amine compounds. When these molecules are associated with 1-naphthol, a different behavior can be evidenced. No proton transfer occurs for methanol until $n = 26$ (Knochenmuss et al. 1988) For water clusters, however, three types of excited state behaviors are observed (Knochenmuss and Leutwyler 1989).

(1) For small clusters ($n \leq 7$), the spectroscopy of clusters is very close to the spectroscopy of bare 1-naphthol. The ionization thresholds of naphthol(H$_2$O)$_{n=2-4}$ are similar to 1-naphthol(H$_2$O)$_1$ (59000 cm^{-1}) Kim et al. 1991b).

(2) For medium sized clusters ($8 \leq n \leq 20$), one observes incremental spectral shifts indicating successive stages of molecular solvation, and the spectra approach the one of 1-naphthol in bulk ice at $n \approx 20$ ($\delta v \approx -200$–250 cm^{-1}).

(3) Larger clusters show an increasing intensity of fluorescence centered at about 25000 cm^{-1}, which is characteristic of the naphtholate anion, with proton transfer occurring for $n \geq 30$. However, this emission is not the fluorescence of the pure naphtholate anion, but can be seen as the superpositions of both naphthol and naphtholate emission. This behavior is different that observed in the liquid water where only naphtholate emission is observed, and also from that observed in ice where only naphthol emission is observed.

These results have been rationalized in the following manner (Knochenmuss and Leutwyler 1989), the key factor being the proton affinity of the solvent: 1-naphthol presents two close-lying excited states, ^{1}L$_a$ and ^{1}L$_b$. In the gas phase or in nonpolar solvents, the ^1L_b state is the lowest state: it is relatively nonpolar and weakly acidic. In constrast, the ^1L_a state is polar, with a large intramolecular charge transfer component, and a large electron density transfer from the –OH group to the ring (Baba and Suzuki; 1961; Nishimoto 1963). This state is more acidic that ^1L_b, and their relative position is solvent sensitive.

Three cases can be considered depending on the dielectric and basicity properties of the solvent.

(1) In the small water clusters with $n < 10$, the ^1L_b state is vertically excited. The absorption spectrum is weakly displaced and the well resolved emission is characterized by the absence of Stokes shift. This behavior has to be compared with that of 1-naphthol in nonprotic solvents with low dielectric constants. In molecular beams, the 1-naphthol–$(H_2O)_{n<10}$ and –$(NH_3)_{n<4}$ clusters fall into this case.

(2) For protic solvents with larger dielectric constants and stronger basicity, the ^1L_a and ^1L_b states are inverted and relaxation from ^1L_b to ^1L_a takes places but there is no proton transfer to the solvent. The fluorescence is then due to the ^1L_a state with a small Stokes shift. The intermediate sized water clusters ($n = 10$–20) belong in this category. The clusters with methanol for any size $n < 10$ (due to a weak basicity or a small dielectric constant) follow this mechanism. From the evaluated proton affinities (see Figure 4-16), it can be seen that for $n \approx 10$ molecules of methanol (PA ≈ 243 kcal mol^{-1} which corresponds to the limit for proton transfer evidence in 1-naphthol complexes with piperidine or ammonia), a proton transfer should be observed. The absence of such a transfer can be related to a cluster structure effect.

Indeed, for large water clusters, the observed emission indicates that some clusters undergo a proton transfer whereas some other ones do not. The proton transfer mechanism implies a reorganization of the solvent around the charges (the proton and the anion) during the lifetime of the excited naphthol. The coexistence of the naphtholate and naphthol emission can be seen as a temperature effect, as can the coexistence in the supersonic expansion of liquid-like clusters and solid ones. In the solid ones, proton transfer does not occur, as in glass, whereas in liquid ones proton transfer occurs. This assumption seems to be confirmed by the lifetime experiments of Knochenmuss et al. (1993), where the appearance lifetime of the naphtholate emission increases with the mean cluster size. This effect has been interpreted as the increase of rigidity of the cluster as the cluster size increases. The above description is simplified since one suspects that, for these finite systems, it can coexist within the same cluster liquid-like (near the surface) and solid-like regions (Torchet et al. 1983). In this case, the observed fluorescence behavior reflects the local environment of the naphthol molecule in the cluster: clusters with the naphthol molecule near the surface lead to proton transfer and those where the molecule is deep inside give naphthol emission. It should be noticed that these systems ought to be very interesting to use a cluster temperature internal probe.

(3) The last case concerns the solvent molecules with large dielectric constants or strong basicity; the ions can be rapidly solvated (in the bulk or in large clusters) and proton transfer occurs. Since the emission arises from the transferred state, the Stokes shift is important (typically around 9000 cm^{-1} with a large bandwidth). The 1-naphtholate fluorescence in neutral water or a mixture of polar solvents

with small amounts of base complexes exhibits this type of behavior (Knochenmuss and Leutwyler 1989). Large phenol or naphthol amine clusters fall into this category.

From these experiments, it is shown that proton transfer is not only dependent on the proton affinity of the solvent or solvent cluster but also on the possibility of interaction of the solvent molecules with the nascent naphtholate anion.

4B.2.3. Comparison of Clusters/Matrices

The proton transfer reaction has been also studied in rare gas matrices, showing some discrepancies with the gas phase (Brucker and Kelley 1989a,b; Crépin and Tramer 1991): in 1- and 2-naphthol with ammonia, the $AH^*-B_n \rightarrow A^-*-HB_n^+$ transition takes place for $AH-(NH_3)_n/Ar$ matrix-embedded clusters with $n \geq 3$ (Brucker and Kelley 1989a,b).

For the phenol–ammonia system, the results can be summarized in the following manner (Crépin and Tramer 1991):

1. As for naphthol–ammonia, phenol$(NH_3)_n$ in the S_1 state presents the proton transfer for $n \geq 3$.
2. In the ground state, the proton transfer occurs only in very large phenol–ammonia clusters. The coexistence of PhOH and PhO^- anions in pure ammonia may result from different environments (sites) rather than a true chemical equilibrium.
3. In the first triplet state, PhO^- phosphorescence is only observed in pure ammonia matrix (the T_1 state is known to be less acidic than the first excited state).
4. The red shift of the 0–0 band induced by formation of a given $PhOH \cdots (NH_3)_n$ cluster is larger in the case of $S_1 \leftarrow S_0$ than in that of the $T_1 \leftarrow S_0$ transitions. Moreover, the red shift of the $S_1 \leftarrow S_0$ transition in matrices is larger than in the free clusters.

These experiments show that the more polar ionic forms seem to be favored in the matrix with respect to the neutral ones, as is expected from well known formulae for the stabilization energy of a system of charges in an apolar solvent. Their leading term,

$$\Delta E = (\mu^2/4\pi\varepsilon_0 r^3)\, \frac{n^2 - 1}{2n^2 + 1},$$

describes the interaction of an electric dipole μ with an infinite continuous dielectric of refractive index, n, with r being the radius of the sphere representing the dimensions of the molecular system. This formula may be used for a rough estimation of the stabilization (solvation) energies of neutral and ionic forms of phenol–ammonia clusters in the argon matrix (Crépin and Tramer 1991).

From the experiments in the gas phase. it has been established that the dependence of the enthalpy of the $PhO^* \cdots (NH_3)_n \rightarrow PhO^-* \cdots H^+(NH_3)_n$ re-action is essentially due to the $n -$ dependence of the proton affinity of $(NH_3)_n$ clusters—PA(n). From the decrease of the size limit of reactivity ($n = 4$ in the gas phase $\rightarrow n = 3$ in the Ar matrix), and as ΔE_{neutr} is negligible as compared with ΔE_{ion}, it is possible to obtain an evolution of the stabilization energy in the argon matrix which is related to the difference in proton affinity between $n = 4$ and $n = 3$ ammonia clusters; this must be of the order of magnitude of 0.34 eV. A good agreement between experiment and this model is obtained for a cavity radius of 4.4 Å, which is reasonable in an argon matrix.

4B.2.4. Proton Transfer in the Ionic State

The dissociative proton transfer of the solvated phenol cation corresponding to process VIII (scheme 1),

$$PhOH^+ - B_n \rightarrow PhO·BH^+(B)_{n-1},$$

has been studied by Syage and coworkers (Steadman and Syage 1991; Syage and Steadman 1992). As dicussed earlier, solvated phenol in the S_1 state can undergo proton transfer depending on the basicity of the solvent. Picosecond experiments have shown that this reaction occurs only for a certain size of solvent clusters with a time constant of about 60–70 ps for phenol–ammonia clusters.

As shown in Figure 4-18, during the reaction one can ionize, at short times, the reactants, and, at longer times, the proton transferred products; then, two different wells in the ionic reaction coordinates can be reached and can be studied by picosecond pump-delay–probe REMPI experiment (Steadman and Syage 1991):

1. If the time delay between both lasers (Δt is zero), PHOH$-^+$B$_n$ clusters are produced in the inner potential well.
2. If $\Delta t > 70$ ps for phenol–ammonia clusters ($n > 4$), the outer potential well can be reached.

The reaction is followed by mass detection of the disappearing PhOH$^+$ signal or the evolution of the H$^+$(NH$_3$)$_n$ signal resulting from dissociative proton transfer as a function of time. The time resolved measurements establish that the dissociative proton transfer in the ionic state occurs in the outer potential well and not in the inner well, but, as shown by Hineman et al. (1993), the cluster size for which such processes occur may be questioned due to possible fragmentation.

These results are confirmed in less basic (CH$_3$OH)$_n$ clusters: in the excited state for moderate cluster sizes, dissociation proton transfer signal was absent on the mass spectra.

Very different proton transfer reaction efficiencies from these two potential wells have been measured, and only the clusters in which the internal energy is sufficient to cross over the barrier will lead to the reaction. The empirical potential surface and the potential barrier between the two wells have been estimated (Steadman and Syage 1991). For ammonia clusters, upper limits for the barriers were estimated to be about 1.5 eV ($n = 1$) to 1.0 eV ($n = 4$).

The internal energy distribution of PhOH$^+$(NH$_3$)$_n$ has been measured by photoelectron spectroscopy (Syage and Steadman 1992) and has shown that the fraction of cluster ions with an internal energy greater than 1 eV is at least 10% for all values of n (not exceeding $n = 6$). Consequently, a lower limit of the proton

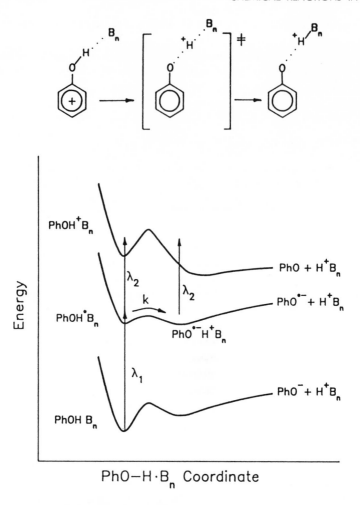

Figure 4-18. Energy diagram for phenol(NH$_3$)$_n$ for a cluster size of $n \cong 5$. The excitation scheme shows how delayed ionization by λ_2 can produce cluster ions in the outer potential well for reactive PhOH*B$_n$ clusters excited by λ_1 (from Steadman and Syage 1991).

transfer barrier height of about 1.0 eV was estimated for the reactive complex, which is in good agreement with the preceeding results.

4B.3.   NUCLEOPHILIC SUBSTITUTIONS

As described in the previous section, a powerful technique to probe most of the chemical or physical properties of molecular clusters is mass spectrometry after ionization of these clusters. Generally, an excess of energy is given by this ionization process and can lead to various dynamical behaviors: from the simplest one—fragmentations of the clusters (evaporation)—to intracluster chemical reactions. This means that the observed distribution of the ionic clusters often does not represent that of the neutral beam. To avoid decay of the ionized clusters, most of the experiments must be performed in particularly "soft" conditions: that is, low laser power (minimization of the "multiphoton" processes) or ionization of clusters near the threshold (evaporation processses are then no longer energetically allowed).

Intracluster ion–molecule reactions have been the subject of numerous investigations (see other chapters of this book). However, in this chapter, we will focus only on aromatic nucleophilic substitutions in substituted benzene/molecule clusters where the two-photon ionization process allows us to obtain spectroscopic information on the parent cluster involved in the reaction process.

The nucleophilic substitution (S_N) of aromatic compounds is one of the fundamental organic reactions in the condensed phase. For example, it is known that aniline can be produced from chlorobenzene under a severe basic condition such as in liquid ammonia (Pine et al. 1980). This bimolecular reaction depends strongly on the solvent. In this respect, the S_N reaction in the gas phase may be inefficient due to the absence of solvent molecules assisting the attack, by the nucleophilic agent, on the aromatic ring. However, the situation is completely different for molecular complexes where the nucleophile can be directly bound to the aromatic ring and then can attack the positively charged carbon atom (Mayeama and Mikami 1988).

4B.3.1. Aromatic Halides with Water or Alcohol Clusters

The nucleophilic substitution reaction in clusters has been observed by Brutschy and coworkers (Awdiew et al. 1990; Brutschy 1989, 1990; Brutschy et al. 1988, 1991, 1992) for the fluorobenzene–methanol system.

The time-of-flight mass spectrum recorded for an expansion of fluorobenzene (FB) with deuterated methanol presents mass peaks corresponding to deuterated anisole$^+$—evidence of an intracluster nucleophilic substitution reaction. In the clusters, the two decay channels of $FB(CD_3OD)_n^+$ are:

$$FB(CD_3OD)_2^+ \xrightarrow{51\%} FB-CD_3OD + CD_3OD \quad (vdWF)$$

$$\xrightarrow{49\%} anisole(d_3)^+ + DF + CD_3OD \quad (S_N)$$

$$FB(CD_3OD)_3^+ \xrightarrow{27\%} FB(CD_3OD)_2^+ + CD_3OD \quad (vdWF)$$

$$\xrightarrow{22\%} anisole(d_3)^+ + CD_3OD + DF + CD_3OD \quad (S_N)$$

$$\xrightarrow{51\%} anisole(d_3)^+ + (CD_3OD)_2 + DF \quad (S_N, vdWF).$$

The interpretation of these results is not obvious since the substitution reaction competes with van der Waals fragmentations (vdWF). The size of the parent cluster ion can be deduced from the R2PI spectra shown in Figure 4-19: the anisole$^+$ ion product is monitored as a function of laser wavelength tuned on the $S_1 \leftarrow S_0$ transition of the precursor $FB(CD_3OD)_n$ clusters. From the position of the different bands on the spectra, the assignment of the neutral precursors has

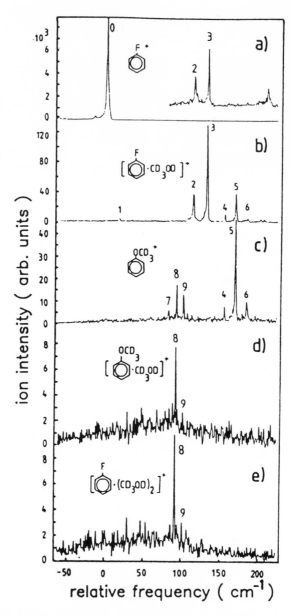

Figure 4-19. Resonance-enhanced two-photon ionization spectra of ions issued from fluorobenzene/methanol–d_4/helium clusters, measured by scanning the laser near the 0_0^0 transition of fluorobenzene [0]. Bands [4–6] are due to the $FB^+(CD_3OD)_2$ precursor which totally fragments, either by evaporation of one methanol–d_4 molecule or reaction leading to anisole + DF + CD_3OD. Bands [7–9] are due to the 1–3 precursor also losing one CD_3OD molecule or reacting. Bands [1–3] are more likely attributed to the 1–1 complex (isomer(?) [1], hot band [2], 0_0^0 [3]) (from Brutschy et al. 1991).

been proposed. One can see from these spectra that the fingerprint (absorption bands) observed at the mass of the 1–1 complex is different from the one observed at the mass of the anisole product; nevertheless, some peaks are seen in both spectra: they are assigned to the van der Waals fragmentation of the 1–2 complex in the 1–1 complex. The 1–2 cluster does not appear in the mass spectrum at its right mass due to a fast reaction. The peaks appearing in this mass spectrum are assigned to the evaporation of the 1–3 complex. One can record the spectra in various expansion conditions (to vary the cluster size) or with two-color R2PI to avoid fragmentations (Brutschy 1990). This has been done by Brutschy and coworkers (Awdiew et al. 1990; Brutschy 1989; Brutschy et al. 1991, 1992): in this case, the appearance potential of fragments is 140 meV lower than the threshold of the chemical reaction but there is no complete switchover to the fragmentation process. The substitution reaction turns out to be almost activationless.

The assigment of the parent ion $[FB(CD_3OD)_2^+]$ as the precursor of the fluoroanisole product has been clearly demonstrated by an infrared UV double-resonance experiment. Here, the product ion signal is monitored as the infrared excitation is scanned. When the IR laser is in resonance with the C–O stretching of the methanol dimer within the 1–2 cluster, a decrease of the product signal is observed since the population of the parent ion decreases by the loss of one methanol molecule through the IR predissociation process (Brutschy et al. 1992).

One should also notice that the fluorescence of the $S_1 \leftarrow S_0$ cluster emission has been recorded, which is good evidence that the reaction does not occur in the first excited state.

Nucleophilic substitution for ionic para-diflurobenzene (PDFB) combined with methanol clusters with formation of para-fluoroanisole$^+$ has been also observed (Brutschy 1989; Brutschy et al. 1991) for $n = 2$.

In water clusters, ionic para-difluorobenzene leads to hydrated fluorophenol$^+$. The most striking difference in this case was the ineffciency of the substitution reaction for the 1–2 cluster. The reaction appears for $n = 3$, leading to para-flurophenol$^+$–H_2O (Martrenchard 1993; Martrenchard et al. 1991). The PDFB$(H_2O)_n$ is a very good example of the competition between van der Waals fragmentation and the substitution reaction. As can be seen on the one-color two-photon excitation spectra of the PDFB$(H_2O)_n$ clusters (see Figure 4-20), at a given mass one observes the superposition of the spectrum corresponding to the larger dissociating clusters. A study of the energetics of the system and a careful analysis of the bands on the spectra are then needed to allow correlations with the products of reaction (see Figure 4-20).

4B.3.2. Aromatic Halides with Ammonia or Amine Clusters

Ammonia is well known to be a strong nucleophile. The nucleophilic substitution reactions in the ionized clusters of X–benzene (X = F, Cl) have been widely studied (Brutschy 1989, 1990; Brutschy et al. 1988, 1991; Mayeama and Mikami 1988, 1990, 1991; Riehn et al. 1992). All these experiments clearly show a specific behaviour not encountered in the hydroxylic solvents: depending on the substituted aromatic and the cluster size, two types of substitution reaction can be detected. The first type leads to ionic amines by substitution of X by the $-NH_2$

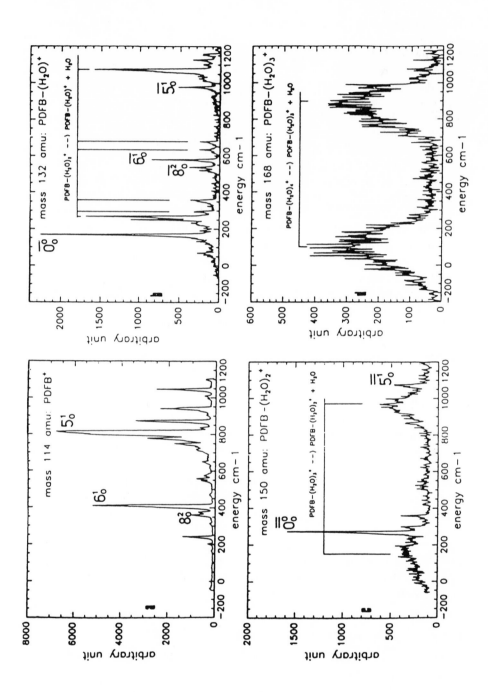

138

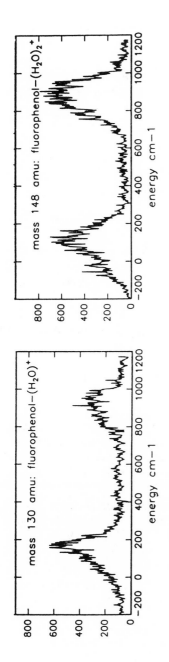

Figure 4-20. One-color resonance-enhanced two-photon ionization spectra of the *para*-difluorobenzene (PDFB)/water clusters (abscissa refers to the energy in the excited state). (a) Mass = 114 amu, corresponds to free PDFB$^+$. This spectrum has been recorded without water. The reference corresponds to the 0_0^0 level of free PDBF*. This level is not observed through one-color two-photon ionization excitation, the energy of $2h\nu_1$ being lower than the ionization threshold of PDBF*. (b) Mass = 132 amu (PDBF–H$_2$O)$^+$: the pointed peaks are due to dissociation of PDFB(H$_2$O)$_2$. (c) Mass = 150 amu (PDBF–(H$_2$O)$_2$)$^+$; the pointed peaks are due to dissociation of (PDFB–(H$_2$O)$_3$)$^+$. (d) Mass = 168 amu (PDBF–(H$_2$O)$_3$)$^+$: the pointed peaks are due to dissociation of (PDFB–(H$_2$O)$_4$)$^+$. The recorded spectra at masses of 130 and 148 amu for (fluorophenol–H$_2$O)$^+$ and (fluorophenol–(H$_2$O)$_2$)$^+$, respectively, are similar to (PDBF–(H$_2$O)$_{n=3,4}$)$^+$ spectra and are assigned to the reaction (PDFB–(H$_2$O)$_{n=3,4}$)$^+$ → (fluorophenol–(H$_2$O)$_{n=1,2}$)$^+$ + H$_2$O + HF.

group and the second one to the substitution of X by the $-NH_3^+$ group. As for hydroxylic aromatic clusters, the experimental procedure is the same: comparison of the excitation spectra of the parent molecule with those of the products.

Fluorobenzene with ammonia leads only to the formation of aniline$^+$. However, the TOF mass spectra also exhibit signals due to protonated and unprotonated ammonia clusters which must be produced by dissociative electron transfer (dET). In this case, direct evidence for a 1–2 precursor for the aniline$^+$ product with three competing channels is provided:

$$C_6H_5F-(NH_3)_2^+ \rightarrow C_6H_5NH_2^+ + NH_3 + HF \quad (S_N) \tag{1}$$

$$C_6H_5F^+-(NH_3)_2 \rightarrow C_6H_5F + (NH_3)_2^+ \quad (dET) \tag{2}$$

$$C_6H_5F-(NH_3)_2^+ \rightarrow C_6H_5F-NH_3^+ + NH_3 \quad (vdW) \tag{3}$$

Using mono-, di-, or trimethylamine, the substitution reactions are not observed, indicating that the barrier to the substitution reaction is probably too high for these molecules, and that the charge transfer process becomes the main exit channel, starting already for the 1–1 complex.

For *chlorobenzene*, the situation is slightly different because the 1–1 complex with ammonia has been found to be reactive (Mayeama and Mikami 1988, 1990) leading to a new class of products—protonated aniline:

$$C_6H_5Cl^+-NH_3 \rightarrow C_6H_5NH_3^+ + Cl^-$$

A similar reaction takes place with methylamine.

For di- and triethylamine, the dissociative electron transfer is again the only exist channel, as for fluorobenzene. These experiments show that the substitution reaction can be observed for 1–1 complexes: the efficiency of reaction of these 1–1 complexes is correlated with a greater reactivity in bimolecular collision experiments.

The presence of a fluorine or a chlorine atom on the aromatic ring clearly leads to two different behaviors, giving two different products—aniline$^+$ or anilinium ion with formation of HX or X$^\cdot$. Apparently, two different reaction coordinates are involved. In disubstituted aromatic compounds, depending on the nature of the substitutents, these two processes are competitive.

For the 1–1 ammonia/*1-chloro,2-fluorobenzene* complex, the S_N reaction leads quantitatively to the 2-fluoroanilinium (Riehn et al. 1992):

This behavior is very similar to that of chlorobenzene, studied previously. No 2-fluoroaniline$^+$ is detected: however, the TOF spectra show the presence of

2-chloroaniline$^+$. The R2PI spectra show that this product is correlated with the 1–2 clusters and is due to the reaction:

$$\text{(F, Cl-benzene)}^+ - (NH_3)_2 \longrightarrow \text{(Cl, NH}_2\text{-benzene)}^+ + HF - NH_3$$

With the *1-chloro,3-fluorobenzene*, the situation is more complicated. In the 1–1 complex, both S_N reaction channels are open:

$$\text{(F, Cl-benzene)}^+ - NH_3 \longrightarrow \text{(F, NH}_2\text{-benzene)}^+ + HCl \qquad [I]$$

$$\text{(F, Cl-benzene)}^+ - NH_3 \longrightarrow \text{(F, NH}_3^+\text{-benzene)} + Cl^{\cdot} \qquad [II]$$

The efficiency of the second channel is lower than in the *ortho*-Cl–F–benzene and it has been shown that in the *meta*–Cl–F–benzene the channel [I]/channel [II] ratio decreases as the energy of the second ionizing laser increases (Martrenchard-Barra et al. 1995).

For the 1–2 complex, the substitution reaction of the florine atom by –NH$_2$, as for the *ortho*-derivative, has been observed:

$$\text{(F, Cl-benzene)}^+ - (NH_3)_2 \longrightarrow \text{(NH}_2\text{, Cl-benzene)}^+ + HF - NH_3$$

The assigment of the parent ion spectroscopic bands to the 1–2 cluster is confirmed by the observation of the completitive electron transfer reaction:

$$\text{(F, Cl-benzene)}^+ - (NH_3)_2 \longrightarrow \text{(F, Cl-benzene)} + (NH_3)_2^+$$

For the *para*-Cl–F–benzene, the efficiency of reaction with ammonia is found to be lower: the fluoroaniline$^+$ product appears in the mass spectrum as a metastable peak. However, both substitution channels are open for the 1–1 complex; the Cl$^{\cdot}$ abstraction is the most efficient—leading to 4-fluoroanilinium.

Substitution in the 1–2 cluster leads to 4–fluoroaniline$^+$ (HCl elimination). This new channel is in competition with the HF elimination:

4B.4. CONCLUSIONS

The results of nucleophilic substitutions in solvated ionic aromatic cations (RX$^+$) are strongly dependent on the nature of the halogen and on the solvent cluster size. The energetics of these systems have been studied and it has been shown that the reactions are thermodynamically allowed for each solvent, even for the 1–1 complexes. The observed different behavior related to cluster size is probably governed by kinetic reasons (existence of barriers to the reactions). Two types of substitution reaction have been identified—leading either to X· radical or to HX molecule formation.

The first reaction, leading to X·, can be understood (Riehn et al. 1992) in terms of the Shaik–Pross model proposed to interpret experiments in the gas phase (Shaik and Pross 1989; Thölmann and Grützmacher 1991). The reaction goes through an intermediate σ complex

followed by elimination of X·. This reaction is observed for chlorobenzene but not for fluorobenzene. The difference between chloro- and fluorobenzene can be understood because the binding energy of the C–F bond is 1.4 eV larger than that of the C–Cl bond.

This model predicts the existence of a barrier to the reaction whose height depends on the dipole moment of the chromophore. The efficiency of this process for distributed benzenes is clearly correlated with their dipole moments: it is larger for *ortho*-disubstituted benzene than for *meta*- and *para*-distributed benzenes. This explains the cluster experiments (Brutschy et al. 1991) as well as the variation of reactivity in the gas phase (Thölman and Grützmacher 1991).

For the reaction leading to the elimination of HX, the situation is not so clear. Two mechanisms have been proposed.

(1) A mechanism is proposed that goes through the same σ complex as before (for X· production), followed by a proton transfer from the NH$_3$ in the σ complex toward the solvent cluster, and a capture of the halogen by the protonated cluster. For example, in the case of fluorine, the C–F bond cannot break alone, but the

barrier to the reaction can be lowered by the concerted proton transfer from NH_3^+ toward the cluster, and the capture of the fluorine atom by the protonated cluster would lead to $C_6H_5NH_2 + HF$.

(2) An electron transfer is proposed from the cluster toward the aromatic ring, leading to the formation of $C_6H_5X\cdots(NH_3)_n^+$ and followed by a rearrangement leading to the formation of $NH_2^. + (NH_3)_nH^+$. In a concerted mechanism, the $NH_2^.$ attacks the phenyl cation leading to aniline$^+$ + HX.

Both mechanisms account for the size effects observed in the reactions within the clusters. In the first case, the proton transfer should depend on the proton affinity of the cluster, which varies with the cluster size (Knochenmuss and Leutwyler 1989):

$$
\begin{array}{ll}
NH_3 & -206 \text{ kcal mol}^{-1} \\
(NH_3)_2 & -227 \text{ kcal mol}^{-1} \\
CH_3OH & -182 \text{ kcal mol}^{-1} \\
(CH_3OH)_2 & -211 \text{ kcal mol}^{-1} \\
H_2O & -167 \text{ kcal mol}^{-1} \\
(H_2O)_2 & -195 \text{ kcal mol}^{-1} \\
(H_2O)_3 & -206 \text{ kcal mol}^{-1}
\end{array}
$$

From these data it can be pointed out that for a given size of the clusters the proton affinity of water is smaller than for the other solvents; consequently, for fluorobenzene/methanol or *para*-difluorobenzene/water systems, a proton affinity of 205/215 kcal mol^{-1} seems to be the limit of the reaction process (it is reached for two molecules of methanol and three molecules of water).

In the second mechanism, the electron transfer from the nucleophile cluster into the aromatic ring should be facilitated by the decrease of the ionization potential (IP) of the solvent clusters as n increases. This mechanism is convincing for the ammonia or methanol clusters which show relatively low IPs when cluster size is increasing; however, for water clusters, the IPs of $n \geq 3$ clusters are not known. The IPs of water and its dimer are 12.6 and 11.2 eV, respectively (Ng et al. 1977). However, these IPs are certainly higher than the one of PDFB (9.2 eV), which is not in favor of a sequential electron transfer followed by a proton transfer mechanism. This mechanism is more likely possible if one assumes, in agreement with Brutschy and coworkers, that the barrier to the reaction is lowered by a concerted electron transfer/proton transfer mechanism (Brutschy 1989, 1990; Brutschy et al. 1988, 1991, 1992, in press).

At first sight, it seems surprising to observe competitive reactions within the same complex. However, it must be noticed that in the ionization processes the internal energy distribution within the ions can be broad since the cluster's geometries in the S_1 excited state and the ionic state can be very different. This can be seen by the ionization threshold measurements which do not exhibit clear onsets. Therefore, the presence of competitive processes can be explained by different barrier heights for the different channels. When the ions are prepared below one barrier and above the other one, only one product will be observed. Due to this broad internal energy distribution, on average, many channels can be detected. Coincidence detection of the zero kinetic electron and the product ions would be the way to measure properly the different barriers (C. Dedonder-Lardeux et al. 1995).

From these experiments on molecular clusters, a new aspect of ion–molecule reactions can be studied as a function of the cluster size and compared with results on the gas phase and in the liquid phase. The role of the solvent in bimolecular reactions has been evidenced and the catalytic effect depends on the number of molecules (which represents a new result compared with experiments in solution) and the nature of the solvent. These systems appear to be very good systems to study the chemistry under microsolvation.

ACKNOWLEDGMENTS: We gratefully acknowledge S. Martrenchard-Barra, C. Dedonder–Lardeux, B. Soep, A. Tramer, and Ph. Millié for stimulating discussions. Special thanks to O. Benoist d'Azy and M. Pellan for their help preparing this chapter. We thank also B. Brutschy, S. Leutwyler, and J. Syage who responded to our request for the figures.

REFERENCES

Abe, H.; Mikami, N.; Ito, M. 1982a J. Phys. Chem. 86:1768

Abe, H.; Mikami, N.; Ito, M.; Ugawada, Y. 1982b Chem. Phys. Lett. 93:217

Abe, H.; Mikami, N.; Ito, M.; Ugawada, Y. 1982c J. Phys. Chem. 86:2567

Awdiew, J.; Riehn, C.; Brutschy, B.; Baumgärtel, H. 1990 Ber. Bunsenges. Phys. Chem. 94:1353

Baba, H.; Suzuki, S. 1961 J. Phys. Chem. 53:1118

Bartok, W.; Lucchesi, P. J.; Snider, N. S. 1962 J. Amer. Chem. Soc. 84:1842

Beens, H.; Grellmann, K. H.; Gurr, M.; Weller, A. H. 1965 Discuss. Faraday Soc. 39:183

Bernier, A.; Millié, P. 1987 Chem. Phys. Lett. 134:379

Bernstein, E. R. 1992 J. Phys. Chem. 96:10105

Bisling, P. G. F.; Rühl, E.; Brutschy, B.; Baumgärtel, H. 1987 J. Phys. Chem. 91:4310

Boivineau, M. 1987, Ph.D. Thèse, Université de Paris-Sud

Boivineau, M.; Le Calvé, J.; Castex, M. C.; Jouvet, C. 1986a Chem. Phys. Lett. 128:528

Boivineau, M.; Le Calvé, J.; Castex, M. C.; Jouvet, C. 19886b Chem. Phys. Lett. 130:208

Boivineau, M.; Le Calvé, J.; Castex, M. C.; Jouvet, C. 19886c J. Chem. Phys. 84:4712

Breckenridge, W. H. 1983 In Reaction in Small Transient Species, A. Fontijn and M. A. A. Clyne, eds., Academic Press, London, and references therein

Breckenridge, W. H.; Duval, M. C.; Jouvet, C.; Soep, B. Chem. Phys. Lett. 119:317

Breckenridge, W. H.; Duval, M. C.; Jouvet, C.; Soep, B. 1987 J. Phys. Chem. 84:381

Breckenridge, W. H.; Jouvet, C.; Soep, B. 1986 J. Chem. Phys. 84:1443

Breckenridge, W. H.; Jouvet, C.; Soep, B. 1996 In Advances in Metal and Semiconductor Clusters, Vol. III, M. Duncan ed., JAI Press, Greenwich in press

Breen, J. J.; Peng, L. W.; Wilberg, D. M.; Heikal, A.; Cong, P.; Zewail, A. 1990 J. Chem. Phys.92:805

Brinckmann, U.; Schmidt, V. H.; Telle, H. 1980 Chem. Phys. Lett. 73:530

Brinckman, U.; Telle, H. 1977 J. Phys. B 10:133

Brucker, G. A.; Kelley, D. F. 1989a Chem. Phys. 136:213

Brucker, G. A.; Kelley, D. F. 1989b J. Chem. Phys. 90:5243

Brumbaugh, D. V., Kenny, J. E.; Levy, D. H. 1983 J. Chem. Phys. 78:3415

Brutschy, B. 1989 Habilitationsscrift, University of Berlin

Brutschy, B. 1990 J. Phys. Chem. 94:8637

Brutschy, B.; Eggert, J.; Janes, C.; Baumgärtel, H. 1991 J. Phys. Chem. 95:5041

Brutschy, B.; Janes, C.; Eggert, J. 1988 Ber. Bunsenges. Phys. Chem. 92:435

Brutschy, B.; Riehn, C.; Lahmann, C.; Wassermann, B. 1992 Communication at XIVth International Symposium on Molecular Beams, Asilomar, USA

Buelow, S.; Noble, M.; Radhakrishnan, G.; Reisler, H.; Wittig, C.; Hancock, G. 1986 J. Phys. Chem. 90:1015

Callear, A. B.; McGurk, J. C. 1972 J. Chem. Soc. Faraday Trans. 2 68:289

Callear, A. B.; Wood, P. M. 1972 J. Chem. Soc. Faraday Trans. 2 68:302

Ceyer, S. T.; Tiedmann, P. W.; Mahan, B. H.; Lee, Y. T. 1979 J. Chem. Phys. 70:14

Cheshnovsky, O.; Leutwyler, S. 1985 Chem. Phys. Lett. 121:1

Cheshnovsky, O.; Leutwyler, S. 1988 J. Chem. Phys. 88:4127

Crépin, C; Tramer, A. 1991 Chem. Phys. 156:281

Dedonder-Lardeux C.; Dimicoli, I.; Jouvet, C.; Martrenchard-Barra, S.; Richard-Viard, M.; Solgadi, D.; Vervldet, M. 1995 Chem. Phys. Lett. 240:97

Docker, M. P.; Hodgson, A.; Simons, J. P. 1986 Faraday Discuss. Chem. Soc. 82:25

Dreiling, T. D.; Setser, D. W. 1984 J. Chem. Phys. 79:5423

Droz, T.; Knochenmuss, R.; Leutwyler, S. 1990 J. Chem. Phys. 93:4520

Dubbov, V. S.; Lapsker, Y. E.; Samoilova, A. N.; Gurvich, L. V. 1981 Chem. Phys. Lett.83:518

Duval, M. C.; Jouvet, C.; Soep, B. 1985 Chem. Phys. Lett. 119:317

Faisal, F. H. M.; Wallenstein, R.; Zacharias, H. 1977 Rev. Lett. 39:1138

Föster, T. 1950a Z. Elektrochem. 54:531

Föster, T. 1950b Z. Elektrochem. Angew, Phys. Chem. 54:42

Foth, H.; Polanyi, J. C.; Telle, H. H. 1982 J. Phys. Chem. 86:5027

Fuke, K.; Kaya, K. 1983 Chem. Phys. Lett. 94:97

Fuke, K.; Saito, T.; Kaya, K. 1984 J. Chem. Phys. 81:2591

Gonohe, N.; Abe, H.; Mikami, N.; Ito, M. 1985 J. Phys. Chem. 89:3642

Goto, A.; Fujii, M.; Mikami, N.; Ito, M. 1986 J. Phys. Chem. 90:2370

Grabner, G.; Köhler, B.; Zechner, J.; Getoff, N. 1977 Photochem. Photobiol. 26:449

Grieneisen, H. P.; Xue Jing, H.; Kompa, K. L. 1983 Chem. Phys. Lett. 82:421

Halberstadt, N.; Soep, B. 1984 J. Chem. Phys. 80:2340

Harria, C. M.; Selinger, B. K. 1980 J. Phys. Chem. 84:1366

Hering, P.; Brooks, P. F.; Curl, R. F; Judson, R. S.; Lowe, R. S. 1980 Phys. Rev. Lett. 44:687

Hineman, M. F.; Brucker, G. A.; Kelley, D. F.; Bernstein, E. R. 1992 J. Phys. Chem. 97:3341

Hineman, M. F.; Kelley, D. F.; Bernstein, E. R. 1993 J. Chem. Phys. 99:4533

Husain, J.; Weinsenfeld, J. R.; Zare, R. N. 1980 J. Chem. Phys. 72:2479

Imre, D. G.; Kinsey, J. L.; Field, R. W.; Katayama, D. H. 1982 J. Phys. Chem. 86:2564

Inoue, G.; Ku, J. K.; Setser, D. W. 1984 J. Chem. Phys. 80:6006

Ishiwata, T.; Tokunaga, A.; Tanaka, I. 1984 Chem. Phys. Lett. 112:356

Johnson, K.; Pease, R.; Simons, J. P.; Smith, P. A.; Kvaran, A. 1986a J. Chem. Soc. Faraday Trans. 2 82:1281

Johnson, K.; Simons, J. P.; Smith, P. A.; Kvaran, A. 1987, J. Chim. Phys. 84:371

Johnson, K.; Simons, J. P.; Smith, P. A.; Washington, C.; Kvaran, A. 1986 Mol. Phys. 57:255

Jouvet, C. 1985 Thèsis, Université de Paris-Sud

Jouvet, C.; Boivineau, M.; Duval, M. C.; Soep, B. 1987 J. Phys. Chem. 91:5416

Jouvet, C.; Lardeux-Dedonder, C.; Richard-Viard, M.; Solgadi, D.; Tramer, A. 1990 J. Phys. Chem. 94:5041

Jouvet, C.; Soep, B. 1981 J. Chem. Phys. 75:1661

Jouvet, C.; Soep, B. 1983 Chem. Phys. Lett. 96:426

Jouvet, C.; Soep, B. 1985a J. Phys. (Paris) 46:C1, 313

Jouvet, C.; Soep, B. 1985b Laser Chem. 5:157

Kamke, W.; Hermann, R.; Wang, Z.; Hertel, I. V. 1988 Z. Phys. D 10:491

Keller, A. 1991 Thèse, Université de Paris-Sud

Kim, S. K.; Hsu, S. C.; Li, S.; Bernstein, E. R. 1991a J. Chem. Phys. 95:3290

Kim, S. K.; Li, S.; Bernstein, E. R. 1991b J. Chem. Phys. 95:3119

Klee, S.; Gericke, K. H.; Comes, F. J. 1986 J. Chem. Phys. 85:40

Kleiber, P. D.; Lyyra, A. M.; Sando, K. M.; Heneghan, S. P.; Stwalley, W. C. 1985 Phys. Rev. Lett. 54:2003

Kleiber, P. D.; Lyyra, A. M.; Sando, K. M.; Zafiropulos, V.; Stwallet, W. C. 1986 J. Chem. Phys. 85:5493

Knochenmuss, R.; Cheshnovsky, O.; Leutwyler, S. 1988 Chem. Phys. Lett. 144:317

Knochenmuss, R.; Holtom, G. R.; Ray, D. 1993 Chem. Phys. Lett. 215:188

Knochenmuss, R.; Leutwyler, S. 1989 J. Chem. Phys. 91:1268

Krause, H. F.; Johnson, S. G.; Datz, S.; Schmidt Bleek, F. K. 1975 Chem. Phys. Lett. 31:577

Krim, L. 1994 Thèsë, Universitë de Paris-Sud

Krim, L.; Qiu, P.; Halberstadt, N.; Soep, B.; Visticot, J. P. 1994 In *Femtosecond Chemistry*, J. Manz and L. Wöste, eds. VCH, Weinheim, Germany

Ku, J. K.; Inoue, G.; Setser, D. W. 1983 J. Phys. Chem. 87:2989

Le Calvé, J.; Castex, M. C.; Jordan, B.; Zimmerer, G.; Moller, T.; Haaks, D. 185 *Photophysics and Photochemistry above 6 eV*, F. Lahmani ed. Elsevier, Amsterdam, p. 639

Legay-Sommaire, N.; Legay, F. 1993 Chem. Phys. Lett. 207:123

Leighton, P. A.; Leighton, W. G. 1935 J. Chem. Education 12:139

Lipert, R. J.; Colson, S. D. 1988 J. Chem. Phys. 89:4579

Macguire, T. C.; Brooks, P. R.; Curl, R. F. 1983 Phys. Rev. Lett. 50:1918

Martrenchard, S. 1993 PhD. Thesis, Université de Paris-Sud

Martenchard, S.; Jouvet, C.; Lardeux-Dedonder, C.; Solgadi, D. 1991 J. Phys. Chem. 95:9186

Martrenchard-Barra, S.; Lardeux-Dedonder, C.; Jouvet, C.; Rockland, U.; Solgadi, D. 1995 J. Phys. Chem. 99:13716

Mataga, N.; Kaifu, Y. 1964 Mol. Phys. 7:137

Mataga, N.; Kawasaki, Y.; Torihashi, Y. 1964 Theor. Chim. Acta 2:168

Mayeama, T.; Mikami, N. 1988 J. Amer. Chem. Soc. 110:7238

Mayeama, T.; Mikami, N. 1990 J. Phys. Chem. 94:6973

Mayeama, T.; Mikami, N. 1991 J. Phys. Chem. 95:7197

Mikami, N.; Okabe, A.; Suzuki, I. 1988 J. Phys. Chem. 92:1858

Mikami, N.; Suzuki, I.; Okabe, A. 1987 J. Phys. Chem. 91:5242

Ng, C. Y.; Trevor, D. J.; Tiedmann, P. W.; Ceyer, S. T.; Kronebusch, P. L.; Mahan, B. H.; Lee, Y. T. 1977 J. Chem. Phys. 67:4235

Nichimoto, K. 1963 J. Phys. Chem. 67:1443

Ogg, R. A.; Martin, H. C.; Leighton, P. A. 1936 Phys. Rev. 58:1922

O'Grady, B. V.; Donovan, R. J. 1985 Chem. Phys. Lett. 122:503

Oikawa, A.; Abe, H.; Mikami, N.; Ito, M. 1983 J. Phys. Chem. 87:5083

Papanikis, J. M.; Vorsa, V.; Nadal, M. E.; Campagnola, P. J.; Buchenau, H. K.; Lineberger, W. C. 1993 J. Chem. Phys. 99:8733, and references therein

Pine, S. H.; Hendrickson, J. B.; Cram, D. J.; Hammond, G. S. 1980 *Organic Chemistry*, 4th Ed., McGraw-Hill, New York

Plusquellic, D. F.; Tan, X. Q.; Pratt, D. W. 1992 J. Chem. Phys. 96:8026

Polanyi, J. C. 1972 Acc. Chem. Res. 5:161

Polyani, M. 1932 *Atomic Reactions*, Williams and Norgate, London

Reichardt, C. 1988 *Solvent Effects in Organic Chemistry*, VCH, Weinheim, FRG

Rettner, C. T.; Zare, R. N. 1982 J. Chem. Phys. 77:2416

Rettner, C. T.; Zare, R. N. 1981 J. Chem. Phys. 75:3636

Rice, S. A. 1986, J. Phys. Chem. 90:3063

Richman, M. K.; Nelson, T. O.; Setser, D. W. 1993 Chem. Phys. Lett. 210:71

Riehn, C.; Lahmann, C.; Brutschy, B. 1992 J. Phys. Chem. 96:3626

Selinger, B. K.; Weller, A. 1977 Austral. J. Chem. 30:2377

Setser, D. W.; Ku, J. K. 1985 *Photophysics and Photochemistry above 6 eV*, F. Lahmani ed. Elsevier, Amsterdam, p. 621

Setser, D. W.; Qin, J. 1991 J. Phys. (Paris) C7:579

Shaik, S. S.; Pross, A. 1989 J. Amer. Chem. Soc. 111:4306

Shea, J. A.; Campbell, E. J. 1984 J. Chem. Phys. 81:5326

Shin, S.; Wittig, C.; Goddard, W. A. 1991 J. Phys. Chem. 95:8048

Smalley, R. E.; Auerbach, D. A.; Fitch, P. S. H.; Levy, D. H.; Wharton, L. 1977 J. Chem. Phys. 66:3778

Soep, B.; Abbès, S.; Keller, A.; Visticott, J. P. 1992 J. Chem. Phys. 96:440

Soep, B.; Whitham, C. J.; Keller, A.; Visticot, J. P. 1991 Faraday Discuss, Chem. Soc. 91:191

Solgadi, D.; Jouvet, C.; Tramer, A. 1988 J. Phys. Chem. 92:3313

Steadman, J.; Syage, J. A. 1990 J. Chem. Phys. 92:4630

Steadman, J.; Syage, J. A. 1991 J. Amer. Chem. Soc. 113:6786

Stephenson, T. A.; Ricem S. A. 1984 J. Chem. Phys. 81:1083

Sur, A.; Hui, A. K.; Tellingshuisen, J. 1979 J. Mol. Spectr. 74:465

Syage, J. A. 1990 Soc. Photo Opt. Inst. Eng. 1209:64

Syage, J. A. 1993 J. Phys. Chem. 97:12523

Syage, J. A., Steadman, J. 1991 J. Chem. Phys. 95:2497

Syage, J. A.; Steadman, J. 1992 J. Phys. Chem. 96:9606

Takayanagi, M.; Hanazaki, I. 1991 Chem. Rev. 91:1193

Tamagake, K.; Kolts, J. H.; Setser, D. W. 1979 J. Chem. Phys. 71:1264

Telle, H.; Brinckmann, U. 1990 Mol. Phys. 39:361

Tellinghuisen, J.; Ragone, A.; Kim, M. S.; Auerbach, D. A.; Smalley, R. E.; Wharton, L.; Levy, D. H. 1979 J. Chem. Phys. 71:1283

Thölmann, D.; Grützmacher, H. F. 1991 J. Amer. Chem. Soc. 113:3281

Torchet, G.; Schwartz, P.; Farges, J.; de Feraudy, M. F.; Raoult, B. 1983 J. Chem. Phys. 79:6196

Vikis, A. C.; le Roy, D. J. 1973 Can. J. Chem. 51:1207, and references therein

Wadt, W. R. 1980 J. Chem. Phys. 72:2469

Wehry, E. L.; Rogers, L. B. 1965 J. Amer. Chem. Soc. 87:4334

Weller, A. 1952 Z. Elektrochem. 56:662

Weller, A. 1958 Z. Physik, Chem. 17:224

Wilcomb, B. E.; Burnham, R. 1981 J. Chem. Phys. 74:6784

Wilkinson, J. P. T.; Kerr, E. A.; Lawley, K. P.; Donovan, R. J.; Shaw, D.; Hopkirk, A.; Munro, I. 1986 Chem. Phys. Lett. 130:213

Yu, Y. C.; Setser, D. W.; Horiguchi, H. 1983 J. Chem. Phys. 87:2209, and references therein

Zewail, A. H. 1991 Faraday Discuss. Chem. Soc. 91:1

5

Intermolecular Dynamics and Bimolecular Reactions

ELLIOT R. BERNSTEIN

5.1. INTRODUCTION

The study of clusters has taken the path that is quite typical in physical chemistry research for a newly discovered system or state of matter: (1) elucidation of energy eigenstates, both experimentally and theoretically, (2) elucidation of structure through experiments and calculations of various degrees of sophistication, (3) exploration of system dynamics, and (4) explorations of chemical reactivity within the new system. Indeed, previous review volumes covering cluster research have dealt mostly with eigenstates and structure, with some attention given to the dynamics and reactions of clusters (Bernstein 1990; Halberstadt and Janda 1990; Jena et al. 1987; Weber 1987).

The study of all aspects of cluster energy levels, structure, and behavior is both important and useful on a number of different levels. First, cluster research is performed because clusters themselves are a fascinating system in which to study intermolecular interactions and solvation behavior. Second, and perhaps more useful, cluster investigations can lead to a far better understanding of condensed phase and surface systems. With regard to condensed phase dynamics and chemical reactions, clusters provide three very important components for the basic data set: (1) clusters can generate the minimum irreducible set required for a dynamical event or reaction to occur, (2) clusters provide an excellent proving ground for the comparison between experiment and theory because both sets of results can be based on exactly the same systems, and (3) an isolated cluster of minimum size with regard to a dynamical event or reaction is an ideal entity in which to investigate a dynamical or reaction coordinate, its dimensionality, and its dependence on system properties.

Of course, clusters are not simply small condensed phase systems: they can express behavior quite different from condensed phase systems. For example, reactions that occur in condensed phases may not occur in clusters; however, these differences can be understood and related to the different properties of the two systems. In clusters, as generated in the gas phase at low temperatures, the structure is controlled mostly by energy considerations, whereas in the condensed phase, both high temperatures and a large number of possible configurations mandate the importance of an entropy term ($\Delta G = \Delta H - T\Delta S$). The density of states in clusters and condensed phases may also be different but, for clusters at least, good estimates of energy levels and their densities are possible. Solvation effects on barriers, transition states and the reaction coordinate or pathway itself can be controlling influences on condensed phase, but not necessarily, cluster chemistry; for clusters, "solvation effects" can be controlled and explored. Thus, the comparison between condensed phase and cluster behavior can be important and informative even when the two systems do not behave similarly. These differences can be understood and employed to model condensed phase dynamics and reactions.

In this chapter, we will discuss various studies that deal with cluster dynamics and cluster chemical reactions. First, the dynamics (intermolecular vibrational energy redistribution—IVR, and vibrational predissociation—VP) of small solute (chromophore)/solvent clusters will be discussed. These data will be analyzed and shown to be of importance for condensed phase modeling. Second, we will discuss various reactions investigated in clusters; electron transfer, proton transfer, and radical addition and generation. In these sections, we will discuss some reactions that do not occur in clusters, even though they do occur in condensed phases. Solvation effects can thus be investigated through such comparisons. Additionally, some preliminary discussion will also be presented for work in progress and systems that can be predicted to be quite interesting from what has already been learned about cluster dynamics and chemistry.

The early experimental work on cluster dynamics was carried out by Levy and coworkers (Brumbaugh et al. 1983; Haynam 1984a,b; Levy 1981; Park and Levy 1984), Rettschnick and coworkers (Heppener et al. 1985; Heppener and Rettschnick 1987; Ramackers et al. 1982, 1983a,b), Beswick, Jortner, and coworkers (Beswick and Jortner 1981; Jortner et al. 1988; Lin 1980; Mukamel and Jortner 1977), Rice and coworkers (Chernoff and Rice 1979; Jacobsen et al. 1988; Stephenson et al. 1984; Stephenson and Rice 1984; Weber and Rice 1988a,b), and the early work on cluster chemistry stems from the experiments of Jouvet and coworkers (Jouvet et al. 1987; Jourvet and Soep 1983; Loison et al. 1991), Wittig and coworkers (Böhmer et al. 1992, 1993; Davis et al. 1993; Ionov et al. 1992, 1993a,b; Shin et al. 1991; Wittig et al. 1988), and Leutwyler and coworkers (Chesnovsky and Leutwyler 1985, 1988; Droz and Knochenmuss 1990; Droz et al. 1988). These original studies will be discussed and referenced in the appropriate sections below as the presentation unfolds. A nice review of the early small molecule clusters chemistry has recently been published (Takayanagi and Hanazaki 1991).

5.2. CLUSTER VIBRATIONAL DYNAMICS—IVR/VP

Almost from the initial studies of cluster fluorescence excitation on systems such as I_2/He, Ne and tetrazine/Ar by the Levy group (Brumbaugh et al. 1983; Haynam 1984a,b; Heppener et al. 1985; Heppener and Rettschnick 1987; Levy 1981; Park and Levy 1984; Ramackers et al. 1982, 1983a,b), vibrational dynamics were clearly known to be a factor in the experimental data set: if a cluster transition is excited for which the vibrational energy is substantially greater than the cluster binding energy in the excited electronic state accessed (usually S_1), emission from the bare chromophore molecule (I_2, aromatic, carbonyl, amine, etc.) is recorded along with some cluster emission. Moreover, even if the cluster does not undergo vibrational predissociation and bare molecule emission is not observed, one often observes, upon cluster vibronic excitation, extensive cluster emission from cluster states other than the cluster state accessed by the laser excitation process. Clearly, clusters readily undergo vibrational dynamics, and both the separate processes of IVR and VP occur on the time scale of the excited state (fluorescence) lifetime. Relaxed cluster emission (say from the cluster $\overline{0^0}$ state upon pumping the cluster $\overline{X^1}$ state) is often broad, as would be expected from a vibrationally hot system.

The question to be addressed is, then, how do the IVR and VP occur? Does VP follow IVR or do they occur simultaneously? Originally, the mechanism for IVR/VP was assumed to be parallel (they occur simultaneously) (Brumbaugh et al. 1983; Haynam et al. 1984a,b; Heppener et al. 1985; Heppener and Rettschnick 1987; Jena et al. 1987; Levy 1981; Park and Levy 1984; Ramackers et al. 1982, 1983a,b), perhaps because I_2/He, Ne results were modeled with a direct VP process (cluster states directly couple to the translational continuum states). For this system, the small cluster binding energy and only three cluster intermolecular vibrational modes mandate that a loss of one quantum of energy in the chromophore mode (ca. 100 cm^{-1}) dissociates the cluster in all three van der Waals degrees of freedom (Levy 1981). The difference then between a serial (IVR from the chromophore modes to the van der Waals intermolecular modes followed by VP) process (Kelley and Bernstein 1986) and a parallel process becomes somewhat blurred for these very weakly bound molecule/atom clusters. Applying this molecule/atom direct VP parallel IVR/VP process model to molecule/molecule clusters, in which the binding energy is large ($\geq 500 \text{ cm}^{-1}$) and five or six van der Waals modes are present, is not obviously correct. In fact, for the cases that have been studied for chromophore/molecule clusters (often referred to as solute/solvent clusters), the serial IVR/VP model is found to be the proper description of the cluster vibrational dynamics (Hineman et al. 1992b, 1993a, 1994; Kelley and Bernstein 1986; Nimlos et al. 1989; Outhouse et al. 1991; Smith et al. 1993).

The conflicting serial/parallel models for IVR/VP are not readily distinguished until time resolved experiments can be performed on the systems of interest. Both models can relate the relative intensities of the emission features to the various model parameters, but the serial process seems more in line with a simple, conventional [Fermi's Golden Rule for IVR (Avouris et al. 1977; Beswick and Jortner 1981; Jortner et al. 1988; Lin 1980; Mukamel 1985; Mukamel and Jortner 1977) and RRKM theory for VP (Forst 1973; Gilbert and Smith 1990; Kelley and Bernstein 1986; Levine and Bernstein 1987; Pritchard 1984; Robinson and Holbrook 1972; Steinfeld et al. 1989)], few parameter approach. Time resolved measurements do distinguish the models because in a serial model the rises and decays of various vibronic states should be linked, whereas in a parallel one they are, in general, unrelated. Moreover, the time dependent studies allow one to determine how the rates of the IVR and VP processes vary with excitation energy, density of states, mode properties, and isotropic substitution.

The studies discussed below in detail are ones from our own laboratory on aniline/Ar, N_2, CH_4 clusters. Comparable studies on indole/Ar, CH_4 by Wallace and coworkers (Outhouse et al. 1991), employing emission and mass resolved techniques, and Knee and coworkers (Smith et al. 1993) on aniline/Ar, CH_4, employing photoelectron techniques, support and substantiate our findings.

The experiment data set and the serial model for IVR/VP, which are discussed below to demonstrate the analysis, yield the following results: (1) a serial IVR/VP mechanism explains all extant data, in which IVR is given by Fermi's Golden Rule and VP is given by a restricted statistical RRKM model, (2) the single most important factor for both IVR and VP is the density of states for the cluster—IVR rates are increased by a high density of states and VP rates are decreased by a high density of states, (3) van der Waals (vdW) modes and low energy chromophore modes contribute equally to the effective density of states for the cluster, with the lower energy modes being most important, (4) a phase space theory can be employed to model final product state distributions of vibrational modes of the chromophore, and (5) a "mode selective" IVR/VP for the final product state distributions is not necessarily inconsistent with this statistical/density of states-based model, as each chromophore mode may have its own coupling matrix element to the van der Waals vibrational density of states for IVR.

5.2.1. Aniline/Ar, N_2, CH_4

We have studied the excited electronic state vibrational dynamics of these clusters by two experimental techniques: time correlated single photon counting (TCSPC) at ca. 50 ps time resolution; and time resolved stimulated emission pumping (TRSEP) at ca. 10 ps time resolution. The experiments are challenging for a number of reasons: concentration of both aniline and the solvent must be maintained at very low levels (ca. 1% of the expansion gas) in order to be certain that only the cluster of interest is present in the expansion, because mass selection of the detected clusters is not employed in these measurements; and cluster origin ($\overline{0^0}$) emission and emission from the cluster state accessed by laser excitation must be observed to have a single exponential decay and thus be free of interference from other species and recurrences from coherent phenomena (Felker and Zewail 1988). If this latter requirement is not met, determination of growths and decays of cluster and bare molecule signals can be quite difficult and confusing. The experimental set up for these determinations is presented in the recent literature (Hineman et al. 1992b, 1993a, 1994; Nimlos et al. 1989).

Beginning with the simplest system, aniline$(Ar)_1$, we present in Figure 5-1 the aniline bare molecule mass resolved excitation spectrum for the $S_1 \leftarrow S_0$ absorption. Figure 5-2 and 5-3 show dispersed emission spectra of aniline$(Ar)_1$ clusters excited at various different vibronic energy levels of S_1. The following observations can be made for this cluster: (1) with $\overline{0^0}$ excitation, only emission from the cluster $\overline{0^0}$ is observed, as is to be expected, (2) at $\overline{X}$ excitation, emission is observed from $\overline{X}$ and $\overline{0^0}$ showing the presence of IVR during the 8 ns cluster excited state lifetime, (3) the cluster does not dissociative with $\overline{X}$ excitation so the excited state cluster binding energy is greater than 450 cm^{-1}, assuming (correctly, see below) that VP is rapid, (4) at $\overline{6a^1}$ excitation, only $\overline{6a^1}$ and 0^0 emission are

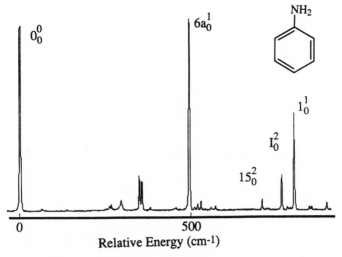

Figure 5-1. Mass resolved excitation spectrum of bare aniline in a molecular jet. Several of the more intense vibronic transitions are assigned.

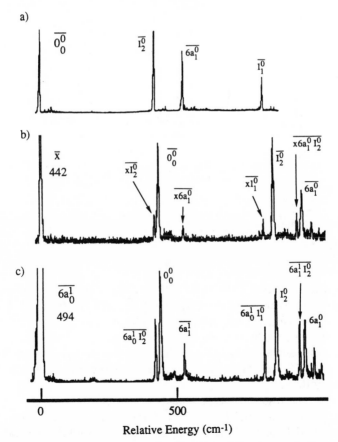

Figure 5-2. Dispersed emission spectra of aniline(Ar)$_1$ following excitation at (a) $\overline{0_0^0}$, (b) 442 cm^{-1}, and (c) $\overline{6a_0^1}$ (494 cm^{-1}). Assignments of the more intense features are indicated. The intense peaks at 0 cm^{-1} are due partially to scattered excitation light.

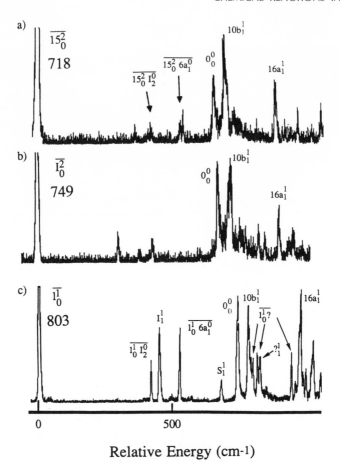

Relative Energy (cm-1)

Figure 5-3. Dispersed emission spectra of aniline (Ar)$_1$ following excitation at (a) $\overline{15^2_0}$ (718 cm^{-1}), (b) $\overline{I^2_0}$ (749 cm^{-1}), and (c) $\overline{I^1_0}$ (803 cm^{-1}). Assignments of the more intense features are indicated. The intense peaks at 0 cm^{-1} are due partially to scattered laser excitation light.

observed, suggesting slow IVR, rapid VP, and an excited state binding energy between 450 and 494 cm^{-1}, and (5) upon $\overline{15^2}$, $\overline{I^2}$, and $\overline{I^2}$ excitation, both cluster and bare molecule emission are observed—all vibronic states of the bare molecule emit that are energetically accessible following vibrational predissociation. The latter point again suggests that IVR is slow and rate controlling and that VP is fast; as soon as IVR has placed enough energy into the van der Waals modes for VP to occur, it does.

While the time resolved measurements (for example, see Figure 5-4) do not, in this case, mandate a serial IVR/VP mechanism, they are consistent with it in terms of the model calculations employing Fermi's Golden Rule for IVR and RRKM theory for VP.

Similar data are presented in Figure 5-5 for aniline(CH$_4$)$_1$ clusters. Note that these spectra are quite different in detail from those for aniline(Ar)$_1$ shown in

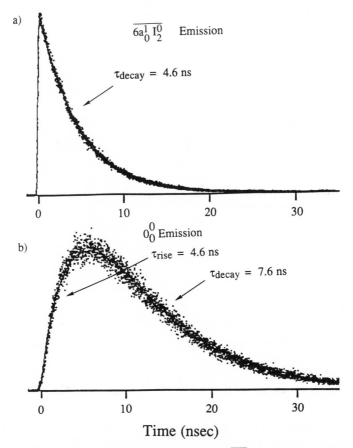

Figure 5-4. Time resolved emission kinetics following $\overline{6a_0^1}$ excitation of aniline(Ar)$_1$. The kinetics of the $\overline{6a_0^1 I_2^0}$ and 0_0^0 transitions are shown. Also shown are calculated curves corresponding to (a) a 4.6 ns decay, and (b) a 4.6 ns rise and a 7.6 ns decay.

Figures 5-2 and 5-3. First, the dispersed emission from the cluster $\overline{0^0}$ contains a good deal of van der Waals mode intensity due to the change in Franck–Condon factor between the two clusters. The difference in Franck–Condon factors probably arises because the Ar/aniline and CH$_4$/aniline intermolecular potentials are somewhat different. Second, excitation of the $\overline{6a^1}$ state yields only $\overline{0^0}$ and 0^0 emission with much more intensity in the cluster emission. This suggests that now IVR is fast, VP is slow, and that the cluster binding energy is close to 494 cm^{-1}. Third, emission from the cluster is now "hot" in that the $\overline{0_x^0}$ features are quite broad. The CH$_4$ cluster emission at $\overline{6a^1}$ excitation is broad, whereas the Ar cluster emission is sharp due to the difference in Franck–Condon factors for the two clusters.

The kinetic data for aniline(CH$_4$)$_1$ $\overline{15_n^2}$ excitation actually reveal the mechanism of relaxation. Since no $\overline{15^1}$ emission is observed we know that the lifetime of the state with regard to IVR is ≤ 100 ps. The rise time of 0^0 emission is 240 ps (Figure 5-6); this shows a delay between the decay of the initial state and the appearance of the final state, and suggests that the $\overline{0^0}$ state is the intermediate state. Recall that for

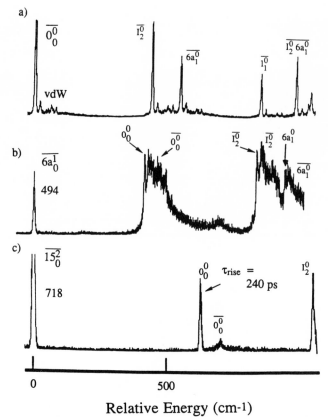

Figure 5-5. Dispersed emission spectra of aniline$(CH_4)_1$ following excitation at (a) 0_0^0, (b) $\overline{6a_0^1}$ (494 cm^{-1}), and (c) $\overline{15_0^2}$ (718 cm^{-1}). Assignments of the more intense features are indicated. The intense peaks at 0 cm^{-1} are due partially to scattered laser excitation light.

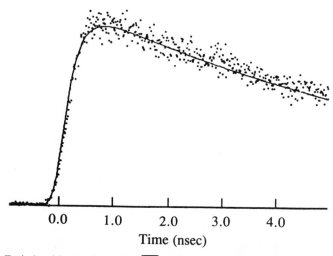

Figure 5-6. Emission kinetics following $\overline{15_0^2}$ excitation of aniline$(CH_4)_1$. The kinetics of the 0^0 transition are shown. Also shown is a calculated curve corresponding to the convolution of the instrument response function with a 240 ps rise and 7.6 ns decay.

$\overline{6a^1}$ excitation, which generates little bare molecule emission, fast IVR for the state pumped is also observed. While concrete proof for a serial mechanism is yet to come in the form of rise and fall times for intermediate states, the inference of this mechanism is quite strong in the results. The discussion below for aniline$(N_2)_1$ clusters, and the simple two parameter serial IVR/VP model based on Fermi's Golden Rule and RRKM theory, will provide the final demonstration for this mechanism.

Due to the nature of the density of states for a chromophore/diatomic molecule cluster, and the pivotal importance of the density of states for cluster vibrational dynamics, the results for the aniline$(N_2)_1$ cluster fall intermediately between those for aniline$(Ar)_1$ and aniline$(CH_4)_1$. This can be seen in emission spectra for various aniline$(N_2)_1$ excitations (Figure 5-7). First, $\overline{6a^1}$ excitation does

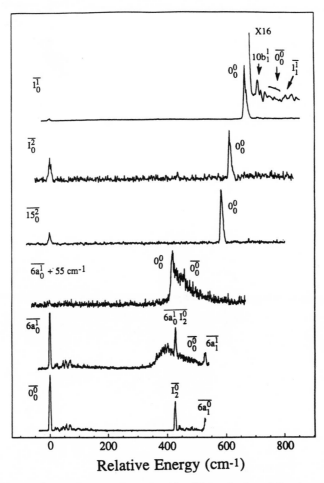

Figure 5-7. Dispersed emission spectra of aniline$(N_2)_1$ clusters following excitation to several vibrational states of S_1. Relative energy is the shift, in wavenumbers, from the excited transition. The top spectrum ($\overline{1_0^1}$ excitation) shows an inset trace for an expanded scale about the 0_0^0 intense feature: $10b_1^1$, $\overline{0_0^0}$, and $\overline{1_1^1}$ emission can be observed. Note that the relaxed cluster emission from $\overline{0_0^0}$ (following IVR) is broad as expected (compare with $\overline{6a_0^1} + 55\ \mathrm{cm^{-1}}$ and $\overline{6a_0^1}$ excitation).

not provide enough excess energy to dissociate the cluster and thus the relaxation time for the $\overline{6a}^1$ state will give the IVR time of this state. Note, too, that the Franck–Condon factors for hot emission are intermediate between those of the other two clusters. Second, excitation of higher vibronic states of the aniline(N_2)$_1$ cluster shows emission from that state pumped, as well as from the aniline 0^0.

Kinetic measurements (Figure 5-8) show exactly what is to be expected for a serial mechanism with relatively slow IVR and faster, but competing, VP. Upon $\overline{1}^T$ excitation the state pumped decays with a roughly 300 ps time constant, the $\overline{0}^0$ rates track the $\overline{1}^T$ kinetics because the rate controlling step is IVR population of the $\overline{0}^0$, and the 0^0 rises with the $\overline{1}^T$ IVR time decay. This is a perfect example of serial kinetics and density of states control of the cluster dynamics.

These decay measurements on the state excited can be repeated by a TRSEP technique (Hineman et al. 1994) to verify the IVR cluster kinetics. This has been done for the aniline(N_2)$_1$ $\overline{1}^T$ vibronic excitation. The experiment involves excitation of the $\overline{1}^T$ state, followed by stimulated emission with a time delayed pulse to deplete the $\overline{1}^T$ population. The total emission from the excited S_1 cluster and bare molecule as a function of time delay between the excitation and dump

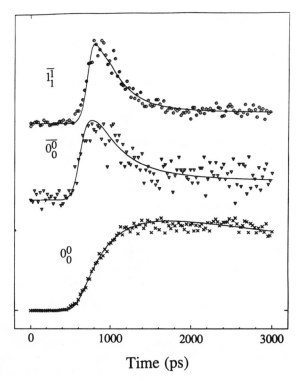

Figure 5-8. Decay curves and fits for several emission bands following $\overline{1}^T$ excitation of the aniline(N_2)$_1$ cluster.

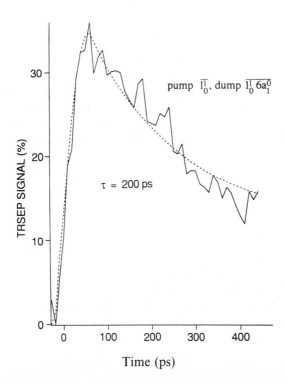

Figure 5-9. TRSEP signal for the aniline$(N_2)_1$ cluster. Excitation laser is tuned to the $\overline{1}_0^1$ transition, and the probe laser is tuned to the $\overline{1}_0^1 6a_1^0$ transition. This plot shows the extent to which the probe pulse diminishes the total fluoresence. The time axis is the difference between the arrival times of the pump and probe pulses. The maximum diminution of the fluorescence is about 30%. The smooth curve is generated using the results of a nonlinear fitting routine. The fast component time constant is 200 ± 50 ps.

pulses $(\Delta\tau)$ will decrease, and then increase as $\Delta\tau$ increases due to depopulation of $\overline{1}^1$ by IVR (see Figure 5-9). Time resolved stimulated emission pumping gives a 200 ± 50 ps lifetime for the $\overline{1}^1$ state of aniline$(N_2)_1$ so $\tau_{IVR} = 200 \pm 50$ ps for this state. The TCSPC time for this decay is 360 ± 100 ps. These two values are in good agreement and show that the observations and analysis are sound.

The above data can be explained nearly quantitatively by a serial IVR/VP mechanism in which energy is transferred from the chromophore vibrational modes to the vdW modes prior to VP (Hineman et al. 1992b, 1993a, 1994; Kelley and Bernstein 1986; Nimos et al. 1989). Once the amount of energy in the vdW modes exceeds the dissociation energy of the complex, VP can proceed or further energy transfer from chromophore modes may occur.

A simplified three-state version of this mechanism can be modeled as

$$A \xrightarrow{k_{IVR}} B \xrightarrow{k_{VP}} C \quad [A = \overline{6a^1} + 55 \text{ cm}^{-1}, \overline{1^1}, \overline{1^2} \text{ or } \overline{15^2}, B = \overline{0^0}, C = 0^0],$$

$$k_f \downarrow \qquad\qquad k_f \downarrow \qquad\qquad k_f \downarrow$$

in which k_{IVR} and k_{VP} are the IVR and VP rates, respectively, and k_f is the fluorescence rate (1/7.8 ns). This mechanism ignores intermediate states populated by IVR (i.e., states with vibrational excitation in both chromophore and vdW modes). As discussed above, these intermediate states are short-lived, and do not significantly alter the kinetics. At $t = 0$ we have $[A] = [A]_0$, $[B] = [C] = 0$, and the concentrations as a function of time are given by

$$[A] = [A]_0 \, e^{-(k_{IVR} + k_f)t},$$

$$[B] = \frac{k_{IVR}[A]_0}{(k_{VP} - k_{IVR})} \, e^{-k_f t} \{ e^{-k_{IVR} t} - e^{-k_{VP} t} \},$$

$$[C] = [A]_0 \, e^{-k_f t} \left\{ 1 + \frac{1}{k_{IVR} - k_{VP}} [k_{VP} \, e^{-k_{IVR} t} - k_{IVR} \, e^{-k_{VP} t}] \right\}. \qquad (5\text{-}1)$$

The IVR rate can be modeled using the Fermi's Golden Rule approximation in which transition probabilities depend on coupling between intitial and final states, and the density of final states. Two major results are generated by this model: (1) a general "energy gap law" for the IVR process under which the rate of IVR decreases as greater numbers of quanta are exchanged between the initial and final states, and (2) the rate of IVR increases with the density of vdW receiving (final) states, $N(E)$. Similar ideas have been suggested by Ewing (1981, 1986, 1987), Jortner (Beswick and Jortner 1981; Jortner et al. 1988; Lin 1980), and Rice (Weber and Rice 1988a,b). The form of the energy gap term depends on assumptions made about the details of the potential surface. For lack of a better alternative, we have used a model describing energy transfer in low temperature matrices. The result is a term proportional to $\exp[-(E_{ij}/hv_{max})^{1/2}]$ in which E_{ij} is the energy transferred in the IVR process upon going from the jth to the ith chromophore energy level and hv_{max} is the energy of one quanta in the highest frequency vdW mode. The ratio E_{ij}/hv_{max} is roughly proportional to the number of quanta deposited into the vdW modes by the IVR transition. (If v_{max} corresponds to the only receiving mode, then the above statement would be exactly correct; however, other vdW modes must also be involved for a Golden Rule expression to be valid.) The final rate constant for IVR is then taken to be

$$k_{IVR} = \frac{A}{v_i} \exp[-(E_{ij}/hv_{max})^{1/2}] N(E_i). \qquad (5\text{-}2)$$

In eq. (5-2), v_i, refers to the frequency of the ith chromophore vibration (populated by the IVR transition), and the proportionality to $1/v_i$ is purely phenomenological. This proportionality reflects the expectation that low frequency chromophore modes will couple most efficiently to the (low frequency) vdW modes. Fermi's Golden Rule expression has two important consequences. First, it predicts that

IVR will be much faster in clusters with more vdW degrees of vibrational freedom and, hence, with higher densities of states. Second, it predicts that the slowest IVR transitions will be from the initially excited state, as subsequent IVR transitions result in more energy in the vdW modes and, hence, a higher density of receiving states. Thus, depopulation of the initially excited state is expected to be the rate limiting IVR step.

The VP rate is calculated by a restricted (to vdW modes) RRKM theory. By assuming a "tight binding" model for the cluster transition state, the resulting expression for the VP rate constant is (Forst 1973; Gilbert and Smith 1990; Kelley and Bernstein 1986; Levine and Bernstein 1987; Pritchard 1984; Robinson and Holbrook 1972; Steinfeld et al. 1989),

$$k_{VP} = \frac{1}{h} \frac{\sum P_V(E - E_0)}{N(E)}, \tag{5-3}$$

in which $\sum P_V(E - E_0)$ is the sum of vdW states above the dissociation threshold, $N(E)$ is the density of vdW states, E is the amount of vibrational energy in the vdW modes, and E_0 is binding energy. In the Marcus–Rice approximation, these components are given by

$$\sum P_V(E - E_0) = \frac{(E - E_0)^{S-1}}{(S - 1)! \Pi' h v_i} \tag{5-4}$$

and

$$N(E) = \frac{E^{S-1}}{(S - 1)! \Pi h v_i}, \tag{5-5}$$

in which S is the number of vdW vibrational modes, the product is over the vdW mode energies, and the prime on the product sign excludes the reaction coordinate. Using the above approximations, we get

$$k_{VP} = \left(\frac{E - E_0}{E} \right)^{S-1} v, \tag{5-6}$$

in which v is the frequency of the vibrational mode corresponding to the reaction coordinate. For a given cluster, k_{VP} always increases as $(E - E_0)$ increases; however, at a given $(E - E_0)$, the cluster with the fewer vdW vibrational modes and, hence, the smaller density of vdW states will have the larger k_{VP}.

The above model makes qualitative and semiquantitative predictions about the IVR and VP rates in different clusters. Of particular relevance is the comparison of the IVR and VP rates for the clusters aniline$(CH_4)_1$, aniline$(N_2)_1$, and aniline$(Ar)_1$: aniline$(Ar)_1$ has only the three vdW modes; aniline$(CH_4)_1$ has six vdW modes; and aniline$(N_2)_1$ has five vdW modes. As a result, the sums and densities of vdW states used in the calculations of k_{IVR} and k_{VP} are quite different in these three clusters. Since the chromophore in these clusters remains the same, however, initial excitation to roughly the same energy in each cluster is possible. Furthermore, we expect that the extent of chromophore–solvent interaction will not be vastly different in these cases; that is, the A term in eq. (5-2) will be of the same order of magnitude for each cluster. The IVR rates for the N_2 cluster are expected to fall below those of the CH_4 cluster but above those of the Ar cluster as the density of vdW states at a given cluster energy increases with the number of vdW vibrational modes. Table 5-1 shows the observed trend: $\tau_{IVR}(Ar) > \tau_{IVR}(N_2) > \tau_{IVR}(CH_4)$. The sharp decrease in decay time of the pumped state for $\overline{6a}^1 + 55$ cm^{-1} compared to $\overline{6a}^1$ for the aniline$(N_2)_1$ cluster can be explained by the increase in vdW state density as the number of quanta in the vdW modes increases.

The VP rates are predicted to order in the opposite manner, as they depend inversely on the density of states. The energy in the vdW modes distributes through phase space in a statistical manner according to RRKM theory, and the greater the number of modes, the smaller is the probability that sufficient energy is in the reaction coordinate. The calculated RRKM VP rate is sensitive to the assumed binding energy E_0. Excitation of the $\overline{15}^2$, $\overline{I}^2$, and $\overline{I}^1$ levels results in about 203, 234, and 288 cm^{-1} of energy in excess of the cluster binding energy (ca. 500 cm^{-1}). If the aniline$(N_2)_1$ cluster has more than about 200 cm^{-1} of excess vibrational energy (i.e., $E > E_0 + 200$ cm^{-1}), then the VP rate for $\overline{0}^0$ given by eq. (5-6) is larger than $\sim(100$ ps$)^{-1}$. This is considerably faster than the IVR rate of the initially accessed state. Simplifying eq. (5-1), we take $k_{VP} \gg k_{IVR}$ and, for times longer than the instrument response function, we take $k_{VP}t \gg 1$. Under these circumstances, we get

$$[B] \sim \frac{k_{IVR}}{k_{VP}} [A]_0 \, e^{-(k_{IVR} + k_f)t}$$

and

$$[C] \sim [A]_0 \, e^{k_f t}\{1 - e^{-k_{IVR}t}\}.$$

Therefore, in such cases, the observed $\overline{0}^0_0$ kinetics [B] are predicted simply to follow the kinetics of the accessed state. Also, the risetime of the 0^0_0 emission (the [C] kinetics) is predicted to match the decay of the $\overline{I}^1$ state and $\overline{0}^0$ states. This is exactly the behavior that is observed and which is shown in Figure 5-4 for the case of $\overline{I}^1$ excitation.

This result is in sharp contrast to what is predicted by a parallel mechanism. The parallel mechanism predicts that similar kinetics should be observed for both $\overline{0}^0_0$ and 0^0_0 emissions: both should rise with the decay of the $\overline{I}^1$ state, and decay with the 7.8 ns emission lifetime.

Table 5-1. Comparison of IVR Lifetimes for Aniline$(X)_1$ Clusters with X = Ar, N$_2$, or CH$_4$[a]

Cluster state excited	Vibrational energy (cm^{-1})	τ_{IVR} X = Ar	X = N$_2$	X = CH$_4$
$\overline{6a}^1_0$	493	11	4.5	<0.1
$\overline{6a}^1_0 + 55$ cm^{-1}	548		0.4	
$\overline{15}^2_0$	718	1.8	0.3	<0.1
$\overline{I}^2_0$	749		0.3	
$\overline{I}^1_0$	803	4.0	0.36	<0.1

[a] Times are calculated by removing the S$_1$ lifetime from the experimental decay times. Measured times (ns) are accurate to +15% or +0.1 ns, whichever is greater.

A quite different situation is obtained for aniline$(CH_4)_1$. In this case, the VP rate of the $\overline{0^0}$ state is much smaller than the IVR rate of the accessed state and, consequently, the observed decay of the $\overline{0^0}$ state is dominated by its own dissociation kinetics.

In the case of aniline$(CH_4)_1$ $\overline{1^1}$ excitation, some $10b_1^1$ emission is also observed, indicating that the above three-state model is only approximately correct. The model can easily be generalized to include the $\overline{10b^1}$ and $10b^1$ states. If the $\overline{10b^1} \rightarrow \overline{0^0}$ IVR rate is taken to be fast [as predicted by eq. (5-2)], then little population ends up in $10b^1$, in agreement with the observed results.

A summary of both experimental and calculated results for aniline clusters studied to date is given in Table 5-2.

The most remarkable and central observation of these experiments is the direct characterization of the emission kinetic curves for the $\overline{1^1}$, $\overline{0^0}$ and 0_0^0 transitions and the observation of the $10b_1^1$ transition following excitation at $\overline{1^1}$ of the aniline$(N_2)_1$ cluster (see Figure 5-7). The pumped state, an intermediate state populated by IVR, and two bare molecule product states populated by the IVR/VP process are observed. These data are consistent with, and therefore provide strong evidence in support of, the serial IVR/VP mechanism applied to clusters containing a polyatomic chromophore. A parallel mechanism simply cannot explain the observed results.

The general conclusions that we derive from these cluster dynamics experiments can be summarized as follows:

1. The experimental results follow the predictions of a serial IVR/VP mechanism in which IVR is described by Fermi's Golden Rule and VP is described by a restricted RRKM model.
2. Aniline$(N_2)_1$ clusters undergo IVR and VP dynamics which are intermediate with respect to those of the aniline$(CH_4)_1$ and aniline$(Ar)_1$ clusters.
3. The most important factor in the determination of IVR and VP rates is the density of vdW states at a given excess vibrational energy.

We have demonstrated these general conclusions by two analytical approaches: first, all the data are fit with the serial IVR/VP model using only one parameter, in the same manner as has been accomplished for tetrazine/Ar complexes; and second, the intermediate cluster state $\overline{0^0}$ ($\overline{1^1} \rightarrow \overline{0^0} \rightarrow 0^0$) dynamics have been measured and the rise and fall times are consistent with those of the nascent and final state dynamics.

5.2.2. 4-Ethylaniline/Ar, N_2, CH_4

With this system of clusters, we are not only able to expand the studies of IVR/VP but we can identify how the final chromophore state is achieved and whether or not a vibrational equilibrium is established in the cluster prior to VP (Hineman et al. 1993a). Clearly, these determinations depend on the relative rates of IVR and VP. In this discussion, the above serial IVR/VP model will serve as a framework within which to interpret the final chromophore product distributions following cluster IVR/VP.

An essential and tacit assumption of the serial IVR/VP model is that the VP

Table 5-2. Comparison of Experimentally Observed and (Calculated) VP Lifetimes (ns) for Aniline(X)$_1$ Clusters where X = Ar, N$_2$ or CH$_4$[a]

Cluster state excited	Vibrational energy	Transition observed	X = Ar Excess energy	τ_{VP}	X = N$_2$ Excess energy	τ_{VP}	X = CH$_4$ Excess energy	τ_{VP}
$\overline{6a}^1$	493	0^0_0	43	<0.1 (0.08)			13	150[b] (40,000)
$\overline{6a}^1 + 55\ cm^{-1}$	548	0^0_0			33	53[b] (60)		
$\overline{6a}^1 + 68\ cm^{-1}$	561	0^0_0			46	19[b] (16)		
$\overline{15}^2$	718	0^0_0	268	<0.1 (0.004)	203		238	0.24 (0.18)
$\overline{I}^2$	749	0^0_0	299		234			
$\overline{I}^1$	803	0^0_0	353	<0.1 (0.003)	288	<0.1 (0.04)		

[a] Binding energies for the clusters are 450, 515, and 480 cm^{-1}, respectively. All energies in cm^{-1}.
[b] Determined by integrated peak ratios because overlapping bands perturb the decay curves for these states.

162

process itself has no effect on the chromophore vibrational state. In this sense, dissociation is considered to be vibrationally adiabatic. This model is complicated by the fact that VP may compete with subsequent energy flow into the vibrational phase space of the van der Waals modes. Such flow results in VP rates that vary as IVR proceeds. Under these circumstances, terms such as "nonstatistical" IVR or VP must be carefully defined. The rates of IVR depend on the couplings between vibrational states and, as such, are inherently nonstatistical. Thereby, even if the VP rates for any specified amount of energy in the van der Waals modes are indeed given by RRKM theory, a complicated distribution of final product states may obtain. This distribution might then be interpreted as "nonstatistical" or "modespecific" VP; however, because the VP rates are given by RRKM theory, we would consider this point of view to be incorrect.

One special case of the serial IVR/VP model which is commonly encountered involves clusters for which IVR is much faster than VP. In these cases, all VP rates may be calculated from statistical theories, ignoring the IVR process completely. For example, this was found to be the case for aniline$(CH_4)_1$ (Hineman et al. 1992b; Nimlos et al. 1989).

Throughout the previous studies of IVR/VP processes, the cluster vibrational phase space has been separated into two distinct regions—the high frequency chromophore modes and the low frequency van der Waals modes. When this is the case, virtually the entire contribution to the total density of vibrational states comes from the van der Waals modes. Thus, in the absence of VP, IVR results in the almost complete and irreversible flow of energy from the chromophore to the van der Waals modes. 4-Ethylaniline(4EA)/Ar, N_2, CH_4 clusters are studied to explore the dynamical consequences of low frequency chromophore modes (Hineman et al. 1993a).

Ring-tail systems (Felker and Zewail 1988; Hopkins et al. 1980a,b,c, 1981; Knee et al. 1987; Powers et al. 1980) (alkylbenzenes and alkylanilines) play an important role in the elucidation of IVR in isolated molecule systems. The addition of an alkyl chain to the aromatic ring molecules typically used to study IVR adds several low frequency modes to the vibrational phase space. These additional modes significantly increase the vibrational density of states and thereby affect the rates of IVR. The effect of these low frequency chromophore modes on the IVR/VP processes in clusters has not been fully explored. The questions which one might ask about these cluster systems include: does the IVR behavior of the cluster favor chromophore mode to chromophore mode IVR when the lowest frequency chromophore vibrations are of similar energy to the van der Waals modes of the cluster; does the addition of low frequency chromophore vibrations to the vibrational phase space of the cluster alter the VP rates; and can the final product distribution of chromophore states be described by statistical theories? We find that while the results of these 4EA cluster studies might be interpreted as being highly mode specific, they can be modeled quantitatively by simple RRKM theory.

The experimental results for these clusters are presented in Figures 5-10 and 5-11. The 4EA(Ar)$_1$ data show that the emission from the cluster/bare molecule systems changes dramatically as a function of vibrational energy in the cluster. At higher energies, only the 0_0^0 of the bare molecule is observed with a 12 cm^{-1} "hot molecule" sequence structure related to the ethyl group bend (toward and away

DE SPECTRA OF 4EA(Ar)₁

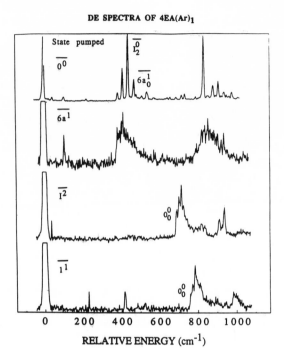

Figure 5-10. Dispersed emission spectra of 4EA(Ar)₁ clusters arising from various excited cluster states. Positive numbers indicate the red shift from the excitation wavelength.

from the ring) which has a 35 cm⁻¹ energy in the S_1 and a 47 cm⁻¹ energy in the S_0 state. The broad $\overline{6a^1}$ excited emission spectrum arises due to hot cluster emission, much like what is characterized for aniline(X)₁ clusters: the cluster, due to IVR, has a great deal of vdW and low energy chromophore mode excitation. Due to this congestion, the mode distribution of the emitting (hot cluster) system cannot be ascertained. Note that the VP does not occur for $\overline{6a^1}$ (435 cm⁻¹) excitation because the cluster binding energy must be larger than the 6a¹ energy.

Upon higher energy excitation at $\overline{I^2}$ and $\overline{I^1}$, the emission is quite different indeed. Figure 5-11 shows an expanded view of the 0⁰ emission following these excitations. The observed sequence structure for these clusters is assigned as a $\Delta v = 0$ transition in the ethyl bend for the CH₄ and N₂ clusters and $\Delta v = 0$ transition in both ethyl bend and torsion modes for the Ar cluster. Up to $v = 4$ in the bend mode must be populated. An equilibrium cluster temperature is clearly established for 4EA/CH₄, N₂ clusters but not the 4EA/Ar cluster.

Time correlated single photon counting studies give the rise and decay times for the various emissions and excitations for these clusters. These data are presented in Table 5-3 and Figure 5-12. For $\overline{6a^1}$ excitation, τ_{IVR} is measured in the rise times of the $\overline{0^0}$ emission and the decay of $\overline{6a^1}$ emission. For $\overline{I^2}$ and $\overline{I^1}$ excitation, the rise times measured involve both IVR and VP because 0⁰ emission is observed from 4EA.

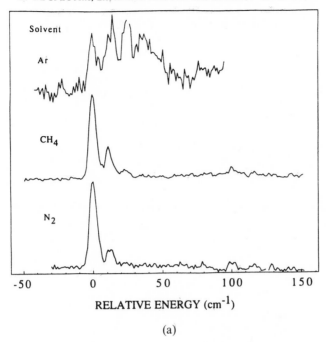

(a)

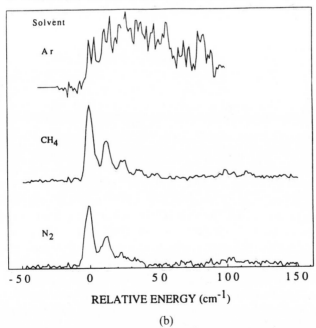

(b)

Figure 5-11. Dispersed emission spectra of 4EA(solvent)$_1$ clusters expanded near the 4EA origin region: (a) $\overline{I}^2$ excitation; (b) $\overline{1}^1$ excitation. Positive numbers indicate a red shift from the 4EA bare molecule origin.

Table 5-3. Results of Fitting Decay Curves for 4-Ethylaniline Clusters

State pumped	Transition observed	Solvent	Rise time (ps)	Decay time (ps)	Component
$\overline{6a}^1$	0_0^0	Ar	200	6000	τ_{IVR}
		CH_4	<100	6200	
		N_2	80	6900	
$\overline{I}^2$	0_0^0	Ar	250	8200	$\tau_{IVR} + \tau_{VP}$
		CH_4	400	8000	
		N_2	585	7200	
$\overline{I}^1$	0_0^0	Ar	185	7200	$\tau_{IVR} + \tau_{VP}$
		CH_4	110	8000	
		N_2	125	7200	

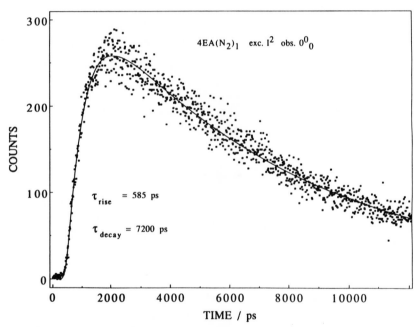

Figure 5-12. A typical fit of an observed TCSPC decay curve. Data correspond to the $4EA(N_2)_1$ cluster pumped at $\overline{I}^2$ and observed at 0_0^0. The fit parameters are rise time 585 ps and decay time 7200 ps.

The data presented show features of the IVR/VP process which any theory of van der Waals molecule dissociation must be able to reproduce. First, only the bare molecule ethyl torsion mode and 0^0 are populated and emit following IVR and VP of 4EA/polyatomic solvent clusters. Second, specific product state distributions in the torsional mode manifold are observed which depend on the cluster and the excitation energy. Third, the rate of dissociation depends on the excitation energy. Fourth, the emission behavior of the three clusters studied is qualitatively different. To rationalize and explain these data, we have applied RRKM theory of unimolecular reactions to the calculation of the numerical data for N_2 and CH_4 clusters, as well as qualitative evaluation of the Ar cluster results.

The model employed to understand the results presented above consists of the following conditions and assumptions. First, IVR and VP are serial processes with IVR occurring first from the excited chromophore modes to the van der Waals modes. Only after the van der Waals modes are populated with sufficient energy will the cluster undergo VP. Second, the rate of dissociation is given by RRKM theory once the van der Waals modes have sufficient energy to cause dissociation. If we consider the case for which all IVR is rapid (4EA/polyatomic clusters) compared with VP, then the vibrational energy in the cluster is partitioned statistically among the chromophore and van der Waals modes. This latter approximation results in considerable simplification of the model. Third, dissociation may occur whenever the amount of energy in the van der Waals modes exceeds the cluster binding energy and the rate of dissociation is then given solely by RRKM theory. Fourth, when the excited clusters dissociate from a given cluster vibrational state, the energy which was (statistically) proportioned into chromophore vibrational modes remains in the bare chromophore molecule after dissociation. These states of the bare chromophore then fluoresce with their characteristic emission wavelengths and times. The relative intensities of these emissions give the population distribution within the manifold of bare 4EA states. These bare solute distributions can then be related to the cluster state distributons and rates of dissociation.

Based on previous results for aniline/Ar, N_2, and CH_4 clusters, the IVR rate among the low frequency modes of 4EA clusters should be very high. Given this rapid flow of energy among the low frequency modes of the cluster, an equilibrium among all these modes should exist. In this case, the intensity of a product state (bare molecule) emission will be given by the rate of dissociation from the parent cluster state multiplied by the probability of populating that cluster state $P_V k(v)$.

The RRKM theory of unimolecular reactions predicts that the rate constant for dissociation will be given by eq. (5-3). The probability of populating a state with energy E_v restricted into the chromophore vibrations is proportional to the ratio of the density of van der Waals states at $E - E_v$ to that at E:

$$P_V = [N(E - E_V)/N(E)]\bigg/\sum_V [N(E - E_V)/N(E)], \qquad (5\text{-}7)$$

in which $N(E)$ is the density of states at energy E. Assuming that the Franck–Condon factors for all $\Delta v = 0$ transitions within the torsional mode manifold are equal, the relative intensities of product state emissions will be the rate of dissociation of the cluster state multiplied by the probability of populating that state:

$$I(v) = P_V k(v). \qquad (5\text{-}8)$$

The total dissociation rate from these equilibrated modes will be given by the sum of the rates out of each state multiplied by the probability of populating each state:

$$k(E) = \sum_V P_V k(v). \qquad (5\text{-}9)$$

This is, of course, equal to the RRKM rate obtained by considering all the vibrational [chromophore + van der Waals) modes (see eq. (5-3)].

Evaluation of eqs. (5-3) and (5-8) requires the calculation of the sums and densities of the van der Waals vibrational states. These are calculated by a direct count method (Forst 1973; Gilbert and Smith 1990; Kelley and Bernstein 1986; Levine and Bernstein 1987; Pritchard 1984; Robinson and Holbrook 1972; Steinfeld et al. 1989), which requires values of the vibrational frequencies. Since eq. (5-3) is restricted to the van der Waals modes, only these frequencies are required. The low frequency chromophore modes which are populated by the IVR/VP processes are not part of this count because the rates for each of the states of these vibrations will be calculated individually, and thus the energy in these modes will be restricted or isolated in the calculation. The van der Waals mode energies are calculated using an atom–atom potential model and normal mode analysis. The density of states is a strongly nonmonotomic function of energy at low energies. It is characterized by peaks and valleys, the energies of which are strongly dependent upon the assumed frequencies. Thus, at low energies (below $100 \, \text{cm}^{-1}$), the required sums and densities are not well calculated by these methods due to inadequacies in both the model potential and the harmonic approximation. Nonetheless, because the RRKM rates are a function of a sum of states divided by a density of states, a significant amount of cancelation of these errors is to be expected.

To complete the RRKM calculations for the cluster dissociation rates and final bare 4EA molecule product distributions, the cluster binding energy E_0 and the energy E_V of the chromophore vibrational state to be populated must be found. These can be estimated from selected fits to the experimental rates and intensities (Hineman et al. 1993a). The results of the rate and product distribution calculations are presented in Table 5-4. The predictions of the model are quite good—less than 30% error for all observations for the $4EA(N_2)_1$ and $4EA(CH_4)_1$ clusters.

At first glance, the emission spectra of 4EA following excitation of the N_2 and CH_4 clusters might appear to be due to some special propensity for dissociation from a particular vibrational mode of 4EA; however, all of the observed data can be reproduced by a statistical approach to VP. Systems for which IVR is fast in comparison with VP should maintain an equilibrium distribution among the low energy van der Waals and chromophore modes. Comparison of the calculated and observed product state distributions in Table 5-4 shows that this expectation is borne out for 4EA(polyatomic solvent)$_1$ clusters. The "special mode" ($35 \, \text{cm}^{-1}$) is observed because of its extremely low energy in S_1. A low energy mode has a high probability of being populated since the system has a high density of states even with population in that mode. Furthermore, the VP rate suffers only a small decrease as quanta are left in the low frequency mode. The model also correctly predicts very little population in the $84 \, \text{cm}^{-1}$ ethyl torsion mode in these clusters.

Quantitative agreement can be obtained for the polyatomic solvent clusters but not for the $4EA(Ar)_1$ cluster using this cluster thermal equilibrium model. While the dispersed emission spectra of $4EA(Ar)_1$ clusters are not sufficiently well resolved to allow quantitative measurement of product state distributions, the model predicts that the 4EA 0_0^0 transition (at $0 \, \text{cm}^{-1}$ in Figure 5-11) should be the

Table 5-4. Comparison of Calculated VP Rates and Product State Intensities with Experimental Results

Pumped state and observable	Solute and binding energy			
	CH_4 (580 cm^{-1})		N_2 (630 cm^{-1})	
	Calculated	Observed	Calculated	Observed
$\overline{I^2}$				
Total rate (s^{-1})	2.8×10^9	2.5×10^9	2.3×10^9	1.7×10^9
Relative intensity				
$v = 0$	1	1	1	1
$v = 1$	0.46	0.38	0.40	0.21
$v = 2$	0.19	0.11	0.10	0.06
$v = 3$	0.06	0.029	—	—
$\overline{I^1}$				
Total rate (s^{-1})	5.8×10^9	9.1×10^9	5.5×10^9	7.9×10^9
Relative intensity				
$v = 0$	1	1	1	1
$v = 2$	0.56	0.51	0.56	0.50
$v = 3$	0.14	0.13	0.10	0.12
$v = 4$	0.06	0.076	—	—

most intense feature in the emission if IVR can establish an equilibrium population distribution within the clusters, prior to VP. Clearly, this is not the case for $4EA(Ar)_1$. This is almost surely due to breakdown of the assumption that all IVR is fast in comparison with VP. This can be clearly seen from Tables 5-3 and 5-4. The IVR time in the N_2 and CH_4 clusters following $\overline{6a^1}$ excitation is $\lesssim 100$ ps (Table 5-3), and is probably faster following $\overline{I^2}$ or $\overline{I^1}$ excitation, due to the increased density of states at these high vibrational energies. The slower product (0_0^0) rise times following $\overline{I^2}$ or $\overline{I^1}$ excitation of N_2 and CH_4 clusters are therefore determined by the VP time, and good agreement between the calculated and observed rates is obtained (Table 5-4). The kinetics of the $4EA(Ar)_1$ cluster are quite different. Slow IVR (200 ps) is obtained following $\overline{6a^1}$ excitation. Calculated VP times for $\overline{I^2}$ or $\overline{I^1}$ excitation are roughly 10–20 ps. Product state rise times for $\overline{I^2}$ and $\overline{I^1}$ excitations are comparable to the $\overline{6a^1}$ IVR time of 200 ps, however, indicating that the dissociation process in $4EA(Ar)_1$ is always limited by the IVR rate. Nonetheless, the model does predict several of the qualitative trends in the $4EA(Ar)_1$ data: the VP rates decrease much less quickly with vibrational energy left in the 4EA molecule for Ar than for polyatomic solvents. For example, at approximately 175 cm^{-1} of energy restricted to 4EA vibrations (five quanta of the ethyl torsion) the VP rates for N_2 and CH_4 clusters decrease by a factor of 15–20 as compared with a factor of 3 for the Ar cluster. This would result in far more intensity in the higher overtone sequence structure of $4EA(Ar)_1$, as is observed. The presence of fewer van der Waals modes in the Ar cluster results in the prediction that the higher energy 84 cm^{-1} ethyl bend mode would be significantly populated. In agreement with this prediction, at least one member of the 16 cm^{-1} progression is observed.

Intermolecular vibrational energy redistribution may populate several different chromophore states with sufficient energy in the van der Waals modes that VP can occur. If VP, which is predicted to be very fast in the Ar cluster, competes with subsequent IVR, then these chromophore states will be populated in the bare molecule. This would give a more crowded spectrum for $4EA(Ar)_1$ with greater intensity away from the 0_0^0 transition, as is observed. These predictions are in qualitative agreement with all cluster data; however, quantitative comparison for the Ar cluster is rendered impossible by crowding of the spectra (see Figure 5-11).

The most important conclusions of these dynamical studies is that van der Waals clusters behave in a statistical manner and that IVR/VP kinetics are given by standard vibrational relaxation theories (Beswick and Jortner 1981; Jortner et al. 1988; Lin 1980; Mukamel and Jortner 1977) and unimolecular dissociation theories (Forst 1973; Gilbert and Smith 1990; Kelley and Bernstein 1986; Levine and Bernstein 1987; Pritchard 1984; Robinson and Holbrook 1972; Steinfeld et al. 1989). One can even arrive at a prediction for final chromophore product state distributions based on low energy chromophore modes. If $\tau_{IVR} \gg \tau_{VP} [4EA(Ar)_1]$, a statistical distribution of cluster states is not achieved and vibrational population of the cluster does not reflect an internal equilibrium distribution of vibrational energy between vdW and chromophore states. If $\tau_{VP} \gg \tau_{IVR}$, and internal vibrational equilibrium between the vibrational modes is established, and the relative intensities of the $\Delta v = 0$ torsional sequence bands of the bare chromophore following IVR/VP can be accurately calculated. A statisticsl sequential IVR/VP model readily explains the data set (i.e., rates, intensities, final product state distributions) for these clusters.

5.2.3. Future Cluster Dynamics Studies

From the results on anilines discussed above (Hineman et al. 1992b, 1993a, 1994; Nimlos et al. 1989) and the referenced work on indoles (Outhouse et al. 1991) and tetrazines (Alfano et al. 1991; Brumbauch et al. 1983; Haynam et al 1984a,b; Morter et al. 1991; Park and Levy 1984), one can be rather confident that IVR/VP dynamics in S_1 is well understood, both in the statistical limit for high densities of states and good coupling between vdW and chromophore modes, and in the limit of sparse densities of states and weak vdW and chromophore mode coupling (e.g., I_2/He, etc.) (Levy 1981).

Two related remaining questions concerning the IVR/VP process are: what are the coupling matrix elements for chromophore to vdW mode IVR, and does the IVR/VP process take on a different character on the ground electronic state surface? Some data for ground state systems [tetrazine $(Ar)_1$ (Alfano et al. 1991; Morter et al. 1991), and perylene $(X)_1$ (Wittmeyer and Topp 1993)], and alkylphenols and alkylanilines (Ebata and Ito 1992) seem to suggest that the IVR/VP times are very long, but a systematic study has not been undertaken. If these results are correct, the difference between S_0 and S_1 IVR/VP kinetics could be due to chromophore to van der Waals mode coupling differences within the two different surfaces. One can suggest that this S_0/S_1 coupling matrix element difference is due to a Herzberg–Teller-like vibronic coupling between the excited electronic states involving cross terms of the nature

$$\left\langle \psi_i \left| \frac{\partial^2 V}{\partial Q\, \partial q} \right| \psi_j \right\rangle, \left\langle \psi_i \left| \frac{\partial V}{\partial Q} \frac{\partial V}{\partial q} \right| \psi_j \right\rangle,$$

or higher order terms in a van der Waals coordinate, q, which would preferentially enhance IVR in the excited electronic states with respect to that in the ground state.

Ground state IVR/VP can be explored by employing fluorescence, stimulated Raman pumping, and stimulated emission pumping (SEP) to populate specific ground vibrational modes of a cluster. The IVR/VP processes can then be followed through fluorescence or ion detection with a subsequent excitation laser pulse. If the preliminary indications are correct, the IVR part of the overall ground state cluster vibrational dynamics can be followed with time resolved nanosecond lasers. The VP part of the dynamics may still require picosecond time resolution at higher vibrational energies and for small vdW phase spaces.

5.3. PROTON TRANSFER REACTIONS—NAPHTHOL AND PHENOL/NH₃

The importance of proton transfer reactions to chemistry and biology is well known and often emphasized. Elucidation of the parameters, reaction coordinates, mechanisms, minimum systems and solvent dependence of proton transfer serves both practical and theoretical needs. Learning the details of the proton transfer reaction will aid in either accelerating or inhibiting such processes in systems, as is appropriate, and studying this relatively simple reaction will serve as a test for our microscopic understanding of chemical reactions in general.

5.3.1. Naphthol/NH₃

The reaction we will explore for 1-naphthol clustered with ammonia was first studied by Leutwyler and coworkers (Chesnovsky and Leutwyler 1985, 1988; Droz and Knochenmuss 1990; Droz et al. 1988) and Zewail and coworkers (Breen et al. 1990). The $pK_a(S_0)$ for 1-naphthol is roughly 10 and the $pK_a(S_1)$ for this molecule is ca. 0 (Ireland and Wyatt 1976). This dramatic change in proton lability is not atypical for aromatic alcohols and it sets up a nearly perfect situation for the study of cluster proton transfer reactions. The reaction is initiated by a photon ($S_1 \leftarrow S_0$ transition) in clusters of the proper size and composition. The original work (Chesnovsky and Leutwyler 1985, 1988; Droz and Knochenmuss 1990; Droz et al. 1988) on this system is not time resolved but does nonetheless conclude that proton transfer occurs for clusters of 1-naphthol$(NH_3)_n$, $n \geq 4$. The determination of an excited state proton transfer reaction is accomplished through many of the same techniques employed in condensed phase studies (Brucker and Kelley 1987a,b, 1988, 1989a,b,c; Brucker et al. 1991; Swinney and Kelley 1991): (1) large red shift and broadening of the mass resolved excitation spectra of $n \geq 4$ clusters relative to those of the bare molecule 1-naphthol and the $1 \leq n \leq 3$ clusters, (2) large red shift and broadening for emission from $n \geq 4$ clusters, and (3) similarity between solution naptholate ion emission and the larger cluster emission. Kim et al. (1991) measured ionization thresholds and found that two $n = 3$ clusters of different geometries exist in the supersonic expansion for naphthol/NH_3, and that one of them has a much lower ionization energy than the other. In fact, the ionization energy of this latter cluster is very similar to that found for the $n = 4$ cluster. This suggests that one geometry of the $n = 3$ cluster also undergoes proton transfer. This observation is confirmed by the time measurements presented by the Zewail group (Breen et al. 1990). A proton transfer time of ca. 100 ps is recorded in this study, and is perhaps the strongest statement that a proton transfer reaction occurs in this system.

Nonetheless, elucidation of the detailed mechanism (e.g., energetics, reaction coordinate, mechanism, potential surface) of the process is not possible with the above qualitative information; to elaborate these details, a systematic variation of isotopic composition, energy in the cluster, cluster size, and even solvent, is necessary. Such a study and mechanistic model determination have been carried out (Hineman et al. 1992a) and are discussed below.

Excited state proton transfer is observed in 1-naphthol(NH_3)$_n$ clusters for $n = 3$ and 4. The experimental technique employed for this investigation is standard pump/probe ionization ($I \leftarrow (S_1^T \leftsquigarrow S_1) \leftarrow S_0$), mass resolved excitation spectroscopy with 10 ps time resolution for the pulses and the delay between them. Because the signal is measured as a function of delay time between the pump and probe ionization pulses, it only depends on neutral S_1 dynamics, not on proton transfer following ionization. As the proton is transferred (or moves along the reaction coordinate or path), the ionization cross section decreases so a signal decay is observed. The change in signal due to proton translocation could equally well have been an increase or, in the worse case scenario, the cross section could have been constant with proton reaction coordinate, meaning proton transfer could not be studied by this approach.

The picosecond time resolved experiments show the following set of results: (1) excited state proton transfer occurs for $n = 3$ and 4 naphthol(NH_3)$_n$ clusters only, (2) for $n = 5$ clusters, the proton is transferred in the ground state and, for $n = 2$ clusters, no proton transfer can be observed, (3) the proton transfer time in the $n = 3$ cluster at the 0_0^0 transition is ca. 60 ps, (4) this time is reduced to ca. 40 ps and ca. 10 ps for 800 cm^{-1} and 1400 cm^{-1} of vibrational energy in S_1, respectively, (5) for the $n = 4$ clusters, these times are approximately 70, 70, and 30 ps for 0, 800, and 1400 cm^{-1} of vibrational energy in S_1, respectively, (6) both $n = 3$ and 4 clusters exhibit a second low amplitude decay component, which is about an order of magnitude slower than the initial decay, and (7) 1-naphthol-d_1(ND_3)$_n$ clusters have a greatly reduced rate constant for the excited state proton transfer dynamics. An example of the data is presented in Figure 5-13 and a summary of these data is given in Tables 5-5 and 5-6. The analysis of the proton transfer dynamics of 1-naphthol(NH_3)$_{3,4}$ discussed below will focus on the proton transfer decay time τ_1. The second long decay time τ_2 can be attributed to a cluster or solvent reorganization (which changes the $I \leftarrow S_1$ cross section) subsequent to the proton transfer event. Similar solvent reorganization times have been suggested for phenol/NH_3 clusters (Steadman and Syage 1991, 1992; Syage 1993; Syage and Steadman 1990, 1991).

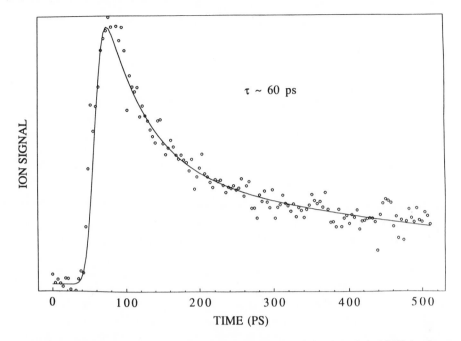

$\tau \sim 60$ ps

Figure 5-13. Time resolved $1 + 1$ ionization signal scan of the 1-naphthol(NH$_3$)$_3$ cluster. Excitation energy, 31,100 cm^{-1}; ionization energy, 29,000 cm^{-1}. 0, data, solid line, fit. The fit parameters are instrument limited rise followed by a biexponential decay $\tau_1 = 60$ ps and $\tau_2 = 500$ ps.

The model employed to rationalize these data and to gain insight into the proton transfer process characterized by τ_1 consists of three essential features: (1) the untransferred and the transferred structures are separated from one another by a potential-energy barrier which can be characterized by a width and a height (Brucker and Kelley 1987a,b, 1988, 1989a,b,c; Brucker et al. 1991; Swinney and Kelley 1991), (2) the barrier width and height are modulated by

Table 5-5. Measured Time Constants for the Non-deuterated Clusters as a Function of Vibrational Energy[a]

Mass (n)	Excitation energy (cm^{-1})	τ_1 (ps)	τ_2 (ps)
3	31,100 (0_0^0)	57 (15)	440 (125)
	32,040	43 (11)	500 (74)
	32,490	12 (3)	119 (10)
4	30,800 (0_0^0)	71 (31)	800 (400)
	32,040	70 (30)	600 (100)
	32,490	33 (20)	1000 (500)

[a] Uncertainties (2σ) are given in parentheses.

Table 5-6. Effect of Dueteration on Measured Time Constants for $n = 3$ as a Function of Vibrational Energy

Mass (H/D)	Excitation energy (cm^{-1})	τ_1 (ps)	τ_2 (ps)
H	31,100	57	440
D	31,100	>1000	—
H	32,490	12	119
D	32,490	75	>1000

vibrational excitation of the intermolecular cluster modes, and (3) vibrational energy is distributed statistically among the vibrational (van der Waals) modes (i.e., $k_{IVR} \gg k_{PT}$).

Two vibrational coordinates take on special significance in this model (Bell 1980; Kebarle 1977). First, the reaction coordinate is taken to be the O–H stretch. The hydroxyl group is hydrogen bonded to the ammonia nitrogen, and the barrier to proton transfer is taken to be along this O–H $\cdots$ N reaction coordinate. Second, the O $\cdots$ N distance in this hydrogen bond is of central importance in determining the height and width of the proton transfer barrier. This distance can be modulated by excitation of the intermolecular "diatomic" stretch mode, for which the 1-naphthol and $(NH_3)_3$ moieties are taken to be the "atoms". Excitation of the O $\cdots$ N stretch can dramatically affect the proton tunneling rate (Borgis et al. 1989; McKinnon and Hurd 1983; Siebrand et al. 1984; Trakhtenberg et al. 1982). Numerical application of this model requires calculation of the proton tunneling rate for each quantum state of the diatomic stretch mode, and calculation of the probability of occupation of each quantum state. These calculations are discussed below. Tunneling rates can be calculated as a function of the O $\cdots$ N separation based upon the WKB approximation for particle penetration through a barrier of assumed functional form.

Figure 5-14 shows the model potential surface for this system in which the untransferred proton is bound in a well and the transferred proton is represented as a free particle. The curve in Figure 5-14 thus represents the hydrogen stretch coordinate. The barrier along the reaction coordinate is taken to be an inverted parabola centered about the top of the barrier: its height and width are measured from the OH vibration zero point energy. The proton transfer reaction is exothermic and the products are expected to have a high density of states: this justifies both the "free particle" product description and the barrier penetration limit. A more complex model with a structured set of product states could readily be constructed. Under these approximations, the tunneling rate constant for the proton transfer reaction reduces to the well known expression (Bell 1980),

$$k = v \exp\left[-\frac{a}{h} \pi (2mE_h)^{1/2} \right], \qquad (5-10)$$

in which v is the zero point frequency for the proton in the potential well, E_h is

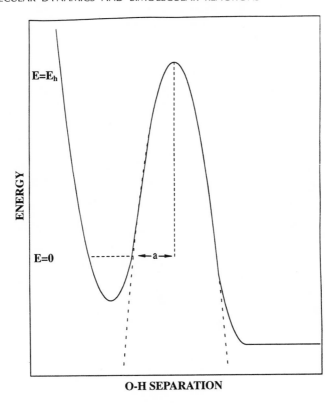

Figure 5-14. Schematic diagram of the potential model used for proton transfer rate calculations. The dashed line is the parabolic barrier using parameters fit to origin $n = 3$ data. The solid curve is the barrier model including a harmonic potential well with an OH vibrational energy of 3000 cm^{-1}. The well is centered at 1.0 Å. The other parameters for the model are $E_0 = 8700$ cm^{-1}, $a_0 = 0.2$ Å, and the van der Waals stretch energy is 110 cm^{-1} for 1-naphthol(NH$_3$)$_3$.

the height of the barrier, m is the mass of the tunneling species (H/D), and a is the barrier half-width at the zero point energy position in the reactant well. This expression is derived from the WKB approximation to the wave function for particle penetration through a potential barrier. Assuming that this barrier is modulated by the stretching motion between the naphthol molecule and the ammonia cluster situated near the OH group, the barrier width and height are a function of the O$\cdots$N distance.

Calculation of the proton transfer or barrier penetration rate as a function of the van der Waals vibrational excitation in the cluster requires an average of the rate constant of eq. (5-8) over the wavefunction for the "diatomic" O$\cdots$N distance. Thus,

$$k(v) = \langle \psi_v | k(a) | \psi_v \rangle, \tag{5-11}$$

in which ψ_v is taken to be a harmonic oscillator wavefunction, and v is the number of quanta in the O$\cdots$N stretch mode. The rate constant can be calculated for the

$O\cdots N$ vibrational levels of cluster S_1 excitation using eq. (5-11). Note that the vibrational energy in the cluster is never sufficient to excite the OH stretch, either directly or indirectly.

This model has only one adjustable parameter: the equilibrium barrier height, E_0. The vibrationless proton (H) transfer rate for the $n = 3$ cluster is fit to the observed rate of 57 ps^{-1} to find a value for E_0 of 8700 cm^{-1}. The transfer rates are a function of the vibrational state can then be obtained.

Cluster vibrational excitation in S_1 is generated by accessing vibrational states of 1-naphthol, which are then assumed to undergo rapid IVR to the cluster van der Waals modes. Rapid naphthol to van der Waals mode IVR for the 1-naphthol(NH$_3$)$_{3,4}$ clusters is a good assumption, based on the following two considerations: high density of cluster vibrational states; and known IVR cluster behavior for other cluster systems [e.g., aniline(Ar)$_1$, (N$_2$)$_1$, (CH$_4$)$_1$ and indole (Ar)$_{1,2}$ and (CH$_4$)$_1$ clusters] (Hineman et al. 1992b, 1993a, 1994; Nimlos et al. 1989; Outhouse et al. 1991). The energy distribution in the van der Waals modes is then taken to be statistical and can be readily calculated. The probability that a specific van der Waals mode (in this case, the $O\cdots N$ "diatomic" stretch) is populated with v quanta is then calculated using the Marcus–Rice approximation (Forst 1973; Gilbert and Smith 1990; Kelley and Bernstein 1986; Levine and Bernstein 1987; Pritchard 1984; Robinson and Holbrook 1972; Steinfeld et al. 1989):

$$P_v = \left[\frac{E - E_v}{E}\right]^{S-2} \Bigg/ \sum_V \left(\frac{E - E_v}{E}\right)^{S-2}, \tag{5-12}$$

in which E is the vibrational energy in the cluster, E_v is the energy of the specific van der Waals mode and S is the number of van der Waals modes for the cluster. The products of $k(v)$ and P_v for the modes of interest are summed to find the total transfer rate for a given S_1 vibrational energy:

$$k(E) = \sum_V P_v k(v). \tag{5-13}$$

Diatomic harmonic oscillator levels ($hv \sim 110$ cm^{-1}) to $v = 3$ for the H cluster are found to contribute to the observed rates with up to 1500 cm^{-1} of vibrational energy in the $n = 3$ cluster. Table 5-7 presents the results of these calculations; agreement between the calculated transfer rates (τ_1^{-1}) based on the above model and the experimental results is excellent.

Isotopic substitution (H/D) affects the proton transfer rate in S_1 for the 1-naphthol(NH$_3$)$_3$ cluster in several ways: (1) the OD stretch has a zero point energy which is much smaller than that of the OH stretch, and therefore both the barrier height and width increase, (2) the van der Waals vibrational reduced mass increases for (ND$_3$)$_3$, and thus the van der Waals vibrational energy decreases, (3) the tunneling (proton) mass is doubled, and (4) the $v = 1, \ldots, 5$ harmonic oscillator levels contribute to the observed rates. The proton mass appears directly in the exponential [eq. (5-10)] and as m increases, the rate decreases. Increasing the barrier height E_h and width a also decreases the rate. Decreasing the van der Waals vibrational energy also decreases the transfer rate but additionally increases

Table 5-7. Comparison of Barrier Penetration Model Calculations with Measured Proton Transfer Time Constants[a]

Mass $(n, H/D)$	Vibrational energy (cm^{-1})	τ_1 calculated[b] (ps)	τ_1 observed (ps)
3, H	0	57 fit	57
3, D	0	2450	>1000
3, H	940	26	43
3, H	1390	12	12
3, D	1390	195	75

[a] See text for parameters used.
[b] H: $E_0 = 8700\ cm^{-1}$, $a_0 = 0.2$ Å, E (OH) $= 3000\ cm^{-1}$, E (vdW) $= 110\ cm^{-1}$.
D: $E_0 = 9140\ cm^{-1}$, $a_0 = 0.216$ Å, E (OD) $= 2120\ cm^{-1}$, E (vdW) $= 104\ cm^{-1}$.

the population of the particular mode. The model predicts that the deuteron transfer rate for the cluster should be small at the origin and should increase more quickly with vibrational energy than does the proton transfer rate for the protonated cluster. These results are clearly shown in Table 5-7 and represent excellent qualitative and quantitative agreement between the model and the experimental data. Both the trends and the details are well described: the effect of vibrational energy on the proton transfer rate of 1-naphthol$(NH_3)_3$ clusters is reproduced, and the decrease of the isotope effect for the deuteron transfer rate of 1-naphthol-$d_1(ND_3)_3$ clusters as a function of vibrational energy is predicted as well.

Other methods of calculating the $O \cdots N$ separation dependent proton transfer rates, such as a Fermi Golden Rule approach (Siebrand et al. 1984), can also be employed. In this approach, two harmonic potential wells (e.g., O–H $\cdots$ N and, O $\cdots$ H–N) are considered to be coupled by an intermolecular term in the Hamiltonian. Inclusion of the van der Waals modes into this approximation involves integration of the coupling term over the proton and van der Waals mode wavefunctions for all initial and final states populated at a given temperature of the system. Such a procedure requires the reaction exothermicity and a functional form for the variation of the coupling as a function of well separation. In the present study, we employ the barrier penetration approach; this approach is calculationally straightforward and leads to a clear qualitative physical picture of the proton transfer process.

The essential features of the results and the mechanistic model can be summarized as follows: (1) proton transfer occurs for $n = 3$ and 4 clusters in the S_1 state and is suggested to occur in the S_0 state for $n \geq 5$, (2) no transfer is found for $n < 3$ clusters, (3) substitution of deuterium for hydrogen in these clusters [i.e., 1-naphthol-$d_1(ND_3)_3$] has dramatic effects on the observed transfer rates, (4) at the origin of the $S_1 \leftarrow S_0$ transition the kinetic isotope effect is at least a factor of 20 (D^+ transfer is slower than H^+ transfer) but at an energy of 1400 cm^{-1} the kinetic isotope effect is about a factor of 6, (5) the proposed model for proton transfer in these clusters equates proton transfer with a simple barrier penetration problem, (6) the height and width of the barrier are modulated by vibrational energy in the cluster intermolecular van der Waals modes, (7) the deuterium isotope effect is related to the OD zero point energy, the reduced van der Waals mode energy for the 1-naphthol-$d_1(ND_3)_3$ cluster, and the H/D masses, and (8) the general trends and detailed rates observed are well described by the model.

5.3.2. Phenol/NH$_3$

Syage and Steadman (Steadman and Syage 1991, 1992; Syage 1993; Syage and Steadman 1990, 1991) have undertaken an extended study of the phenol/NH$_3$ cluster system to investigate proton transfer in much the same way as described above, using both nanosecond and picosecond time resolved techniques and mass detection of the cluster ions. Unlike the situation for naphthol clusters, the phenol/NH$_3$ system displays some quite complex behavior due to phenol/NH$_3$ cluster ion fragmentation.

The data for the phenol(NH$_3$)$_n$ clusters as obtained by Syage and Steadman (Steadman and Syage 1991, 1992; Syage 1993; Syage and Steadman 1990, 1991) show that for clusters with $n = 4, \ldots, 7$, some dynamical behavior occurs. They also characterize cluster fragmentation of the form PhOH$^+$(NH$_3$)$_n$ → PhO + NH$_4^+$(NH$_3$)$_{n-1}$. In each instance, $\lambda_{excite} = 266$ nm and $\lambda_{ion} = 355$ or 532 nm. High concentrations ($\sim 10\%$) of ammonia are employed in these expansions. The ionization step is possibly multiphoton and, considering the 37,590 cm^{-1} excitation, leaves the cluster with at least ca. 10^4 cm^{-1} of excess energy. This is certainly enough energy to cause extensive cluster ion fragmentation.

Our studies on this system (Hineman et al. 1993b) are carried out with lower concentrations of ammonia (ca. 1%), lower and variable excitation energy, and typically an ionization energy of 29,850 cm^{-1} ($\lambda_{ion} = 335$ nm). Where the experimental conditions overlap for the two studies (Steadman and Syage 1991, 1992; Syage 1993; Syage and Steadman 1990, 1991) (i.e., 10% ammonia concentration, 266 nm excitation and 355 nm ionization), the results are the same. The data for this cluster system can be summarized as follows: (1) spectra for phenol(NH$_3$)$_{1,2}$ are relatively sharp and well defined and shifted 600–800 cm^{-1} red of the phenol origin, (2) spectra of phenol(NH$_3$)$_3$ are broad and fall in the same region, (3) a continuous background of phenol(NH$_3$)$_{1,2}$ signal is due to fragmentation of phenol(NH$_3$)$_n$, $n \geq 3$, clusters throughout the λ_{excite} range accessed ($285 \leq \lambda \leq 266$ nm), (4) extensive fragmentation of phenol/NH$_3$ clusters to yield (NH$_3$)$_m$H$^+$ (m not necessarily equal to n) is found, (5) for origin excitation, phenol(NH$_3$)$_{2,3,4}$ clusters show fast decays which are dramatically slowed upon deuteration, (6) these decays are all concentration dependent, (7) with 266 nm excitation phenol(NH$_3$)$_n$, $n = 1, \ldots, 4$, clusters all show fast decays and (NH$_3$)$_{3,4}$H$^+$ clusters show fast rises, and (8) lower energy ionization could not be obtained with this system due to the poor Franck–Condon factor for near threshold ionization.

The observation of a fast time decay for the $n = 2$ mass channel excited at the cluster origin is surprising. Isotopic substitution experiments show that the decay is due to proton tunneling, yet naphthol, which is a stronger excited state acid, does not exhibit proton transfer until it is clustered with three ammonia molecules. A simple calculation of the thermodynamics of the proton transfer, using the observation that naphthol transfers at $n = 3$ and assuming that binding energies are roughly the same for the two systems, predicts that phenol should have a cluster size threshold for proton transfer at $n = 4$ or 5. Most likely, the clusters which give rise to the proton transfer dynamics observed in the phenol$(NH_3)_2^+$ mass channel are larger $(n \geq 4)$ clusters which fragment to the lower mass channel. These same clusters probably give rise to the broad background observed to be associated with the low mass channels $(n < 3)$ in the excitation spectra.

The concentration dependence of this time resolved signal also suggests that the signal is due to fragmentation of larger clusters. Changing the ammonia concentration changes the distribution of cluster sizes formed in the jet, and thereby changes the relative importance of various parent cluster contributions to a given mass channel signal. Thus, if both the fast and slow components of the $n = 2$ mass channel signal are from $n = 2$ parent clusters, the relative amplitudes would not be expected to vary with ammonia concentration. The decay times of the fast component also vary slightly with concentration, although the low concentration results are relatively noisy. This indicates that more than one parent cluster may fragment into a given mass channel and, thus, ammonia loss for each cluster is not fixed.

The ionization energy could be lowered in order to minimize the amount of fragmentation in the cluster ion; however, the cross section for ionization drops rapidly below the employed ionization energy, making pump–probe experiments difficult. In the experiments reported here, nearly 10,000 cm^{-1} of excess energy is placed in the large cluster ions $(n \geq 4)$. This extra energy may account for the extensive cluster fragmentation into many different daughter ions.

The phenol$(NH_3)_1$ mass channel shows only a slow decay following origin (280 nm) excitation, but shows a fast component following 266 nm excitation. Excitation at 266 nm provides the cluster with an additional ~ 2200 cm^{-1} of vibrational energy. Isotopic substitution experiments confirm that the fast decay time observed in the $n = 1$ mass channel following 266 nm excitation is indeed due to proton transfer. Certainly, this signal is due to larger clusters fragmenting into the $n = 1$ mass channel, as the concentration dependence demonstrates. We suggest that 266 nm excitation provides sufficient energy to fragment larger clusters (which undergo proton transfer) into the $n = 1$ mass channel. Moreover, the spectrum of phenol$(NH_3)_1$ is sharp and structured, and may have little absorption at 266 nm. This would contribute to the observed dynamics being dominated by fragmentation of larger clusters.

The model of proton transfer proposed above for the naphthol/ammonia system predicts a substantial cluster vibrational energy effect on proton transfer in the cluster size range $n < 6$. Nonetheless, the observed time constants for 266 nm excitation (~ 2200 cm^{-1} of vibrational energy in the large clusters $n \geq 3$) do not decrease significantly. As mentioned above, deuteration studies show that the observed decays are indeed due to proton transfer. One possible explanation for the apparent absence of S_1 vibrational energy dependence of the proton transfer rates is that the cluster distribution which contributes to a given mass channel for origin excitation may not be the same as that which contributes to the same mass channel with $2200 \, cm^{-1}$ of additional vibrational energy in S_1. In fact, the increased $(NH_3)_n H^+$ signals at the higher excitation energy suggest that the fragmentation patterns are indeed different at different excitation energies.

Because of the uncertainty in fragmentation pathways and cluster ion parentage, the proton transfer model used to interpret the naphthol/ammonia results quantitatively (or any model which predicts cluster size dependent proton transfer rates) cannot be applied to the present data set for phenol/ammonia. Qualitatively, increased vibrational energy is expected to increase the proton transfer rate; however, the cluster distribution giving rise to the proton transfer dynamics observed in any mass channel cannot be sorted out at either of the excitation energies used. Furthermore, no correlation or correspondence between the two fragmentation distributions at the two energies can be achieved. As a result, not even qualitative expectations for the observed rates can be generated.

This study can be summarized in the following manner: (1) S_1 dynamics of the phenol/ammonia cluster system are determined, by isotopic substitution, to be due to a proton tunneling event in the neutral cluster [i.e., $PhOH(NH_3)_n \rightarrow PhO^-(NH_3)_n H^+$], (2) significant cluster fragmentation is observed following ionization with ca. $10,000 \, cm^{-1}$ of excess energy, (3) dynamics observed in a specific mass channel cannot be assigned to a specific parent cluster as the fragmentation pathways are a function of excess energy in the cluster ion, and (4) the phenol/ammonia cluster system, therefore, cannot be used as a test for models of proton transfer. The naphthol/ammonia cluster system serves as a test case for proton transfer models because it does not exhibit significant cluster ion fragmentation at the ionization energies employed. A qualitatively and quantatively accurate proton transfer model can only be based on data for known cluster size.

5.3.3. Proton Transfer Models

The model put forward above for proton transfer by a tunneling mechanism based on naphthol/NH_3 cluster data is both simple and sufficient to explain (nearly quantitatively) the observed cluster behavior. This same model can be used to explain cluster matrix isolation behavior as well (Brucker and Kelley 1987a,b, 1988, 1989a,b,c; Brucker et al. 1991; Swinney and Kelley 1991).

The model used for naphthol clusters maintains that at $T = 0$ K the S_1 proton transfer is exothermic such that

$$ROH(NH_3)_n[S_0] \xrightarrow{hv} ROH(NH_3)_n[S_1] \rightarrow RO^-(NH_3)_n H^+[S_1] + [\Delta H < 0].$$

Thus, the general curve of E vs q_H, as schematically represented in Figure 5-14, can be constructed without recourse to more complicated cluster rearrangements. This suggests that the tunneling motion is simple and does not involve solvent reorganization or fluctuations.

These tunneling concepts can be compared with the Marcus theory for electron transfer (Closs and Miller 1988; Kang et al. 1990; Kim and Hynes 1990a,b; Marcus and Sutin 1985; McLendon 1988; Minaga et al. 1991; Sutin 1986). In the normal regime, solvent fluctuation is required to equilibrate the reactant and product energetics so that the electron transfer can occur with $\Delta G = 0$ through a barrier. Thus, solvent motion and fluctuation control the transfer dynamics because the rate for transfer through the barrier at the equilibrated or resonance condition is much faster than the solvent fluctuation rate for energetic equilibration. In this instance, if solvent fluctuation does not occur (e.g., $T = 0$ K), proton transfer is thermodynamically uphill and not favored. In the Marcus inverted regime, transfer is thermodynamically favored without solvent reorganization and one needs only to consider the barrier height and width or Franck–Condon factor for the transfer process.

While the present cluster data do not demand or require a more sophisticated or elaborate model for the proton transfer tunneling mechanism, a number of possible improvements can be considered in light of the above discussion (Syage 1993). First, a more realistic, less idealized barrier (e.g., nonparabolic) could be employed. Second, the importance of anharmonicity could be recognized and included in the calculation of energies and densities of states. Third, the product "potential" could be considered as a well and not as a continuum of states based on "free particle" high density final proton states. This latter reduction seems reasonable only if large exothermicities are appropriate for the transfer reaction, so that the product well is deep and proton tunnels from a low density of states to a very high one. Fourth, a more realistic approach to IVR in the cluster could be taken that would include the dynamics associated with both chromophore and vdW modes as required. Finally, the actual cluster geometry could be included in the well structure and the model, rather than simply the present "diatomic" $O–H \cdots N''$ reduced q_H reaction coordinate, so that more dynamical details (e.g., mode structure, etc.) of the process might be explored. Note that these augmentations would still be within the context of the "Marcus inverted region," because the cluster is cold and solvent fluctuations do not occur on the experimental time scales. Solvent reorganization is observed (τ_2 in Tables 5-4 and 5-5) following proton transfer and is enabled by the ΔH of the transfer process (i.e., the cluster is now hot).

Proton transfer in a room temperature liquid would certainly depend on solvent dynamics and may well be controlled by solvent fluctuations, as predicted and measured by many authors (e.g., Swinney and Kelley 1993). A recent paper by Syage reviews these notions quite thoroughly (Syage 1993), although we suggest, as stated above, that the temperature is too low for solvent fluctuation to bring reactant and product proton transfer systems into resonance prior to the proton transfer event. If ΔH for the transfer reaction is greater than 1 kcal mol^{-1} (~ 350 cm^{-1}), we can expect the density of cluster states in the product well to be sufficiently large that the continuum model is reasonable and the conditions for resonant transfer are always met.

5.3.4. ROH/H$_2$O Clusters

The general result for naphthol and phenol clustered with up to 30 to 40 water molecules is that excited state proton transfer does not occur in cold clusters (Knochenmuss et al. 1993; Knochenmuss and Smith 1994). Proton transfer occurs in clusters only if the solvent is a good proton acceptor and if ions generated in the proton transfer process (RO^- and H^+) are well solvated within the excited state lifetime. The reaction can be detected, as pointed out above, by red-shifted fluorescence from (solvated) RO^-, by the appearance of $(NH_3)_n H^+$ ions due to fragmentation of the cluster ion, by a greatly reduced ionization potential for the cluster following proton transfer (Kim et al. 1991), by a red-shifted absorption spectrum for the cluster which undergoes proton transfer, and by picosecond time resolved dynamics within the S_1 state which changes dramatically upon H/D substitution. The reaction is found to depend both on the basicity of the proton acceptor ($> 243 \, kcal \, mol^{-1}$) and the cluster geometry (Aue and Bowers 1979; Ebata and Ito 1992; Kebarle 1977; Knochenmuss et al. 1993; Knochenmuss and Smith 1994; Mautner 1984, 1986; Payzant et al. 1973). A water cluster with $20 \leq n \leq 30$ should have a proton affinity which is greater than $\sim 245 \, kcal \, mol^{-1}$ and, thus, should generate an excited state proton transfer reaction for naphthol and phenol/H_2O clusters with $n \geq 30$. In room temperature water, proton transfer readily occurs.

The difference between the ROH/NH_3 and H_2O behavior with regard to proton transfer is probably related to cluster structure and cluster rigidity. The interaction between ammonia and an aromatic π-system (benzene and naphthalene) is between $600\text{--}900 \, cm^{-1}$ based on atom–atom potential energy calculations (Menapace and Bernstein 1987a,b; Wanna et al. 1986). The hydrogen bonding interaction between ammonia and ROH is roughly $1000 \, cm^{-1}$ (Hineman et al. n.d.; Li and Bernstein 1991) and $(NH_3)_2$ energy is also ca. $1000 \, cm^{-1}$ (Klemperer 1992; Klemperer et al. 1988; Miller 1988; Nelson et al. 1987; Nesbitt 1988; Nesbitt and Naaman 1989; Weber and Rice 1988a,b). Water has roughly the same interaction energy as ammonia with the π-system, but it has a ca. $1500 \, cm^{-1}$ hydrogen bonding energy with the aromatic alcohol and a ca. $2000 \, cm^{-1}$ interaction energy with itself (Li and Bernstein 1991). Given these interaction energies, one can suggest that a large ROH/H_2O cluster would resemble a drop or lattice of water, hydrogen bonded to the ROH, and that an ROH/NH_3 cluster would have a much more distributed structure in which the NH_3 more or less equally interacts with the π-system, the OH moiety, and itself. The water structure is also expected to more rigid. Consequently, both RO^- and H^+ are better solvated in an NH_3 cluster than an H_2O cluster, and whatever contribution solvent dynamics might make to the transfer process would be more effective in NH_3 clusters. These ideas are qualitatively supported by what little structural data exist on small clusters of NH_3 and H_2O with aromatic alcohols (Felker 1992; Hartland et al. 1992; Plusquellic et al. 1992).

5.4. ELECTRON TRANSFER REACTIONS

5.4.1. Intramolecular-Solvent Stabilized

During the last 6 years, we have studied solvent stabilization of the excited state electron transfer reaction for dimethylaminobenzonitrile (DMABN) (Grassian et al. 1989, 1990; Shang and Bernstein 1992; Warren et al. 1988). Investigation of the cluster behavior of this reaction has led to an important finding that pertains to both condensed phase behavior and the mechanism for the electron transfer, as we shall see below. Since these studies have been reviewed previously (Bernstein 1992; Grassian 1989, 1990; Warren et al. 1988), we only present an outline of the results to provide a demonstration of the main point.

The spectrum of DMABN is quite sharp, although somewhat complex for the cold bare molecule. The origin region is composed of a series of sharp lines (ca. 12) which can be attributed to the compound motion of the dimethylamino group (inversion and rotation) and possibly rotation of the individual methyl groups about the C–N methylamine bond (Bernstein 1992; Grassian et al. 1989, 1990). A potential surface for this motion can be evaluated (Bernstein 1992; Grassian et al. 1989, 1990) and the dimethylamino group rotates ca. 25° and moves into the plane of the benzene ring upon $S_1 \leftarrow S_0$ excitation.

This origin 0_0^0 structure is maintained in clusters of DMABN with nonpolar and hydrogen bonding solvents (e.g., CH_4, H_2O). In these systems, even for larger clusters ($n \geq 3$ solvent molecules), the local excited state is the lowest singlet excitation. The local excited $\pi\pi^*$ state is recognized by sharp absorption and emission, which follow those of the bare molecule very closely. If a polar solvent such as $(CH_3)_2CO$, CH_2Cl_2, CH_3CN is clustered with DMABN, the spectrum changes dramatically: large red shifts for absorption and emission are observed, both are thousands of cm^{-1} broad, and the red shifts correlate with solvent dipole movement. The important point here is that all these changes are for one-to-one clusters; that is, DMABN (solvent)$_1$ clusters. In this small cluster limit, cluster conformation or solvent orientation with respect to the S_1 (near) molecular plane matters a great deal for the lowering of the electron transfer state below the local excited state. If the solvent does not assume a dipolar orientation with respect to the ring (i.e., "head to tail" with the DMABN dipole) lying above the ring plane, the local excited state is lower in energy than the electron transfer state. Since different cluster conformations can be observed for these clusters, both electron transfer and local excited state type clusters can be identified in the spectra.

The potential surfaces can be qualitatively expressed for these clusters through a model quite similar to that employed by Levy and coworkers (Tubergen et al. 1990; Tubergen and Levy 1991) for substituted indole clusters. Polar solvation in this case lowers the polar L_a state below the local excited L_b state, which is the first excited singlet state of the bare molecule.

Thus, in discussing the degree of solvation needed to stabilize the excited state intramolecular electron transfer for DMABN in condensed phases, one must realize that only one polar solvent molecule, properly placed, will lower the energy of this state below that of the local $\pi\pi^*$ excited state (see Grassian et al. 1989, 1990; Shang and Bernstein 1992; Warren et al. 1988 for more discussion of this point).

5.4.2. Intermolecular Electron Transfer in Neutral Systems

Rydberg states are readily perturbed by their environment (Miladi et al. 1975; Robin 1974, 1975; Robin and Kuebler 1970). In the presence of an inert gas (e.g., N_2, rare gases, etc.) at medium pressures, the Rydberg transition experiences a blue shift and asymmetric broadening to higher energy. This phenomenon can be attributed to an exchange repulsion effect between the Rydberg electronic configuration of the "solute" and the closed shells of the inert solvent. In fact, this observation has become the empirical technique employed to distinguish Rydberg from valence states. Cluster studies of the Rydberg states of solute/solvent clusters give a microscopic and detailed picture of these phenomena. In fact, these investigations have shown that Rydberg states of molecules experience three different categories of intermolecular interaction: exchange repulsion between the electron in the extended Rydberg orbital and the solvent closed shells, a dipole-induced dipole interaction with polar solvents, and an electron transfer interaction with solvents which have available virtual orbitals of similar energy to the half-filled Rydberg orbital. With regard to the study of electron transfer chemistry, the use of Rydberg states is ideal because they have low ionization energies and the solvents with available vacant orbitals have good electron affinities.

We have investigated a variety of clusters with ethers and amines as the Rydberg donor systems and polar solvents as the acceptors. The donors we have studied include dioxane ($C_4H_8O_2$), azabicyclooctane (ABCO), diazabicyclooctane (DABCO), hexamethylenetetramine (HMT), and others (Moreno et al. 1992; Shang and Bernstein 1994; Shang et al. 1993a,b,c, 1994a,b,c). The acceptors include ethers, amines, and aromatics.

One can observe the different types of interactions as they dominate the static and dynamic behavior of the various clusters. For example, the DABCO(Ar)$_n$ $^1R_1 \leftarrow S_0$ transitions show a ca. 100 cm^{-1} blue shift/Ar atom for the coordination site between the ethylene bridges connecting the two nitrogen atoms and a 290 cm^{-1} blue shift for each argon atom coordinated near the nitrogen atoms of DABCO. This demonstrates the exchange repulsion or Pauli exclusion interaction for the Rydberg state. This interaction can even be larger than the binding energy for the cluster and can cause some larger ($n = 2, 3, 4, \ldots$) clusters to dissociate upon $^1R_1 \leftarrow S_0$ excitation (Shang et al. 1993c). The dipole-induced dipole interaction is well demonstrated by the pair of solvents propane and dimethylsulfide (CH_3SCH_3). Propane has $\mu = 0$ D, a polarizability of 6.3 Å^3, and generates $a + 267$ cm^{-1} blue shift for the $^1R_1 \leftarrow S_0$ electronic transition. On the other hand, dimethylsulfide has $\mu = 1.5$ D, a polarizability of 7.4 Å^3, and generates a blue shift for the transition of $+72$ cm^{-1}. The dipole-induced dipole interaction should be attractive and generate a red shift; this would reduce the close shell exchange interaction blue shift—in this instance by almost 200 cm^{-1}. These shifts can be compared with those of solvents with available, equi-energy, virtual Rydberg, or other orbitals to accept an electron transfer from DABCO. The amines DABCO, ABCO, and triethylamine yield transition red shifts for ABCO and DABO clusters of more than 500 cm^{-1}. Ethers give red shifts ca. 500 cm^{-1} as well, and aromatics (C_6H_6, $C_6H_5CH_3$) give shifts ca. -350 cm^{-1}. The polarizabilities of these solvents are all ca. 5–10 Å^3 and these red shifts do not scale well with dipole movements.

If these shift data really do represent the onset of an intermolecular electron transfer reaction in DABCO, ABCO, HMT clustered with amine, either, and aromatic solvents, one ought to be able to observe the reaction kinetics or dynamics. Consider the specific instance of DABCO. The singlet Rydberg state lifetime for DABCO (and all the other Rydberg molecules studied for this determination (Shang et al. 1993c, 1994a) is ca. 2 µs for the isolated molecule and ca. 1.2 µs for the nonpolar rare gas, hydrocarbon, and fluorocarbon solvents. This reduction in lifetime is due to intersystem crossing and the exact dynamics depends on the coordination site (Shang et al. 1994c). For clusters with large red shifted Rydberg transitions, the excited state lifetime is quite different: an initial decay of ca. 100 ns is measured for the R_1 excited state, and the R_1 state decays to another long-lived ($\tau > 100$ μs) excited state (see Figures 5-15 and 5-16). This excited state also has an evolving ionization threshold: at $t \sim 0$ μs, the cluster Rydberg state can be ionized with about 650 cm^{-1} less energy than at $t > 500$ ns. Thus, the electron is transferred in ca. 100 ns and the electron transfer state (DABCO$^+ \cdot$solv$^-$) is ca. 650 cm^{-1} lower in energy than the $^1R_1$ state of DABCO (see Figure 5-17). The other systems behave in a similar fashion. Thus, both static and dynamic consequences of the electron transfer interactions can be characterized in these cluster systems.

The study of Rydberg states in clusters is just at its beginning, especially with respect to Rydberg state chemistry. The behavior of cluster Rydberg states is apparently much different than that of cluster valence states. The Rydberg states

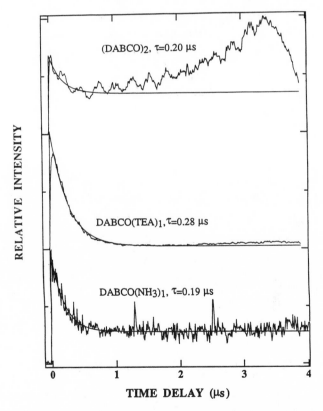

Figure 5-15. State decay of DABCO/amine cluster (2p3s) Rydberg states observed through origin excitation. The ionization energy is 22,222 cm^{-1}. Note that the decays end in a constant signal for each case. These curves have been fit to single exponential function plus a constant. The constant signal is assigned as due to long-lived charge-transfer intermediate. The signal rise at later time is probably due to solvent reorganization accompanied by an ionization efficiency enhancement.

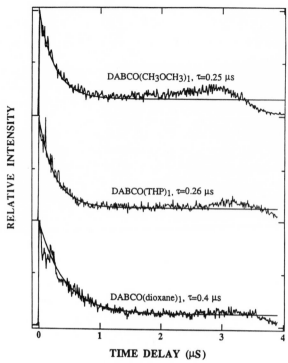

Figure 5-16. State decay of DABCO/ether cluster (2p3s) Rydberg state observed through origin excitation. The conditions and conclusions are the same as described for Figure 5-15.

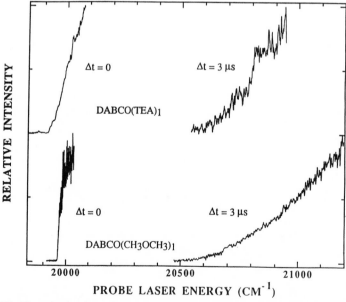

Figure 5-17. Photoionization threshold energy spectra for DABCO(CH$_3$OCH$_3$)$_1$ and DABCO(TEA)$_1$ (2p3s) Rydberg state. Excitation is at the cluster first observed origin transition. The values of Δt represent the time delay between the excitation laser and the ionization laser.

are highly sensitive to their environments. Only 3s Rydberg states have been thus far studied in this manner and higher states may have directional properties and may be even more environmentally sensitive than those (2p3s) states studied. Reactivity is expected to increase as the Rydberg state energy (n) increases: both ion- and radical-like behavior can be anticipated from Rydberg state chemistry. Electron transfer reactions are anticipated to be more facile for higher Rydberg states as well.

The theory for this intermolecular electron transfer reaction can be approached on a microscopic quantum mechanical level, as suggested above, based on a molecular orbital (filled and virtual) approach for both donor (solute) and acceptor (solvent) molecules. If the two sets of molecular orbitals can be in resonance and can physically overlap for a given cluster geometry, then the electron transfer is relatively efficient. In the cases discussed above, a barrier to electron transfer clearly exists, but the overall reaction in certainly exothermic. The barrier must be coupled to a nuclear motion and, thus, Franck–Condon factors for the electron transfer process must be small. This interaction should be modeled by Marcus inverted region electron transfer theory and is well described in the literature (Closs and Miller 1988; Kang et al. 1990; Kim and Hynes 1990a,b; Marcus and Sutin 1985; McLendon 1988; Minaga et al. 1991; Sutin 1986).

5.5. REACTIVE INTERMEDIATES—RADICALS, CARBENES, NITRENES, AND CLUSTER IONS

The reactions discussed above deal with IVR/VP of clusters (the breaking of a van der Waals bond), excited state intermolecular proton transfer, and excited state intra- and intermolecular electron transfer. These reactions are relatively easy to study because they are adiabatic; that is, the reactants and products remain on the "same" excited electronic state potential energy surface. The reason that such reactions are readily accessible is that products can be observed though emission and/or further excitation to the ion and subsequent mass analysis. Radical reactions also can often be adiabatic and thus can be easily accessible for study. Diabatic reactions in clusters are certainly accessible but can generate highly vibrationally excited products, which, for large systems, can be more difficult to investigate. Thus, radical containing clusters are another category of system that can be explored with regard to excited state chemistry by present techniques. Even if such reactions are diabatic, products can be reexcited and fluorescence and mass resolved excitation spectra (MRES) can be collected to analyze the product state cluster or molecule nascent energy level distribution. Studies of this latter nature were pioneered by the groups of Wittig (Böhmer et al. 1992, 1993; Davis et al. 1993; Ionov et al. 1992, 1993a,b; Shin, et al. 1991; Wittig et al. 1988) and Crim (Crim 1984, 1987), particularly for the generation of small radicals like OH and NO. Their work has been previously reviewed in some depth and will not be discussed in detail below.

Radicals are just the simplest member of the family of reactive intermediates which also includes the carbenes (RĊR′) and nitrenes (RN:). With the exception of triatomic carbenes (Jacox 1984, 1988, 1990) and diatomic nitrenes (Huber and

Herzberg 1979), very few studies of these latter two gas phase species have appeared; nonetheless, they are reported to be important for condensed phase chemistry (Platz 1990; Schuster 1986). Characterization of their unsolvated, solvated, and reacting properties and energy levels is certainly a useful endeavor. Parallel calculations at a high *ab initio* level can also shed light on the properties (i.e., geometry, electronic and vibrational energy levels, solvation structures and reactions) of reactive intermediates.

The radicals that have been investigated in the gas phase are listed by Jacox (1984, 1988, 1990), Miller (Foster and Miller 1989), and Huber (Huber and Herzberg 1979). Solvated systems have also been studied and have been reviewed by Heaven (Heaven 1992) and Lester (Giancarlo et al. 1994). The cluster chemistry studies, as such, are discussed below for benzyl ($C_6H_5CH_2 \cdot$) and methyl radicals. Work in progress on small radicals, carbenes, and nitrenes will be briefly mentioned at the end of the section.

5.5.1. Benzyl Radical/Ethylene Reaction—Experiments

The fluorescence spectrum of the cold benzyl radical has been acquired by a number of groups, and has been recently reviewed by various authors (Disselkamp and Bernstein 1993; Foster and Miller 1989). Figure 5-18 shows the MRES of the benzyl radical $D_2/D_1 \leftarrow D_0$ [$2^2B_2/1^2A_2 \leftarrow 1^2B_2$] transition. A comparison

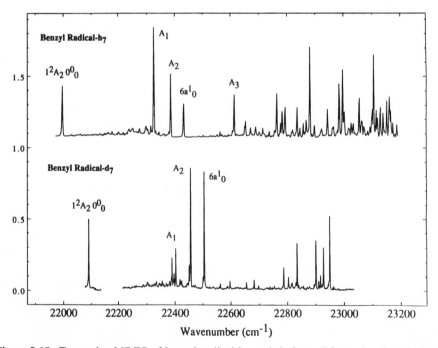

Figure 5-18. Two-color MRES of benzyl radical-h_7 and d_7 formed from the photolysis of benzyl chloride and cooled in the supersonic expansion. The 0_0^0 transitions are at 21,998 cm^{-1} (h_7) and 22,090 cm^{-1} (d_7).

between the d_7 and h_7 substituted radical spectra make the vibronic coupling between the two excited electronic states quite obvious. Spectra of a number of nonreactive clusters have also been obtained (Disselkamp and Bernstein 1993). The solvents include Ar, N_2, and C_nH_{2n+2} ($n = 1, 2, 3$). These show that the cluster binding energies are ca. 500 cm^{-1} and that solvation generates a red-shifted electronic transition (Figure 5-19). The nature of the radical/solvent interactions for this system is quite similar to that expected for aromatic molecular clusters (e.g. C_6H_6 or $C_6H_5CH_3$/Ar, N_2, C_nH_{2n+2}).

Consider now the very different situation that arises for the benzyl radical$(C_2H_4)_n$ clusters as shown in Figure 5-20 (Disselkamp and Bernstein 1994). The clusters with one or two C_2H_4 solvent molecules now have very different spectra. The results can be summarized as follows: (1) the binding energy for the BR(C_2H_6) cluster in D_0, D_1 is ca. 500 cm^{-1}—for BR$(C_2H_4)_1$ the binding energy is greater than 15,000 cm^{-1} (43 kcal mol^{-1}) in the D_1 state but only ca. 600 cm^{-1} for the D_0 state, (2) BR$(C_2H_4)_1$ ionization energy is ca. 1000 cm^{-1} less than that for BR$(C_2H_6)_1$ and the threshold is sharp for BR$(C_2H_6)_1$ but very broad for BR$(C_2H_4)_1$, (3) the D_1 lifetime is ca. 1 μs for Br$(C_2H_6)_1$ but ca. 0.5 μs for the high energy BR$(C_2H_4)_1$ absorption, (4) the spectrum for BR$(C_2H_6)_1$ is sharp, but for BR$(C_2H_4)_1$ it is very broad with only a few discernible features, (5) deuteration has no effect on lifetime or nature of BR$(C_2H_4)_1$ spectrum, and (6) further solvation with another C_2H_4 has no obvious affect on the cluster spectrum. Data and overall spectral features suggest an excited state reaction has occurred

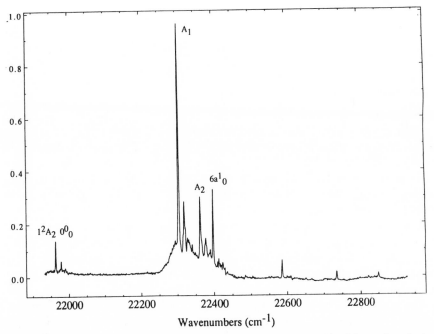

Figure 5-19. Two-color MRES of benzyl radical$(C_2H_6)_1$ cluster obtained from photolyzing benzyl chloride and expanding with 1% ethane in helium. Note the loss of intensity beyond 22,600 cm^{-1}.

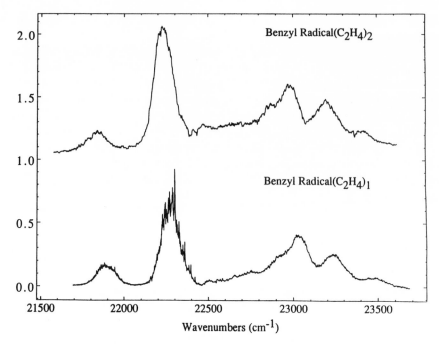

Figure 5-20. Two-color MRES of benzyl radical$(C_2H_4)_n$, $n = 1, 2$, obtained from photo-lyzing benzyl chloride and expanding with 1% C_2H_4 in helium.

for $BR + C_2H_4$ within the cluster ("half-collision" environment). Calculations discussed below are consistent with this expectation.

5.5.2. Benzyl Radical/Ethylene—Theory

Theoretical studies of this excited state reaction are first explored for a simpler system, $CH_3(C_2H_4)_1$, in order to determine the appropriate level of calculation required to reproduce experimental results through a potential energy surface/reaction path generation. The heat of reaction for the system in the ground state is known to be -25.5 kcal mol^{-1} with an activation energy (barrier) of 6.4 kcal mol^{-1} (Disselkamp and Bernstein 1994; Takayanagi and Hanazaki 1991). In order to achieve these numbers, a 5 electron CASSCF plus MRCI (S + D) with a Davidson correction for higher order electron correlation and a DZV basis set must be employed. Within this framework, we find the ground state has a $\Delta E_{react} = -23$ kcal mol^{-1} and a $\Delta E_{act} = 7.0$ kcal mol^{-1}. We find these results very informative for the reaction coordinate in the ground state. Excited state results achieved at the same calculational level show the reaction is exothermic (-30 kcal mol^{-1}), has no activation energy, and the minimum on the excited state surface falls at the same reaction coordinate value as the maximum on the ground state surface. This latter result implies that the two potential surfaces undergo an anticrossing: our calculations make this clear because of "root switching" that occurs near the crossing point. We learn from these results that in order to calculate a cluster radical reaction on the ground and excited state surfaces, one must include extensive electron correlation through CI, the full virtual space for the CASSCF, a multireference (many Slater determinant configuration state functions) description of the state prior to additional configuration interaction, and as large a basis set as it acceptable for the calculational time and expense. These results are so encouraging that they are a main impetus for studying this cluster reaction experimentally in the future.

Based on this success, we have performed calculations for the $BR(C_2H_4)_1$ cluster; however, a 9 or 13 electron CASSCF/MRCI (S + D) system cannot be treated at the current available level of computers. We thus rely on a projected MP2 perturbation approach which can be calibrated against the CH_3/C_2H_4 system experiments and theory. Once this calibration is known, a 9 electron CASSCF approach is taken to the BR/C_2H_4 reaction and the appropriate corrections are applied to obtain acceptable results. These calculations show the same basic results for the BR/C_2H_4 and CH_3/C_2H_4 systems: (1) the ground state reaction for BR/C_2H_4 has a negative heat of reaction ($-21\,kcal\,mol^{-1}$) and a large activation energy ($11\,kcal\,mol^{-1}$), and (2) the excited state reaction has a small barrier ($<2\,kcal\,mol^{-1}$) and $-7\,kcal\,mol^{-1}$ heat of reaction. In this instance, a level anticrossing for the D_0 and D_1 surfaces is not apparent.

This combination of experiment and theory has enabled us to elucidate the BR/C_2H_4 cluster reaction and has encouraged us with further studies of small radical cluster chemistry. Clearly, the synergy between theory and experiment within our own laboratory has afforded this progress and enabled this new work to be conceived and planned.

5.5.3. Work in Progress—Carbenes, Nitrenes, Ions, and Radicals

The gas phase spectroscopy of carbenes ($R\ddot{C}R'$, R, R' = H, Cl, Br, F, CH_3, C_6H_5) has recently become more active and productive due to supersonic jet techniques and the advent of different types of supersonic nozzles [e.g., pyrolysis (Clouthier and Karolczak 1989), discharge (Schlachta et al. 1991), photodissociation (Suto and Steinfeld 1990), and ablation (Li and Bernstein n.d.)]. Carbenes can thus be generated in high concentrations and both fluorescence and mass detected spectroscopy can be carried out for these interesting reactive intermediates. Our efforts have focused on aromatic substituted carbenes (in particular, diphenylcarbene— $C_6H_5\ddot{C}C_6H_5$). Spectra obtained for this system generated by laser photolysis at 1064, 532, 355, 266 and 193 nm of $(C_6H_5)_2CN_2$ are composed of emission from C_2, CN and NCO. We are now pursuing thermal decomposition pathways to the generation of gas phase carbenes (Minsek and Chen 1993; Clauberg and Chen 1991; Kohn et al. 1992). We plan cluster studies of this carbene and will also pursue cluster reaction, such as,

$$RR'C\colon (C_2H_4)_1 \quad \xrightarrow{h\nu} \quad \overset{\textstyle\diagup\!\!\!\!\diagdown}{\underset{R \quad R'}{\times}} \cdot$$

One of the important concerns for this type of chemistry is that singlet and triplet state carbene reactions are supposed to be different (Platz 1990; Schuster 1986). This can be explored by changing the R, R' groups, which will vary the singlet/triplet separation and eventually give both ground state singlet and triplet carbenes. Moreover, the singlet/triplet relative energy is governed by the dihedral angle value: 180° implies a triplet ground state and 90° implies a singlet ground state. The singlet/triplet energy ordering and splitting varies smoothly with this angle from 90 to 180°. For these systems, solvation and structure play an essential role in chemistry and clusters offer an ideal situation for the elucidation of the properties and behavior of both simple and complex carbenes.

Nitrenes (RN:, R = H, F, CH_3, C_6H_5, ...) are another important organic reactive intermediate; they are even more reactive than carbenes in many instances and share some of the properties of carbenes with regard to singlet/triplet state behavior (Platz 1990; Schuster 1986). Phenylnitrene is known to rearrange, for example, to the cyanocylopentadienyl radical through an excited singlet state (Cullin et al. 1990a,b; Im and Bernstein 1991). If electron withdrawing groups are present on the phenyl ring (e.g., 2,6-difluoro- or pentafluorophenylitrene), such rearrangements are inhibited. Nonetheless, C_2, CN, and NCO are the major emissive laser photolysis products from these molecules. Again, thermal chemistry is being pursued to generated these species (Clauberg and Chen, 1991; Kohn et al. 1993; Minsek and Chen 1993). The cluster chemistry of these and other nitrenes should be fascinating and should shed light on the nature of their condensed phase behavior.

Reactions are known to be highly dependent on solvents and the nature of solvent-reactive intermediate interactions: solvents can affect the reaction coordinate, the activation energy, and the overall reaction thermodynamics. Clusters, especially ionic clusters, show this behavior as well. The systems we have studied are α-substituted toluenes; phenol is known to transfer a proton upon $S_1 \leftarrow S_0$ excitation, but what happens for excited states of α-substituted benzyl alcohols ($C_6H_5CH_2OH$)? The results, which are presented in detail by Li and Bernstein (Bernstein 1992; Li and Bernstein 1992a,b) are unique and quite informative. They are different than those discussed by Jouvet and Solgadi in chapter 4 of this volume.

When toluene, benzyl alcohol, and α,α-dimethylbenzyl alcohol are solvated with either water or ammonia and ionized ($I \leftarrow S_1 \leftarrow S_0$), a number of reactions occur. Toluene$^+$ will transfer a proton to water or ammonia, and thereby generate a benzyl radical and a solvated proton. The solvated proton and the benzyl radical are stable enough to drive this unlikely reaction in even small clusters ($3 \leq n \leq 6$). For benzyl alcohol, proton transfer occurs not from the OH moiety but from the C_α position—again, a solvated proton and a substituted benzyl radical are formed. Benzyl alcohol/ammonia clusters undergo the same reaction but, as cluster size increases, OH proton transfer to the ammonia solvent begins to occur. Additionally, in ammonia clusters of α,α' dimethylbenzyl alcohol, methyl radical and substituted benzyl radical formation are observed, rather than OH proton transfer.

Studies of this nature are just at their inception, with solvent variation and different chromophoric species to be explored. This form of ionic fragmentation chemistry is quite interesting and is highly dependent on solvation; structure dependence of these fragmentations can also be investigated in small clusters ($n \leq 6$). The opportunity for more experiments and theory here is quite clear. These are perhaps the most solvent rich and dependent processes thus far characterized in clusters.

In the past, the radicals that have received the most experimental and theoretical attention are, of course, the small systems with only a few heavy atoms. Radicals such as OH, NO, CH_3, CH_3O, C_3H_5, NO_3, SH, CH_3S, and others, are both environmentally important and readily accessible for investigation (Huber and Herzberg 1979; Jacox 1984, 1988, 1990). Studies of cluster chemistry involving these species with many different reactants (e.g., alkanes, alkenes, N_2, chlorofluorocarbons, NO_2, etc.) would have several advantages: ready identification of reactant and product state energy levels and populations, experimental results could be supported by very high levels (MRCI, CASSCF, large basis sets, etc.) of theory, a large background of isolated molecule information is available, and a great deal is known about the reaction chemistry of these species. A number of groups (e.g., Bernstein, Lester) are presently studying one or more of the cluster reactions of these small radicals.

5.6. CONCLUSIONS

Cluster chemical reactions offer a wide variety of systems to be studied by an ever-expanding range of experimental and theoretical methods. The criteria needed for a reliable study with current techniques include: a barrier to reaction on the ground electronic state potential energy surface, accessible excited state surfaces upon which reactions occur, accessible product states for subsequent interrogation, and clusters and molecules and reactive intermediates small enough to be theoretically modeled or calculated on a microscopic potential energy surface/ reaction path (Dunning Harding 1985; Kuntz 1985; Levine 1985; Truhlar 1981; Truhlar et al. 1985) level. In this chapter, we have demonstrated that cluster dynamics, electron transfer, proton transfer, radical addition, ion fragmentation, and reactive intermediate reactions fall into this general category. Certainly, many other unimolecular and bimolecular reactions are also accessible by these techniques.

We can project a very rich and productive future for the study of chemical reactions in clusters.

ACKNOWLEDGMENTS: The work presented herein would not have been possible without the financial support of the Office of Naval Research, National Science Foundation, and the U.S. Army Research Office. The studies were done by my very able coworkers Drs. H. S. Im, S. K. Kim, M. Nimlos, M. A. Young, R. Disselkamp, M. Hineman, S. Li, Q. Y. Shang, P. O. Moreno, and Mr. C. F. Dion. My colleague, Professor D. F. Kelley, has been a major contributor to IVR/VP and proton transfer studies and indeed all of our cluster reaction studies. I wich to thank Dr. T. Selegue for his very careful reading of this manuscript and helpful comments.

REFERENCES

Alfano, J. C.; Martinez, S. J.; Levy, D. H. 1991 J. Chem. Phys. 94:1673, and references therein

Aue, D. H.; Bowers, M. T. 1979 In *Gas Phase Ion Chemistry*, Vol. 2, M. T. Bowers, ed., Academic Press, New York, 1979

Avouris, P.; Gelbart, W. M.; El-Sayed, M. A. 1977 Chem. Rev. 77:793

Bell, R. P. 1980 *The Tunnel Effect in Chemistry*, Chapman and Hall, New York

Bernstein, E. R. (ed.) 1990 *Atomic Molecular Clusters*, Elsevier, Amsterdam

Bernstein, E. R. 1992 J. Phys. Chem. 96:10105

Beswick, J. A.; Jortner, J. 1981 Adv. Chem. Phys. 47:363, and references therein

Böhmer, E.; Mikhaylichenko, K.; Wittig, C. 1993 J. Chem. Phys. 99:6545

Böhmer, E.; Shin, S. K.; Chen, Y.; Wittig, C. 1992 J. Chem. Phys. 97:2536

Borgis, D. C.; Lee, S.; Hynes J. T. 1989 Chem. Phys. Lett. 162:19

Breen, J. J.; Peng, L. W.; Willberg, D. M.; Heikal, A.; Cong, P.; Zewail, A. H. 1990 J. Chem. Phys. 92:805

Brucker, G. A.; Kelley, D. F. 1987a J. Phys. Chem. 91:2856

Brucker, G. A.; Kelley, D. F. 1987b J. Phys. Chem. 91:2862

Brucker, G. A.; Kelley, D. F. 1988 J. Phys. Chem. 92:3805

Brucker, G. A.; Kelley, D. F. 1989a J. Phys. Chem. 93:5179

Brucker, G. A.; Kelley, D. F. 1989b J. Chem. Phys. 90:5243

Brucker, G. A.; Kelley, D. F. 1989c Chem. Phys. 136:213

Brucker, G. A.; Swinney, T. C.; Kelley, D. F. 1991 J. Phys. Chem. 95:3190

Brumbaugh, D. V.; Kenny, J. E.; Levy, D. H. 1983 J. Chem. Phys. 78:3415, and references therein

Chernoff, D. A.; Rice, S. A. 1979 J. Chem. Phys. 70:2521, and references therein

Chesnovsky, O.; Leutwyler, S. 1985 Chem. Phys. Lett. 121:1

Chesnovsky, O.; Leutwyler, S. 1988 J. Chem. Phys. 88:4127

Clauberg, H.; Chen, P. 1991 J. Am. Chem. Soc. 113:1445

Closs, G. L.; Miller, J. R. 1988 Science 240:441

Clouthier, D. J.; Karolczak, J. 1989 J. Phys. Chem. 93:7542

Crim, F. F. 1984 Annu. Rev. Phys. Chem. 35:657

Crim, F. F. 1987 In *Molecular Photodissociation Dynamics*, M. N. R. Ashfold and J. E. Baggott, eds., Royal Society of Chemistry, London, p. 177

Cullin, D. W.; Soundararajan, N.; Platz, M. S.; Miller, T. A. 1990a J. Phys. Chem. 94:8890

Cullin, D. W.; Yu, L.; Williamson, J. M.; Platz, M. S.; Miller, T. A. 1990b J. Phys. Chem. 94:3387

Davis, H. F.; Ionov, P. I.; Ionov, S. I.; Wittig, C. 1993 Chem. Phys. Lett. 215:214

Disselkamp, R.; Bernstein, E. R. 1993 J. Chem. Phys. 98:4339, and references therein to older work

Disselkamp, R.; Bernstein, E. R. 1994 J. Phys. Chem. 98:7260

Droz, T.; Knochenmuss, R.; Leutwyler, S. 1990 J. Chem. Phys. 93:4520

Dunning, T. H.; Harding, L. B. 1985 In *Theory of Chemical Reactions Dynamics*, Vol. I, M. Baer, ed., CRC Press, Boca Raton, p. 1

Ebata, T.; Ito, M. 1992 J. Phys. Chem. 96:3224

Ewing, G. E. 1981 In *Potential Energy Surfaces and Dynamics Calculations*, D. G. Truhlar, ed., Plenum Press, New York, p. 75

Ewing, G. E. 1986 J. Phys. Chem. 90:1790

Ewing, G. E. 1987 J. Phys. Chem. 91:4662

Felker, P. M. 1992 J. Phys. Chem. 96:7844

Felker, P. M.; Zewail, A. H. 1988 Adv. Chem. Phys. 70 (pt. 1):265

Forst, W. 1973 *Theory of Unimolecular Reactions*, Academic Press, New York

Foster, S. C.; Miller, T. A. 1989 J. Phys. Chem. 93:5986

Giancarlo, L. C.; Randall, R. W.; Choi, S. E.; Lester, M. I. 1994, J. Chem. Phys. 101:2914

Gilbert, R. G.; Smith, S. C. 1990 *Theory of Unimolecular and Recombination Reactions*, Blackwell, Oxford

Grassian, V. H.; Warren, J. A.; Bernstein, E. R. 1989 J. Chem. Phys. 90:3994

Grassian, V. H.; Warren, J. A.; Bernstein, E. R. 1990 J. Chem. Phys. 93:6910

Halberstadt, N.; Janda, K. C. (eds.) 1990 *Dynamics of Polyatomic van der Waals Clusters*, NATO ASI Series, Plenum Press, New York

Hartland, G. V.; Henson, B. F.; Venturo, V. A.; Felker, P. M. 1992 J. Phys. Chem. 96:1164

Haynam, C. A.; Brumbauch, D. V.; Levy, D. H. 1984a J. Chem. Phys. 80:2256

Haynam, C. A.; Young, L.; Morter, C.; Levy, D. H. 1984b J. Chem. Phys. 81:5216

Heaven, M. C. 1992 Annu. Rev. Phys. Chem. 43:283

Heppener, M.; Kunst, A. G. M.; Bebelaar, D.; Rettschnick, R. P. H. 1985 J. Chem. Phys. 83:5314

Heppener, M.; Rettschnick, R. P. H. 1987 In *Structure and Dynamics of Weakly Bound Molecular Complexes*, A. Weber, ed., D. Reidel, Dordrecht, p. 533

Hineman, M.; Bernstein, E. R.; Kelley, D. F. 1993a J. Chem. Phys. 98:2516

Hineman, M.; Bernstein, E. R.; Kelley, D. F. 1994 J. Chem. Phys. 101:850

Hineman, M. F.; Brucker, G. A.; Kelley, D. F.; Bernstein, E. R. 1992a J. Chem. Phys. 97:3341

Hineman, M.; Kelley, D. F.; Bernstein, E. R. 1993b J. Chem. Phys. 99:4533

Hineman, M. F.; Kim, S. K.; Bernstein, E. R.; Kelley, D. F. 1992b J. Chem. Phys. 96:4904

Hineman, M.; Li, S.; Bernstein, E. R. n.d. unpublished results

Hopkins, J. B.; Powers, D. E.; Mukamel, S.; Smalley, R. E. 1980b J. Chem. Phys. 72:5049

Hopkins, J. B.; Powers, D. E.; Smalley, R. E. 1980b J. Chem. Phys. 72:5039

Hopkins, J. B.; Powers, D. E.; Smalley, R. E. 1980c J. Chem. Phys. 73:683

Hopkins, J. B.; Powers, D. E.; Smalley, R. E. 1981 J. Chem. Phys. 74:745

Huber, K. P.; Herzberg, G. 1979 *Molecular Spectra and Molecular Structure, IV. Constants of Diatomic Molecules*, Van Nostrand Reinhold, New York

Im, H. S.; Bernstein, E. R. 1991 J. Chem. Phys. 95:6326

Ionov, S. I.; Brucker, G. A.; Jaques, C.; Chen, Y.; Wittig, C. 1993a J. Chem. Phys. 99:3420

Ionov, S. I.; Brucker, G. A.; Jaques, C.; Valachovic, L.; Wittig, C. 1992 J. Chem. Phys. 97:9486

Ionov, S. I.; Brucker, G. A.; Jaques, C.; Valachovic, L.; Wittig, C. 1993b J. Chem. Phys. 99:6553

Ireland, J. F.; Wyatt, P. A. H. 1976 Adv. Phys. Org. Chem. 12:131

Jacobsen, B. A.; Humphrey, S.; Rice, S. A. 1988 J. Chem. Phys. 89:5624

Jacox, M. 1984 J. Phys. Chem. Ref. Data 13:945

Jacox, M. 1988 J. Phys. Chem. Ref. Data 17:269

Jacox, M. 1990 J. Phys. Chem. Ref. Data 19:1387

Jena, P.; Rao, B. K.; Khanna, S. N. (eds.) 1987, *Physics and Chemistry of Small Clusters*, Plenum Press, New York

Jortner, J.; Levine, R. D.; Rice, S. A. 1988 Adv. Chem. Phys. 70 (pt. 1):1

Jouvet, C.; Boivineau, M.; Duval, M. C.; Soep, B. 1987 J. Phys. Chem. 91:5416

Jouvet, C.; Soep, B.; 1983 Chem. Phys. Lett. 96:426

Kang, T. J.; Jarzeba, W.; Barbara, P. F. 1990 Chem. Phys. 149:81

Kebarle, P. 1987 Annu. Rev. Phys. Chem. 28:445

Kelley, D. F.; Bernstein, E. R. 1986 J. Phys. Chem. 90:5164

Kim, H. J.; Hynes, J. T. 1990a J. Phys. Chem. 94:2736

Kim, H. J.; Hynes, J. T. 1990b J. Chem. Phys. 93:5211

Kim, S. K.; Li, S.; Bernstein, E. R. 1991 J. Chem. Phys. 95:3119

Klemperer, W. 1992 Science 257:887

Klemperer, W.; Yaron, D.; Nelson, D. D. 1988 Faraday Discuss. Chem. Soc. 86:261

Knee, J. L.; Khundkar, L. R.; Zewail, A. H. 1987 J. Chem. Phys. 87:115

Knochenmuss, R.; Chesnovsky, O.; Leutwyler, S. 1988 Chem. Phys. Lett. 144:317

Knochenmuss, R.; Holtom, G. R.; Ray, D. 1993 Chem. Phys. Lett. 215:188, and references therein to older work

Knochenmuss, R.; Smith, D. E. 1994 J. Chem. Phys. 101:7327

Kohn, D. W.; Robles, E. S. J.; Logan, C. F.; Chen, P. 1993 J. Phys. Chem. 97:4936

Kuntz, P. J. 1985 In *Theory of Chemical Reactions Dynamics*, Vol. I, M. Baer, ed., CRC Press, Boca Raton, p. 71

Levine, R. D. 1985 In *Theory of Chemical Reactions Dynamics*, Vol. IV, M. Baer, ed., CRC Press, Boca Raton, p. 1

Levine, R. D.; Bernstein, R. B. 1987 *Molecular Reaction Dynamics and Chemical Reactivity*, Oxford University Press, Oxford

Levy, D. H. 1981 Adv. Chem. Phys. 47:323

Li, S.; Bernstein, E. R. 1991 J. Chem. Phys. 95:1577

Li, S.; Bernstein, E. R. 1992a J. Chem. Phys. 97:804

Li, S.; Bernstein, E. R. 1992b J. Chem. Phys. 97:7383

Li, S.; Bernstein, E. R. n.d. unpublished results

Lin, S. H. 1980 *Radiationless Transitions*, Academic Press, New York

Loison, J. C.; Dedonder-Lardeau, C.; Jouvet, C.; Solgadi, D. 1991 J. Phys. Chem. 95:9192

Marcus, R. A.; Sutin, N. 1985 Biochim. Biophys. Acta 811:265

Mautner, M. J. 1984 J. Amer. Chem. Soc. 106:1257, 1265

Mautner, M. J. 1986 J. Amer. Chem. Soc. 108:6189

McKinnon, W. R.; Hurd, C. M. 1983 J. Phys. Chem. 87:1283

McLendon, G. 1988 Acc. Chem. Res. 21:160

Menapace, J. A.; Bernstein, E. R. 1987a J. Phys. Chem. 91:2533

Menapace, J. A.; Bernstein, E. R. 1987b J. Phys. Chem. 91:2843

Miladi, M.; Le Falher, J.-P.; Roncin, J.-Y.; Damany, H. 1975 J. Mol. Spectrosc. 55:81

Miller, R. E. 1988 Science 240:447

Minsek, D. W.; Chen, P. 1990 J. Phys. Chem. 94:8399

Moreno, P. O.; Shang, Q.-Y.; Bernstein, E. R. 1992 J. Chem. Phys. 97:2869

Morter, C. L.; Wu, Y. R.; Levy, D. H. 1991 J. Chem. Phys. 95:1518

Mukamel, S. 1985 J. Phys. Chem. 89:1077

Mukamel, S.; Jortner, J. 1977 In *Excited States*, Vol. III, E. C. Lim, ed., Academic Press, New York, p. 57

Nelson, D. O.; Fraser, G. T.; Klemperer, W. 1987 Science 238:1670

Nesbitt, D. J. 1988 Chem. Rev. 88:843

Nesbitt, D. J.; Naaman, R. 1989 J. Chem. Phys. 91:3801

Nimlos, M. R.; Young, M. A.; Bernstein, E. R.; Kelley, D. F. 1989 J. Chem. Phys. 91:5268

Outhouse, E. A.; Bickel, G. A.; Dremmer, D. R.; Wallace, S. C. 1991 J. Chem. Phys. 95:6261

Park, Y. D.; Levy, D. H. 1984 J. Chem. Phys. 81:5527

Payzant, J. D.; Cunningham, A. J.; Kebarle, P. K. 1973 Can. J. Chem. 51:3242

Platz, M. S. (ed.) 1990 *Kinetics and Spectroscopy of Carbenes and Biradicals*, Plenum Press, New York

Plusquellic, D. F.; Tan, X. Q.; Pratt, D. W. 1992 J. Chem. Phys. 96:8026

Powers, D. E.; Hopkins, J. B.; Smalley, R. E. 1980 J. Chem. Phys. 72:5721

Pritchard, H. O. 1984 *The Quantum Theory of Unimolecular Reactions*, Cambridge University Press, London

Ramackers, J. J. F.; Krijnen, L. B.; Lips, H. J.; Langelaar, J.; Rettschnick, R. P. H. 1983a Laser Chem. 2:125

Ramackers, J. J. F.; Langelaar, J.; Rettschnick, R. P. H. 1982 In *Picosecond Phenomena III*, K. Eisenthal, R. M. Hochstrasser, W. Kaiser, and A. Laubereau, eds., Springer, Berlin, p. 264

Ramackers, J. J. F.; van Dijk, H. K.; Langelaar, J.; Rettschnick, R. P. H. 1983b Faraday Discuss. Chem. Soc. 75:183

Robin, M. B. 1974 *Higher Excited States of Polyatomic Molecules*, Vol. I, Academic Press, New York

Robin, M. B. 1975 *Higher Excited States of Polyatomic Molecules*, Vol. II, Academic Press, New York

Robin, M. B.; Kuebler, N. A. 1970 J. Mol. Spectrosc. 33:274

Robinson, D. J.; Holbrook, K. A. 1972 *Unimolecular Reactions*, Wiley, New York

Schlachta, R.; Lask, G.; Stangassinger, A.; Bondybey, V. E. 1991 J. Phys. Chem. 95:7132

Schuster, G. B. 1986 Adv. Phys. Org. Chem. 22:311

Shang, Q.-Y.; Bernstein, E. R. 1992 J. Chem. Phys. 97:60

Shang, Q.-Y.; Bernstein, E. R. 1994 Chem Rev. 94:2015–2025

Shang, Q.-Y.; Dion, C.; Bernstein, E. R. 1994a J. Chem. Phys. 101:118

Shang, Q.-Y.; Moreno, P. O.; Bernstein, E. R. 1994b J. Amer. Chem. Soc. 116:302

Shang, Q.-Y.; Moreno, P. O.; Bernstein, E. R. 1994c J. Amer. Chem. Soc. 116:311

Shang, Q.-Y.; Moreno, P. O.; Disselkamp, R.; Bernstein, E. R. 1993a J. Chem. Phys. 98:3703

Shang, Q.-Y.; Moreno, P. O.; Li, S.; Bernstein, E. R. 1993b J. Chem. Phys. 98:1876

Shang, Q.-Y.; Moreno, P. O.; Dion, C.; Bernstein, E. R. 1993c J. Chem. Phys. 98:6769

Shin, S. K.; Chen, Y.; Nickolaisen, S.; Sharp, S. W.; Baudet, R. A.; Wittig, C. 1991 Adv. Photochem. 16:249

Siebrand, W.; Wildman, T. A.; Zgierski, M. Z. 1984 J. Amer. Chem. Soc. 106:4083

Smith, J. M.; Zhang, X.; Knee, J. L. 1993 J. Chem. Phys. 99:2550

Steadman, J.; Syage, J. A. 1990 J. Chem. Phys. 92:4630

Steadman, J.; Syage, J. A. 1991 J. Amer. Chem. Soc. 113:6786

Steadman, J.; Syage, J. A. 1992 J. Phys. Chem. 96:9606

Steinfeld, J. I.; Francisco, J. S.; Hase, W. L. 1989 *Chemical Kinetics and Dynamics*, Prentice Hall, New York

Stephensen, T. A.; Radloff, P. L.; Rice, S. A. 1984 J. Chem. Phys. 81:1060

Stephenson, T. A.; Rice, S. A. 1984 J. Chem. Phys. 81:1083

Sutin, N. 1986 Prog. Inorg. Chem. 30:441

Suto, O.; Steinfeld, J. 1990 Chem. Phys. Lett. 168:181

Swinney, T. C.; Kelley, D. F. 1991 J. Phys. Chem. 95:2430

Swinney, T. C.; Kelley, D. F. 1993 J. Chem. Phys. 99:211, and references therein

Syage, J. A. 1993 J. Phys. Chem. 97:12523

Syage, J. A.; Steadman, J. 1991 J. Chem. Phys. 95:1326, 2497

Takayanagi, M.; Hanazaki, I. 1991 Chem. Rev. 91:1193, and references therein to earlier reviews and works

Tominaga, K.; Walker, G. C.; Jarzeba, W.; Barbara, P. F. 1991 J. Phys. Chem. 95:10475

Trakhtenberg, L. I.; Klochikhin, L.; Pshezhetsky, S. Ya. 1982 Chem Phys. 69:121

Truhlar, D. G. (ed.) 1981 *Potential Energy Surfaces and Dynamical Calculations*, Plenum Press, New York

Truhlar, D. G.; Isaacson, A. D.; Garrett, B. C. 1985 In *Theory of Chemical Reactions Dynamics*, Vol. IV, M. Baer, ed., CRC Press, Boca Raton, p. 65

Tubergen, M. J.; Cable, J. R.; Levy, D. H. 1990 J. Chem. Phys. 92:51

Tubergen, M. J.; Levy, D. H. 1991 J. Phys. Chem. 95:2175

Wanna, J.; Menapace, J. A.; Bernstein, E. R. 1986 J. Chem. Phys. 85:1795

Warren, J. A.; Bernstein, E. R.; Seeman, J. I. 1988 J. Chem. Phys. 88:871

Weber, A. (ed.) 1987 *Structure and Dynamics of Weakly Bound Molecular Complexes*, D. Reidel, Dordrecht

Weber, P. M.; Rice, S. A. 1988a J. Chem. Phys. 88:6082, 6107, 6120

Weber, P. M.; Rice, S. A. 1988b J. Phys. Chem. 92:5470

Wittig, C.; Sharpe, S.; Beaudet, R. A. 1988 Acc. Chem. Res. 21:341

Wittmeyer, S. A.; Topp, M. R. 1993 J. Phys. Chem. 97:8718

6

Reaction Dynamics in Femtosecond and Microsecond Time Windows: Ammonia Clusters as a Paradigm

S. WEI AND A. W. CASTLEMAN JR

6.1. INTRODUCTION

The last decade has seen tremendous growth in the study of gas phase clusters. Some areas of cluster research which have received considerable attention in this regard include solvation (Lee et al. 1980), (Armirav et al. 1982), and reactivity (Dantus et al. 1991; Khudkar and Zewail 1990; Rosker et al. 1988; Scherer et al. 1987). In particular, studies of the dynamics of formation and dissociation, and the changing properties of clusters at successively higher degrees of aggregation, enable an investigation of the basic mechanisms of nucleation and the continuous transformation of matter from the gas phase to the condensed phase to be probed at the molecular level (Castleman and Keesee 1986a, 1988). In this context, the progressive clustering of a molecule involves energy transfer and redistribution within the molecular system, with attendant processes of unimolecular dissociation taking place between growth steps (Kay and Castleman 1983). Related processes of energy transfer, proton transfer, and dissociation are also operative during the reorientation of molecules about ions produced during the primary ionization event required in detecting clusters via mass spectrometry (Castleman and Keesee 1986b), providing further motivation for studies of the reaction dynamics of clusters (Begemann et al. 1986; Boesl et al. 1992; Castleman and Keesee 1987; Echt et al. 1985; Levine and Bernstein 1987; Lifshitz et al. 1990; Lifshitz and Louage 1989, 1990; Märk 1987; Märk and Castleman 1984, 1986; Morgan and Castleman 1989; Stace and Moore 1983; Wei et al. 1990a,b).

The real-time probing of cluster reaction dynamics is a facilitating research field through femtosecond pump–probe techniques pioneered by Zewail and coworkers (Dantus et al. 1991; Khundkar and Zewail 1990; Rosker et al. 1988; Scherer et al. 1987). Some real-time investigations have been performed on metal, van der Waals, and hydrogen-bonded clusters by employing these pump–probe spectroscopic techniques. For example, the photoionization and fragmentation of sodium clusters have been investigated by ion mass spectrometry and zero kinetic energy photoelectron spectroscopy in both picosecond (Schreiber et al. 1992) and femtosecond (Baumert et al. 1992, 1993; Bühler et al. 1992) time domains. Studies have also been made to elucidate the effect of solvation on intracluster reactions. Investigations of the dissociation, recombination, and vibrational energy redistribution of iodine in rare gas clusters have been made in real time on the picosecond (Baumert et al. 1993; Gutmann et al. 1992a,b; Willberg et al. 1992) and femtosecond (Liu et al. 1993; Potter et al. 1992) time scales using laser-induced fluorescence. The ultrafast dynamics of those processes also has been studied by mass spectrometry for iodine solvated by carbon dioxide (Papanikolas et al. 1991, 1992; Roy et al. 1989).

Revealing the dynamics of ionization, proton transfer, and fragmentation of clusters has begun to attract the interests of several groups (Breen et al. 1990; Hineman et al. 1992, 1993; Steadman and Syage 1990, 1991). For example, reactions of naphthol–ammonia and phenol–ammonia clusters, and proton transfer of naphthol clustered with other solvents such as water, methanol, and trimethylamine, have been investigated (Syage and Steadman 1991). Until recently, however, little attention has been paid to the competitive processes of ionization and predissociation in hydrogen-bonded clusters. As the body of knowledge on cluster dissociation continues to increase, the complex issue of interpreting the processes arises. Employing multiphoton ionization (Wei et al. 1992, 1993) and time-of-flight mass spectrometry, our group has commenced a series of studies to elucidate the ionization mechanisms of protonated (Purnell et al. 1993) and unprotonated ammonia clusters (Buzza et al. submitted). In the first part of this chapter, we review the application of femtosecond pump–probe techniques in elucidating the mechanisms involved in the ionization of clusters, using ammonia as an example.

After the ionization event, the ionic clusters are normally formed with enough energy to undergo a number of first-order processes, in particular, those of evaporative dissociation that extends to the microsecond time scale to "cool down" the "hot" ionic cluster species. The cooling of a body which accompanies evaporation from its surface is a familiar phenomenon, and readily explicable in terms of a phenomenological approach. Recent progress in the field of cluster beam research now enables the details of these evaporation processes to be quantitatively investigated at the molecular level (Wei et al. 1990a,b).

Several theoretical models (Engelking 1986, 1987; Klots 1985, 1987b, 1991a,b; Stace 1986) have been presented in order to account for these unimolecular evaporative processes. Klots (1985, 1987, 1991a,b) has elucidated various important factors regarding the formation and energetics of dissociating clusters, or, as they are now termed, evaporative ensembles. From the theoretical models, the relative binding energies of the evaporating cluster ions can be derived from studies of the extent of the evaporation processes over a given period of time, but accurate measurements of decay fractions together with kinetic energy release values are required in order to obtain absolute binding energies.

Until now, most studies of dissociation dynamics of metastable cluster ions have been made using a double-focusing mass spectrometry method (Lifshitz et al. 1990; Lifshitz and Louage 1989, 1990; Stace 1986). As discussed herein, the novel technique of reflectron time-of-flight mass spectrometry is a valuable alternative approach to more standard methods. With carefully designed experiments, it is possible to derive both kinetic energy releases and decay fractions for clusters undergoing unimolecular decomposition following multiphoton ionization. The precision which can be obtained with the method is demonstrated for the studies of ammonia clusters $(NH_3)_nH^+$, $n = 4$–17. By applying both Klots' evaporative ensemble model and Engelking's modified RRK/(quasi-equilibrium theory), it is demonstrated that binding energies of ammonia cluster ions can be derived.

6.2. EXPERIMENTAL

6.2.1. Reflection Time-of-Flight Mass Spectrometry

The apparatus used in these experiments is a reflectron time-of-flight (RETOF) mass spectrometer coupled with a femtosecond laser system, described in detail (Wei et al. 1992). In brief, Figure 6-1 is a schematic of the pulsed molecular beam reflectron time-of-flight mass spectrometer. The femtosecond laser beam described in the next section, or the third harmonic output of a Nd/YAG laser, is directed into the source chamber where it intersects a collimated (through a 1 mm skimmer) molecular beam containing neutral ammonia clusters within the time-of-flight acceleration region (between U_1 and U_2). The ions formed, by the multiphoton ionization (MPI) process, are accelerated in a double electric field and directed towards a reflectron. After traversing a second field-free region, the reflected ions are detected by a microchannel chevron plate (MCP). The ion signals detected by the MCP are directed into a digital oscilloscope coupled to a personal computer.

Several different reflectron operational modes (Wei et al. 1990a,b) are employed in this work. When the reflectron is operated in a hard reflection mode,

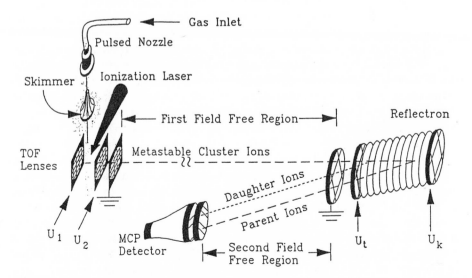

Figure 6-1. Schematic of pulsed molecular beam reflectron time-of-flight mass spectrometer.

where $U_t > U_0$ and $U_k = 0$, both the parent ions and their respective metastable daughter ions are reflected, and are detected simultaneously. Second, if the voltage on U_t is decreased below the parent ion birth potential, U_0, only the daughter ions, D^+, will be reflected through the second field-free region and detected by the MCP. This is commonly referred to as the daughter-only reflection mode, and results in a daughter-only TOF spectrum. A third operational mode, or soft reflection mode, exists when voltages are placed across U_t and U_k. Depending upon the potentials on these two reflecting grids and the kinetic energy of the ions, particular ones will penetrate some distance into the reflecting field. Since the kinetic energy of any specific daughter ion is smaller than its respective parent, the distance into the reflecting field will be smaller. This is reflected by a temporal separation of parent and daughter ions in the resulting TOF spectrum. Often, it is necessary to compare parent and daughter intensities to elucidate dissociation dynamics. Proper comparison has to take into account differences in ion trajectory and collection efficiency of the MCP. In order to correct for these differences, both the parent and daughter ions must be reflected at approximately the same point within the reflectron. This is accomplished by appropriately setting the reflecting potentials and monitoring the arrival times for the parent and daughter ions that are to be compared (Wei et al. 1990a,b).

6.2.2. Femtosecond Pump–Probe Techniques

Figure 6-2 is a schematic of the femtosecond laser system used in the current experiments. Briefly, femtosecond pulses are generated by a colliding pulse mode-locked ring dye laser (CPM), typically centered around 624 nm, approximately 120 fs in duration, with pulse energies in the order of 100–150 pJ per pulse. The pulse energies are too weak for the high photon flux necessary for multiple photon absorption; therefore, three amplification stages are employed. All amplifiers are pumped by the second harmonic (532 nm) of a 30 Hz, injection seeded, Nd:YAG laser (Spectra Physics GCR-5). The pulses are first amplified within a six pass or "bowtie" amplifier, increasing the average pulse energy by five orders of magnitude (average pulse energy, 5–10 μJ per pulse). The second and third stages are 6 and 12 mm bore Bethune cells, respectively. For the majority of this work, the gain dye in all three stages is sulforhodamine 640 dissolved in a 50/50 mixture of methanol and water. Alternate dyes were used to amplify various wavelengths. Amplified pulses have the following characteristics: average pulse energy up to 2 mJ per pulse, spectral bandwidth (10 nm at all employed wavelengths), and pulse duration < 350 fs.

After amplification, the laser beam characteristics are modified for specific experimental requirements. In the investigations into the origin and the subsequent metastable decay of the unprotonated series, the femtosecond beam is immediately directed into the RETOF mass spectrometer. For all other experiments, the beam is split into pump and probe beams. For the Ã state experiments, the pump beam is frequency-tripled and the probe beam is frequency-doubled. This gives pump wavelengths of 214 nm ($v = 0$), 211 nm ($v = 1$), and 208 nm ($v = 2$), and probe wavelengths of 321 nm, 316.5 nm, and 312 nm, respectively. Using a 45° high reflector, coated for the pump wavelength, the beams are separated by reflecting

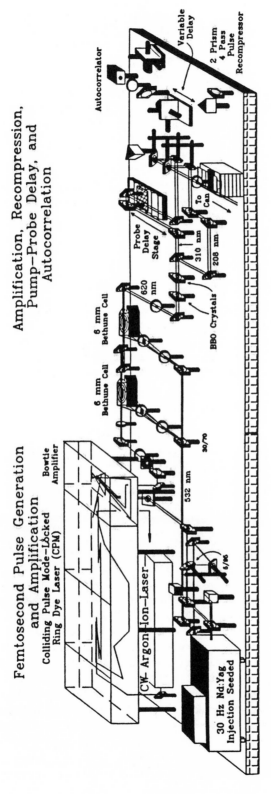

Figure 6-2. Schematic of femtosecond laser system.

201

the pump and transmitting the probe. The probe beam is sent through a delay stage which can be varied from 0.1 μm to 30 cm. Thereafter, the beams are recombined using another 45° high reflector. For the C′($\tilde{B}$) state experiments, the laser beam is split into identical pump and probe beams at a wavelength of 624 nm. A Michelson interferometric arrangement is used to set the time delay between the pump and probe beams. After recombination, the laser beams are focused, with a 50 cm lens, into the interaction region where they intersect the molecular beam containing the neutral ammonia clusters.

The ammonia (NH_3 and ND_3, with minimum purity of 99.98%) used in these experiments was obtained from Union Carbide and Cambridge Isotope Laboratories. These gases were used without further purification.

6.3.   REACTION DYNAMICS ON THE FEMTOSECOND TIME SCALES

6.3.1.   Ionization Mechanisms of Ammonia Clusters

Two mechanisms have been proposed to account for the formation of protonated ammonia clusters under multiphoton resonant ionization conditions. They are absorption–ionization–dissociation (AID) (Echt et al. 1984, 1985; Shinohara and Nishi 1987; Tomoda 1986) and absorption–dissociation–ionization (ADI) (Cao et al. 1984). The absorption–ionization–dissociation mechanism is expressed as follows:

$$(NH_3)_n + h\nu_1 \rightarrow (NH_3)_n^* \tag{1}$$

$$(NH_3)_n^* + h\nu_2 \rightarrow (NH_3)_n^+ + e^- \tag{2}$$

$$(N_3)_n^+ \rightarrow (NH_3)_{n-2} \cdot NH_4^+ + NH_2 \tag{3}$$

The alternative absorption–dissociation–ionization mechanism is expressed as:

$$(NH_3)_n + h\nu_1 \rightarrow (NH_3)_n^* \tag{4}$$

$$(NH_3)_n^* \rightarrow (NH_3)_{n-2}NH_4 + NH_2 \tag{5}$$

$$(NH_3)_{n-2}NH_4 + h\nu_2 \rightarrow (NH_3)_{n-2}NH_4^+ + e^- \tag{6}$$

The ADI mechanism was initially proposed based on theoretical calculations (Cao et al. 1984), and supported (Gellene and Porter 1984) by findings that hydrogenated ammonia clusters can have lifetimes of a few microseconds following neutralization of protonated cluster cations. Recent nanosecond pump–probe studies by Misaizu et al. (1993) also provided some evidence for the ADI mechanism for the case of large clusters ionized through the $\tilde{A}$ state. The fact that protonated ammonia clusters are formed under electron impact and single photon ionization conditions provides evidence that the AID mechanism must be operative at least in some situations. As discussed in the following subsection, the femtosecond pump–probe studies provide a detailed and complete picture for the formation of protonated ammonia cluster cations produced by the ionization of clusters through the $\tilde{A}$ and C′ state of an ammonia molecule. (The various schemes of the pump–probe

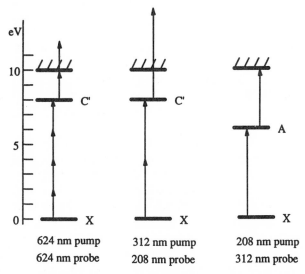

Figure 6-3. Various schemes of the pump–probe experiments relative to the states of the ammonia monomer.

experiments relative to the states of the ammonia monomer are shown in Figure 6-3; it should be noted that the dynamics attributed to the $\tilde{C}$ state may involve the $\tilde{B}$ state as well, since their energy levels are close and lie within the laser bandwidths.)

6.3.2. Femtosecond Reaction Dynamics Through C′ States

The dynamics of ionization of clusters through the C′ state of an ammonia monomer will be considered first. This investigation was carried out using a femtosecond pump–probe technique (Dantus et al. 1991; Khundkar and Zewail 1990; Rosker et al. 1988; Scherer et al. 1987) at 624 nm, where the time delay between the pump and probe beams was set using a Michelson interferometric arrangement. Since ammonia clusters are easily ionized with 1 mJ of light at 624 nm, it was necessary to determine if the observed signals were from a resonant or nonresonant process. For this determination, we carried out studies of the ionization signal versus the laser power dependence. In contrast with the case in typical nanosecond experiments (Wei et al. 1990a,b), fragmentation is not found to be affected by laser fluence, even when varied over two orders of magnitude. It was found (Wei et al. 1992) that a linear relationship between the logarithm of laser power and that of ion intensity is maintained for all cluster ions studied, and for laser powers ranging from the minimum power for observable ionization up to the maximum power obtainable in the current setup. Each measurement revealed a slope of 4 ± 0.2, corresponding to a four-photon process involved in the excitation of the clusters to the resonantly excited electronic state which serves as the intermediate in the ionization process. It is known that the ionization potential (Ceyer et al. 1979) of ammonia is 10.17 eV and the energy (Glownia et al. 1980) corresponding to the C' state of the ammonia monomer is 8.04 eV. (It should be noted that it is also possible that the excitation leads to some population of the $\tilde{B}$ state due to the broad spectral bandwidth of the femtosecond laser. However, for simplicity of the presentation, only the C' state will be mentioned in the text.) Since each photon contains an energy of 2 eV, giving a total of 8 eV, a resonant process is involved. This corresponds to a 4 + 1 or 4 + 2 ionization of the clusters through the C' state, depending on cluster size.

The pump beam excites the clusters to the C' state while the probe beam, at various time delays, ionizes the electronically excited clusters. Figure 6-4 shows a typical pump–probe spectrum of ammonia clusters extending to $n = 4$. The pump and probe beams are of identical wavelength, and in all cases the curve is seen to be symmetrical about zero. Also, although not shown, the baseline of the leading edge (probe before pump) is at the same level as the trailing edge. In order to ascertain the origin of the ion signal existing at long delay times between the pump and probe laser beams, the ion intensities were carefully measured with, and without, the laser beams being individually blocked. The nonzero baseline signal is established to be due to the sum of the ions arising from the pump and probe lasers acting independently. Importantly, as seen from Figure 6-4, the data reveal that the response curves for all ions, at least up to the tetramer, are identical. This suggests that the lifetime of the intermediate state leading to the formation of both the unprotonated and the protonated clusters is the same within the pulse width limit.

It should be noted that the findings of short-lived intermediates for the C' state are also supported by the observation that the leading edge of the data

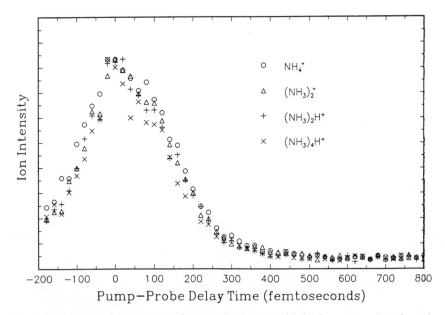

Figure 6-4. Pump–probe spectrum of ammonia clusters with both pump and probe pulses at 624 nm; C' ($v = 1$) of the ammonia molecule.

obtained for studies done through the Ã state, discussed in detail later, does not change with the laser power at a wavelength of 312 nm. With regard to the Ã state experiments shown in Figure 6-5, at low laser power, ionization of the ammonia clusters is achieved through the Ã state only. The leading edge should reflect the extent of the temporal overlap between the pump (208 nm) and the probe (312 nm) pulses. However, at high laser power, ionization can also be achieved through the C' state (two photons of 312 nm laser pulses acting as the pump, 208 nm pulses acting as the probe; see Figure 6-3). In this situation, the leading edge of Figure 6-5 would reflect the convolution of the laser overlap and the dynamics of the C' state. The fact that the leading edge does not change with laser power at 312 nm indicates that the dynamics involved in the C' state are faster than the pulse width limit, in agreement with the 624 nm pump–probe measurements. This finding establishes the fact that failure to observe a long-time tail in the C' state is not due to the inability to ionize any possible NH_4 residing in the cluster with 624 nm pulses.

These findings, shown in Figure 6-4, that lifetimes for the formation of both unprotonated ammonia clusters and protonated ammonia clusters are very short through the C' state can only be explained by the AID mechanism. In considering the ADI mechanism, the neutral species $(NH_3)_nH$, if present, would be formed by the predissociation of ammonia; it would be a long-lived species (microsecond lifetime) (Gellene and Porter 1984). Hence, for the ADI mechanism, the lifetime of the C' state would be expected to be equivalent to the lifetime of the intermediate $(NH_3)_nH$. However, we observe a lifetime of less than 100 fs which could be instrumentally limited. Direct ionization is certainly responsible for the sharp

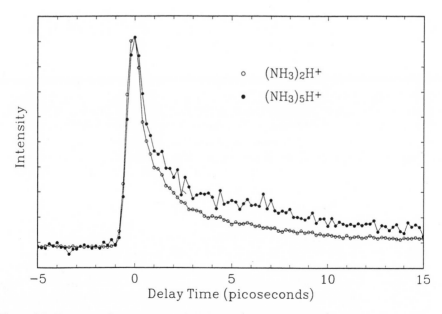

Figure 6-5. Pump–probe spectrum of $(NH_3)_2H^+$ and $(NH_3)_5H^+$ with pump pulses at 208 nm and probe pulses at 312 nm; Ã ($v = 2$) of the ammonia molecule.

intensity peak at $t = 0$, and an initial drop would be expected to be observed irrespective of the mechanism. Predissociation of ammonia would cause the signal intensity to display at least an initially diminishing trend with time. The failure to observe any ionization attributable to that of NH_4 incorporated in the cluster via the predissociation of NH_3 to NH_2 and H, and the subsequent reaction of H and NH_3, eliminates this as the major contributor to the formation mechanism of protonated clusters in the C′ state. This conclusion is reinforced by observations that there is no ionization attributable to NH_4 in the high fluence studies when the pump and probe delays are reversed (leading edge of Figure 6-5, as discussed earlier), establishing a difference between the $\tilde{A}$ and C′ state mechanisms. There are several possible factors that could result in failure to observe the NH_4 species. One possibility is the presence of dissociated channels other than NH_2 + H, such as NH + H_2 as revealed by Simons and coworkers (Quinton and Simons 1982) at higher excitation energy. The other possibility is the formation of high kinetic energy H atoms, due to the large amount of excess energy available at such a high energetic state (Biesner et al. 1989), which results in the low capture rate of the fast H atom and the inability to form the NH_4 species.

6.3.3. Femtosecond Reaction Dynamics Through Ã States

Next, we consider in detail the results for the $\tilde{A}$ state. A series of femtosecond pump–probe experiments were performed at wavelengths corresponding to the Rydberg states $\tilde{A}$ ($v = 0, 1, 2$) of an ammonia monomer (Herzberg 1960; Ziegler 1985). (Note throughout the text that the vibrational levels denote those of unclustered ammonia molecules.) The wavelengths used to access these vibrational levels (in the monomer) were 214 nm, 211 nm, and 208 nm for the pump laser, and 321 nm, 316.5 nm, and 312 nm for the probe laser. For each experiment, the probe beam was appropriately delayed, as discussed earlier.

Figures 6-5 to 6-7 show typical pump–probe spectra of protonated clusters, $(NH_3)_2H^+$ and $(NH_3)_5H^+$, through different vibrational levels of the $\tilde{A}$ state, that is, $v = 2, 1, 0$, respectively. All of these spectra have some features in common, namely, they display a large increase in intensity at $t = 0$ (maximum temporal overlap between the pump and probe pulses), and thereafter a subsequent rapid intensity drop. However, the various spectra do display some noticeable differences with regard to the shape of the falloff region following the initial substantial peak. Except possibly for $v = 0$, when the vibrational energy of the $\tilde{A}$ states increases, the long-time intensity level of all cluster ions increases. More important, the difference between $(NH_3)_2H^+$ and $(NH_3)_5H^+$ becomes evident at higher vibration levels, that is, $v = 1$ and $v = 2$.

At first glance, the results obtained by pumping through the $v = 0$ level (Figure 6-7) seem to be similar to the C′ state results which would seem to eliminate the possibility of NH_4, coming from the predissociation of ammonia, contributing to the mechanism of multiphoton ionization. Yet, careful examination of Figure 6-7 suggests a slight enhancement of the signal above the uncorrelated photon behavior, at long times. This is clearly seen to be the case for $v = 2$, with an even larger contribution for the case of $n = 5$ compared with $n = 2$; see Figure 6-5. It

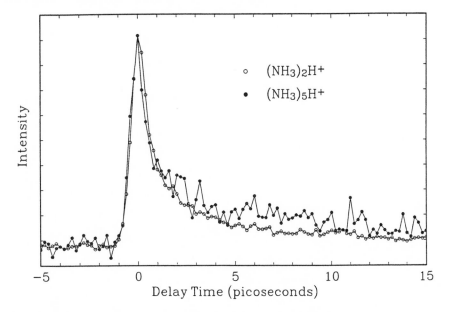

Figure 6-6. Pump–probe spectrum of $(NH_3)_2H^+$ and $(NH_3)_5H^+$ with pump pulses at 211 nm and probe pulses at 317 nm; $\tilde{A}$ ($v = 1$) of the ammonia molecule.

should be noted that due to the pulse width used in these experiments, the leading edge of the pulse profiles may be instrumentally limited. However, the important point is that the presence or absence of the long tail is indicative of the ADI or AID mechanism, respectively.

It is known (Herzberg 1960; Ziegler 1985) that ammonia in the $\tilde{A}$ state predissociates via formation of NH_2 and H. Verification of the predissociation of NH_3 to $NH_2 + H$ being a contributor to the mechanism of protonated cluster formation through the $\tilde{A}$ state is seen from the data at $v = 2$; Figure 6-5. Here, the channel due to predissociation and subsequent ionization becomes readily observable. In the case of $v = 0$, predissociation in the cluster may be endothermic, as revealed by the appearance potential measurements reported in the literature (Misaizu et al. 1993).

In order to account for the time response features observed for the $\tilde{A}$ state, we propose the following dynamical processes.

1. The neutral clusters are excited to the $\tilde{A}$ state through absorption of the first photon,

$$(NH_3)_n + hv_1 \rightarrow (NH_3)_n^+. \tag{7}$$

The excited clusters undergo intracluster reactions as follows.

2. Predissociation of the excited ammonia moiety:

$$(NH_3)_n^* \rightarrow (NH_3)_{n-2} \cdot H_3N \cdot (H \cdots NH_2). \tag{8}$$

3. The intermediate species can lead to the loss of H or NH_2 (not detectable), or

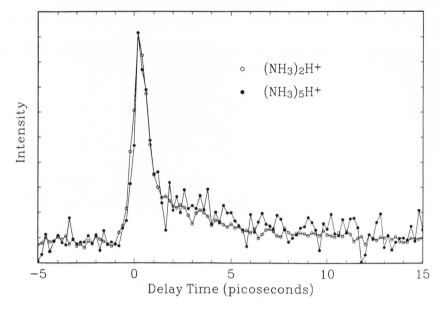

Figure 6-7. Pump–probe spectrum of $(NH_3)_2H^+$ and $(NH_3)_5H^+$ with pump pulses at 214 nm and probe pulses at 321 nm; $\tilde{A}$ ($v = 0$) of the ammonia molecule.

reaction of the H to form NH_4 which is detectable through one-photon ionization with the probe laser:

$$(NH_3)_{n-2} \cdot H_3N \cdot (H \cdots NH_2) \rightarrow (NH_3)_{n-2}(NH_4) + NH_2. \tag{9}$$

4. Ionization of either $(NH_3)_n^*$ or radicals $(NH_3)_{n-2}(NH_4)$ leads to formation of protonated cluster ions:

a. $$(NH_3)_n^* + hv_2 \rightarrow (NH_3)_n^+ + e^- \rightarrow (NH_3)_{n-2}(NH_4)^+ + NH_2 + e^- \tag{10}$$

b. $$(NH_3)_{n-2}(NH_4) + hv_2 \rightarrow (NH_3)_{n-2}(NH_4)^+ + e^-. \tag{11}$$

It should be noted that the rapid intensity drop observed for all protonated cluster ions when $n \geq 2$ is attributed to reaction step 2 [eq. (8)], where the NH_2 or H containing species cannot be readily ionized. Reaction step 3 [eq. (9)] leads to formation of long-lived radicals in accordance with the findings of nonzero ion intensity values at long pump–probe delays observed in the data for the $\tilde{A}$ state. The relative importance of the ionization of the NH_4, for different cluster sizes, in the overall ionization of ammonia through the $\tilde{A}$ state at different vibrational levels is seen by comparing the data for the trimer and hexamer, detected as the protonated dimer and pentamer cluster ions; see Figures 6-5 to 6-7. The overall dependence of the decaying signal intensity on the vibrational levels is indicative of the influence of the energetics on the predissociation and reaction forming NH_4, while the trend in the long-time tail reflects the effects of solvation and retainment of NH_4. Work is in progress to quantitatively model the reaction dynamics involved in the ionization processes of ammonia clusters.

Our results, which show a rapid decay and leveling off to a nonzero value of intensity, suggest that two processes are operating simultaneously in the $\tilde{A}$ state. Since it is known (Herzberg 1960; Ziegler 1985) that ammonia clusters rapidly predissociate into $NH_2 + H$, the rapid decay that we observe would suggest that a similar predissociation is taking place for the clusters. It is also known (Gellene and Porter 1984) that the radicals $(NH_3)_nNH_4$ have long lifetimes (greater than 1 μs), and evidence suggests that formation of these radicals is taking place through intracluster reactions between H and NH_3. This is seen in the leveling off to a nonzero value which persists for longer than 1 ns. Unlike the C′ state, which follows only the AID mechanism, it is evident that the $\tilde{A}$ state competes between both the AID and ADI mechanisms. The AID mechanism, which is the dominant process, is seen when the pump and probe pulses are overlapped ($t = 0$), while the ADI mechanism occurs when the probe photon is absorbed at long time delays.

6.4. REACTION DYNAMICS ON THE MICROSECOND TIME SCALES

6.4.1. Metastable Dissociation Dynamics: Decay Fractions

a. *General considerations*
Following ionization, cluster ions typically undergo extensive evaporative dissociation in time windows of several to tens of microseconds. The excess energy available for this process generally comes from several sources including rearrangement of the ions following the instantaneous ionization of the stable neutral structure, as well as photoabsorption into the ion state during MPI, and possible thermal fluctuations among the many modes of clusters which relay internal energy during the formation process.

The general procedure for measuring the decay fractions is given in subsection 6.2.1. Consider the general reaction for cluster A comprised of n units undergoing evaporative dissociation:

$$A_n^+ \rightarrow A_{n-1}^+ + A. \tag{12}$$

The decay fractions, $D = I_d/(I_d + I_p)$ where I_d and I_p are the daughter and parent ion intensities, respectively, can be measured with high precision by integrating the parent and daughter ion peaks.

b. *Evaporative ensemble model for decay fractions*
The evaporative ensemble model (Klots 1985, 1987b, 1991a,b) assumes that each cluster ion has suffered at least one evaporation before entering the field-free region of the TOF mass spectrometer. For each cluster ion, t_0 is defined as the flight time that the parent ion spends from the ionization region to the last TOF lens, whereas t is the flight time that the parent ion reaches the first grid of the reflectron unit. The value of t_0 can be calculated using the geometry of the TOF lenses and t can be obtained from the TOF spectrum. The evaporative ensemble predicts that at time t the normalized population of daughter ions is given (Klots 1985, 1987b) by

$$D = (C_n/\gamma'^2)\ln\{t/[t_0 + (t - t_0)\exp(-\gamma'^2/C_n)]\} \tag{6-1}$$

$$\gamma'^2 = \gamma^2/[1 - (\gamma/2C_n)^2], \tag{6-2}$$

where C_n is the heat capacity of the cluster ion (in units of Boltzmann constant k_B), and γ is the Gspann parameter. Studies of clusters containing many thousands of atoms by using electron diffraction methods (Farges et al. 1981; Torchet et al. 1983; Valente and Bartell 1983, 1984) suggest that $\gamma \approx 25$, usually independent of cluster size (Klots 1987a).

Using eqs. (6-1) and (6-2) with experimental values of t, t_0, and D, the Gspann parameter γ and heat capacity C_n may be determined. Since the cluster (inter-molecular) modes are much more important than the internal molecular motions in the evaporative dissociation process of a cluster containing n molecules, C_n is chosen to be proportional to $n - 1$.

When one considers that the cluster ions of sizes n and $n + 1$ have different binding energies, as denoted by ΔE_n and ΔE_{n+1}, the daughter ion population can be calculated from the following equation (Klots 1991a,b):

$$D = 1 - (\alpha W_n)^{-1} \ln\{1 + [\exp(\alpha W_n) - 1]t_0/t\}, \tag{6-3}$$

where

$$\alpha W_n = \gamma^2 (W_n/\Delta E_n)/\{C_n[1 - \gamma/C_n + (\gamma/C_n)^2/12\ldots]^2\} \tag{6-4}$$

$$W_n/\Delta E_n = 1 + [(dE/d\Delta E_n)_k^{-1}](\Delta E_n - \Delta E_{n+1})/\Delta E_n \tag{6-5}$$

$$(dE/d\Delta E_n)_k = (C_n/\gamma)[1 - \gamma/2C_n + (\gamma/C_n)^2/12\ldots]. \tag{6-6}$$

By fitting the calculated decay fractions to the measured ones, the values of $(\Delta E_n - \Delta E_{n+1})/\Delta E_n$ are readily obtained.

c. Considering the ammonia system

Ammonia clusters have been studied extensively and provide an excellent test case for these considerations. As discussed later in this chapter, comparison of the deduced thermochemical values is possible by employing data, available in the literature, that was obtained by other methods.

Upon multiphoton ionization of the neutral ammonia clusters, protonated ions are formed via the following internal ion–molecule reactions (Echt et al. 1985; Morgan and Castleman 1989; Wei et al. 1990a,b):

$$NH_3^+ + NH_3 \rightarrow NH_4^+ + NH_2 \qquad \Delta H^0 = -0.74 \text{ eV}. \tag{13}$$

The excess energy within the cluster ion owing to the above exothermic reaction and the relaxation around the newly formed ion, as well as possible further multiphoton excitations, contributes to heating of the cluster ions and concomitant evaporative dissociation. As the cluster ions cool evaporatively, the dissociation extends to longer times, and the present study is directed to an investigation of the metastable dissociation processes of cluster ions $(NH_3)_n H^+$ in the field-free region (where the time window is about one to a few tens of microseconds). In the case of the ammonia cluster ion system, the dissociation process can be expressed as

$$(NH_3)_n H^+ \rightarrow (NH_3)_{n-x} H^+ + x\, NH_3, \tag{14}$$

where $x \geq 1$. Under our usual experimental conditions, and in the absence of collision-induced dissociation processes, we have found (as discussed later) that the dissociation process involves losing only one molecule [i.e., $x = 1$ in reaction (14)] for cluster ions of all sizes ($n = 4$–22).

d. Measurement of decay fractions

A typical hard reflection time-of-flight mass spectrum of ammonia clusters is shown in Figure 6-8(a). When the reflecting voltage is reduced, only the daughter ions are reflected, as shown in Figure 6-8(b). To precisely measure the metastable decay fractions, the soft reflection mode is used for a number of reasons discussed elsewhere (Wei et al. 1990a,b).

As an example, Figure 6-9 displays soft reflection TOF spectra taken at (a) $U_t = 390$ V and $U_k = 3901$ V, and (b) $U_t = 420$ V and $U_k = 4200$ V (the parent ion peaks are labeled as "P" and the corresponding daughter peaks are labeled as "D"). Here, the dissociation process with $n = 14$ in reaction (11) is used as an example. First, the U_t and U_k potentials are set to 420 V and 4200 V, where the overall ion intensity is optimized. As shown in Figure 6-9(b), the arrival time of the parent ion $(NH_3)_{14}H^+$ is 52.22 µs. Since the daughter ions $(NH_3)_{13}H^+$ have an energy of $U_d = (M_d/M_p)U_0$ as a result of metastable decomposition, different trajectories for parent and daughter ions in the reflectron are expected. Corrections for this effect are made as discussed earlier. It is clear that the arrival time (52.22 µs) of the daughter ions $(NH_3)_{13}H^+$ in Figure 6-9(a) is equal to that of their parent ions $(NH_3)_{14}H^+$ in Figure 6-9(b). The integrated intensities of these two peaks are then used to compute the decay fraction, $D = I_d/(I_p + I_d)$ where I_d and I_p are the daughter and remaining parent ion intensity of $(NH_3)_nH^+$, $n = 14$.

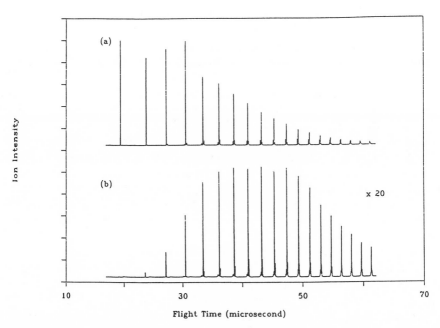

Figure 6-8. Time-of-flight mass spectrum of $(NH_3)_nH^+$, $n = 2$–20, $U_0 = 1625$ V. (a) A "hard reflection" TOF spectrum at $U_t = 2400$ V and $U_k = 0$, (b) a "daughter only" TOF spectrum at $U_t = 1625$ V and $U_k = 0$.

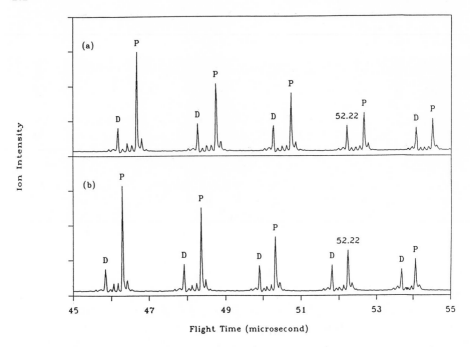

Figure 6-9. Reflectron TOF spectra of $(NH_3)_nH^+$, $n = 11$–15 (P = parent, D = daughter). (a) $U_t = 390$ V and $U_k = 3901$ V, (b) $U_t = 420$ V and $U_k = 4200$ V.

Similar procedure follows for all cluster species of various size, n. The experimentally measured values of the metastable cluster ions $(NH_3)_nH^+$, $n = 4$–23 are plotted as a function of cluster size, as shown in Figure 6-10 (the data points are plotted as $\square$). The experimental uncertainties for each measured point are less than 5%.

The solid line in Figure 6-10 represents the decay fraction of metastable cluster ions $(NH_3)_nH^+$, $n = 4$–23, calculated using eqs. (6-1) and (6-2) with $\gamma = 24.5$ and $C_n = 6(n - 1)$. An alternative way involves: (a) using the bulk heat of evaporation of liquid ammonia and a first estimate of $\gamma = 25$ to estimate Gspann temperature, $T_g = \Delta E_n/(\gamma k_B)$, (b) obtaining the bulk heat capacity of liquid ammonia at T_g, (c) applying this value as C_n to get γ, and (d) repeating processes (a)–(c) if necessary. It was found that values of $\gamma = 24$ and $C_n = 5.5(n - 1)$ give a line which can be superimposed on the solid line in Figure 6-10. The fact that these two methods are consistent indicates that both sets of estimated C_n and γ values are quite plausible.

The general trend agrees well with the experimental measurements; however, it is evident that the predictions from eq. (6-1) are in discrepancy with the experiments for small cluster sizes. As discussed in the following section, this discrepancy results from different binding energies of ammonia cluster ions at various sizes; however, the binding energy difference was neglected in eq. (6-1).

Using eqs. (6-3) to (6-6), it is possible to determine relative bond energies for the clusters from measurements of the decay fractions. The deduced values of

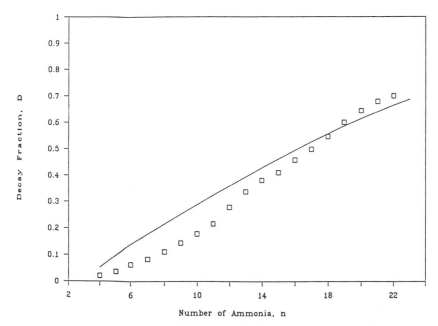

Figure 6-10. A plot of measured decay fractions as a function of cluster size, n: ($\square$) measured points, (——) eq. (6-1).

$(\Delta E_n/\Delta E_{n+1})$ for ammonia cluster ions $(NH_3)_n H^+$, $n = 4$–22, are shown in Figure 6-11. The points designated as " + " are obtained by assuming that $C_n = 6(n - 1)$ and $\gamma = 24.5$. It should be noted that the values of $(\Delta E_n/\Delta E_{n+1})$ are very sensitive to the choice of γ for small cluster sizes; however, the values are insensitive to γ for large ones.

By using the best known literature value of the binding energy of $(NH_3)_5 H^+$ as a starting point, binding energies of $(NH_3)_n H^+$, $n = 4$–22, were calculated and are shown in Figure 6-12 (labeled as "Δ") with $\gamma = 24.5$ and $C_n = 6(n - 1)$. The literature values (designated as "$\bullet$") are also shown in the figure for comparison. It is seen that values from the present approach are in good agreement with the available thermochemical data for $n = 4$–7. An abrupt decrease in the binding energy of $(NH_3)_n H^+$ from $n = 5$–6 is observed. This indicates that $(NH_3)_4 NH_4^+$ is a particularly stable ion and can be pictured as a complete solvation shell formed by four NH_3 molecules bound to a central ammonium ion.

6.4.2. Metastable Dissociations Dynamics: Kinetic Energy Release

a. *Experimental measurements*
The RETOF technique has applications far beyond mass analysis and determination of metastable dissociation fraction. In particular, it provides a very valuable approach to determining kinetic energy release (KER) during evaporative dissociation. As shown in this section, these data also find application in determining absolute values of cluster bond energies for systems without a barrier to

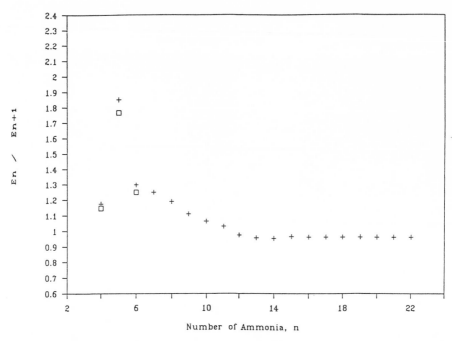

Figure 6-11. Relative binding energies plotted as a function of n. The ($+$) is from eqs. (6-3)–(6-6), the ($\square$) is the literature value (Keesee and Castleman 1986).

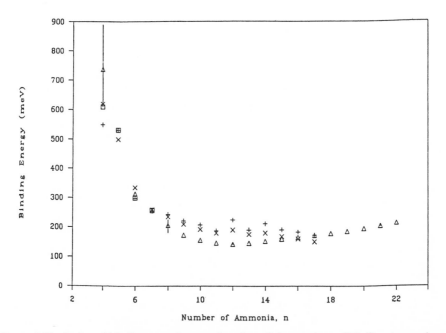

Figure 6-12. A plot of binding energies as a function of cluster size n: ($\square$) literature values (Keesee and Castleman 1986); (Δ) Klots' model, decay fractions; ($\times$) Klots' model, KER; ($+$) Engelking's model, KER.

evaporative metastable dissociation. The technique depends on accurately measuring the profile of the parent and daughter clusters.

In studies of ammonia clusters, all parent and cluster ion peaks are observed to display a Gaussian shape in the TOF spectra. As an example, the experimental data points (10 ns apart from each other) of the normalized peak of the parent ion $(NH_3)_{11}H^+$ (indicated as " + ") and daughter ion $(NH_3)_{10}H^+$ (indicated as " $\bigcirc$ ") are fitted to the pure Gaussian curves (indicated as solid lines) as shown in Figure 6-13. Clearly, the daughter ion peak is broader than the parent ion peak, due to the kinetic energy release involved in the decomposition process. A small tail on the later time side of the parent cluster ion peak is observed, resulting from rapid ion fragmentation in the TOF acceleration zone. This tail does not affect our KER measurements for cluster size up to 17 because the width used to compute the average kinetic energy release value is measured at 22% of the peak height. However, it was found that the tail became more noticeable for peaks of larger parent cluster ions, due to more dissociation in the TOF lens region.

With a measured parent ion birth potential U_0, an estimated daughter ion traveling distance L (distance from the position where the daughter ion is born to the ion detector) in the field-free drift region, and a broadening width W_t observed in the TOF spectra, the average kinetic energy release is given (Wei et al. 1990a,b)

$$\langle E_r \rangle = 1/\langle L_2 \rangle [U_0 W_t][M_d/(2M_p M)], \tag{6-7}$$

where M_p, M_d, and M are the masses of the parent ion, daughter ion, and neutral

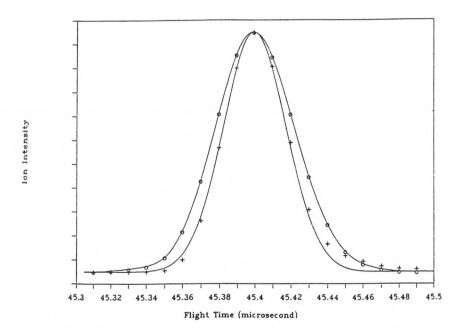

Figure 6-13. Peak shapes of (+) parent $(NH_3)_{11}H^+$ and ($\bigcirc$) daughter $(NH_3)_{10}H^+$ fitted by (——) Gaussian functions.

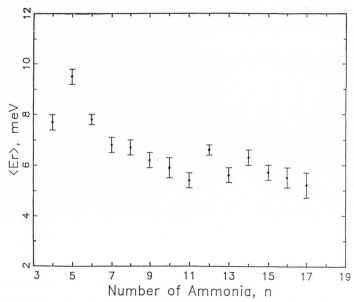

Figure 6-14. A plot of the measured average KER as a function of n.

fragment, respectively. The values deduced by this method are shown in Figure 6-14.

b. *Evaporative ensemble model for kinetic energy release*

The evaporative ensemble approach suggests that the evaporation energy can also be calculated from the measured kinetic energy release with a properly chosen heat capacity and Gspann parameter. It follows that the binding energy of a molecule in a cluster ion of size n can be calculated (Schreiber et al. 1992) using the following equation:

$$\Delta E_n = \gamma \langle E_r \rangle / [1 - (\gamma/2C_n)]. \tag{6-8}$$

Figure 6-12 displays the binding energies of $(NH_3)_nH^+$, $n = 4–17$, as a function of cluster size n (labeled as " $\times$ ") using the measured KER and $\gamma = 24.5$, $C_n = 6(n - 1)$, and compares the data with prior measurements based on high pressure mass spectrometry and values deduced from decay fractions.

c. *QET/RRK theory and approach*

Engelking (1986, 1987) has also proposed a modified QET/RRK statistical model to determine the binding energy of a molecule within a metastable cluster ion, and he has suggested that it can be determined from the evaporative lifetime and the average kinetic energy release. In the present work, we adopt this simple theoretical model, and, by employing the measured average kinetic energy release values, we obtain the binding energies of the protonated ammonia cluster ions $(NH_3)_nH^+$, $n = 4–17$. The binding energy of $(NH_3)_nH^+$ can be calculated using the following equation:

$$E_n^* = A(s - 1)[C/(s - 1)\langle E_r \rangle(s - 2)/(s - 1) - \langle E_r \rangle], \tag{6-9}$$

where $\langle E_r \rangle$ is the measured kinetic energy release, s is the cluster mode (number of internal motions), A is 0.5 (model scaling parameter), given by Engelking (1986, 1987) and

$$C = 16\pi v^3 \mu g S / \Gamma_n.\qquad(6\text{-}10)$$

In order to calculate the constant C, one needs to obtain information on v (vibrational frequency of cluster mode), μ (reduced mass), g (remaining channel degeneracy of the ejected neutral fragment), S (geometrical cross section for forming cluster ion), and Γ_n (unimolecular dissociation rate). From eq. (6-3), it is known that the dissociation rate and the kinetic energy release are the only required experimental values for obtaining the binding energy of a metastable cluster ion. However, the metastable process can be observed only if the lifetimes of the dissociating cluster ions fall within the experimental time window. Since the decay rate depends upon a high power $(s - 2)$ of the internal energy, even a very broad range of experimental lifetimes can imply a narrow range of internal energies of the metastable ions. This point can be further visualized in eq. (6-9). The lifetime is included in the constant C which is taken to a large root. Therefore, the variation of dissociation rate constant does not greatly affect the determination of the binding energy.

In computing the binding energies of $(NH_3)_n H^+$, the cluster modes including rotations and translations of each monomer are treated actively, providing $6(n - 1)$ degrees of freedom. The channel degeneracy g is set to $n - 1$ and the vibrational frequencies are set to $100\ \text{cm}^{-1}$ for $(NH_3)_n H^+$, and $50\ \text{cm}^{-1}$ for $(NH_3)_n H^+$, $n = 6\text{–}17$, considering that the cluster ion $(NH_3)_5 H^+$ forms a closed solvation shell. The geometrical cross section S is chosen to be $100\ \text{Å}^2$. The lifetimes $(1/\Gamma_n)$ of the metastable cluster ions are estimated from the observed ion arrival times, assuming that the ion dissociates in the region midway between the last TOF lens and the first plate of the reflectron unit in our time-of-flight mass spectrometer. They are in the range of $8\text{–}20\ \mu\text{s}$ for $(NH_3)_{17} H^+$, $n = 4\text{–}17$.

It is necessary to choose a single proper model scaling parameter (A) such that the computed binding energies of ammonia cluster ions match the literature values. Figure 6-12 shows the determined binding energies of $(NH_3)_n H^+$, $n = 4\text{–}17$, as a function of cluster size n for $A = 1$ (shown as " + " in the figure). All formula parameters remain the same for cluster ions of all sizes, except the reduced mass, lifetime, channel degeneracy, cluster mode, and measured average kinetic energy release. The deduced binding energy values are scaled to match the best determined value for $(NH_3)_5 H^+$ in the literature (Keesee and Castleman 1986). The binding energies of the protonated ammonia cluster ions, determined using eq. (6-3), are relatively insensitive to the choices of the vibrational frequency and lifetime. A 50% variation in the chosen vibrational frequencies gives rise to about $10\text{–}15\%$ change in the calculated binding energy. In addition, the binding energies change about $5\text{–}10\%$ when lifetime is varied from the earliest to the latest time window: $0.6\text{–}18\ \mu\text{s}$ for 4-mer and $1.2\text{–}36\ \mu\text{s}$ for 17-mer.

As seen from Figure 6-12, it is evident that binding energies determined from different methods are in good agreement with the known thermochemical data for $n = 4\text{–}7$. However, more important, the present approach enables binding energies to be determined for larger clusters, which are very difficult to obtain using the more traditional high pressure mass spectrometry technique.

6.5. SUMMARY AND PERSPECTIVES FOR THE FUTURE

Based on the studies presented, it is clear that ultrafast lasers, in conjunction with supersonic expansion, time-of-flight reflectron methods, reveal a new dimension in the dynamics of intracluster reactions in hydrogen-bonded systems. As shown in this chapter, femtosecond pump–probe techniques have been able to resolve controversies on the mechanisms of ionization of ammonia clusters, in particular. Prior to the ultrafast laser studies, the formation of protonated ammonia clusters, following ionization of the neutral cluster species, remained largely the subject of speculation. Through pump–probe experiments in the C′ state, the present work has established lifetimes of less than 100 fs for the species $(NH_3)_n^+$ and $(NH_3)_nH^+$ ($n = 1$–4), and also established that the AID mechanism is the solely operative one. By contrast, the $\tilde{A}$ state ($n = 2$–5) data display two unique features with respect to the pump–probe delays: a fast decay process is observed, followed by a leveling off to a nonzero value of ion intensity which is a function of the vibrational state excited during the excitation, as well as a function of the size of the cluster. Hence, ionization through the $\tilde{A}$ state occurs by the AID mechanism at short times, and the ADI mechanism for longer time delays between the pump and probe lasers.

The importance of using a reflectron time-of-flight mass spectrometer in the studies of metastable unimolecular dissociation dynamics is that it enables simultaneous determination of the decay fractions and quantitative values of kinetic energy release. Here, we present a precise method for measuring both kinetic energy releases and decay fractions of metastable cluster ions, with corrections concerning instrumental artifacts and the ion trajectory of the parents and daughters. The experimental data are used to derive the Gspann parameter and heat capacity of clusters as described in Klots' evaporative ensemble model. With these two parameters, we apply the evaporative ensemble model to obtain binding energies of ammonia cluster ions $(NH_3)_nH^+$. The deduced binding energy values are found to be in very good agreement with each other and with the thermochemical data.

For small cluster sizes, the values of relative binding energies employing the approach using measurements of decay fractions are very sensitive to choices of the Gspann parameter and heat capacity. It is therefore evident that the method based on kinetic energy release measurement is the preferred one for small cluster sizes. By contrast, the method using decay fractions is more applicable to the larger cluster sizes, where it is comparably insensitive to the choice of parameters which must be employed in deducing the values from the theoretical model. Most important, the present approach enables binding energies to be determined for larger clusters, which are very difficult to obtain using traditional high pressure mass spectrometry techniques. In addition to the above, the reflectron TOF technique enables studies of other processes such as delayed reaction, delayed ionization, and the existence of isomeric structures in cluster ions. It can be expected that these processes will be the subject of continuing investigations, in view of their fundamental importance and the prospects for their study offered by recent experimental advances.

In the future, we can expect many further advances in an understanding of reactions in clusters, including details of the real-time dynamics of proton transfer and the energy disposal following the Coulomb explosion of highly charged systems, through femtosecond pump–probe experiments in conjunction with reflectron time-of-flight mass spectrometry techniques. Also, the origin of un-protonated clusters, and the energy input accompanying ionization over the spectral region associated with the small temporal pulse widths of these experiments, is a subject warranting further study. Evidently, the region of the ionic potential energy accessed during vertical ionization does not change appreciably, which raises interesting questions for the ionization of a wide range of hydrogen-bonded systems.

REFERENCES

Amirav, A.; Even, U.; Jortner, J. 1982 J. Phys. Chem. 86:3345

Baumert, T.; Rüttgermann, C.; Rothenfusser, C.; Thalweiser, R.; Weiss, V.; Gerber, G. 1992 Phys. Rev. Lett. 69:1512

Baumert, T.; Thalweiser, R.; Gerber, G. 1993 Chem. Phys. Lett. 209:29

Begemann, W.; Meiwes-Broer, K. H.; Lutz, H. O. 1986 Phys. Rev. Lett. 56:2248

Biesner, J.; Schneider, L.; Ahlers, G.; Xie, X.; Welge, K. H.; Ashfold, M. N. R.; Dixon, R. N. 1989 J. Chem. Phys. 91:2901

Boesl, U.; Weinkauf, R.; Schlag, E. W. 1992 Int. J. Mass Spectrom. Ion Phys. 112:121

Breen, J. J.; Peng, L. W.; Willberg, D. M.; Heikal, A.; Cong, P.; Zewail, A. H. 1990 J. Chem. Phys. 92:805

Bühler, B.; Thalweiser, R.; Gerber, G. 1992 Chem. Phys. Lett. 188:247

Buzza, S. A.; Wei, S.; Purnell, J.; Castleman, A. W., Jr. 1995 J. Chem. Phys. 102:4832

Cao, H.; Evleth, E. M.; Kassab, E. 1984 J. Chem. Phys. 81:1512

Castleman, A. W., Jr.; Keesee, R. G. 1986a Acc. Chem. Res. 19:413

Castleman, A. W., Jr.; Keesee, R. G. 1986b Chem. Rev. 86:589

Castleman, A. W., Jr.; Keesee, R. G. 1987 In Structure/Reactivity and Thermochemistry of Ions, NATO ASI Series, P. Ausloos and S. G. Lias, eds. D. Reidel, Dordrecht, Boston, Lancaster, Tokyo, pp. 185–217

Castleman, A. W., Jr.; Keesee, R. G. 1988 Science 241:36

Ceyer, S. T.; Tiedemann, P. W.; Mahan, B. H.; Lee, Y. T. 1979 J. Chem. Phys. 70:14

Dantus, M.; Janssen, M. H. M.; Zewail, A. H. 1991 Chem. Phys. Lett. 181:281

Echt, O.; Dao, P. D.; Morgan, S.; Castleman, A. W., Jr. 1985 J. Chem. Phys. 82:4076

Echt, O.; Morgan, S.; Dao, P. D.; Stanley, R. J.; Castleman, A. W., Jr. 1984 Ber. Bunsen-Gesellschaft Phys. Chem. 88:217

Engelking, P. C. 1986 J. Chem. Phys. 85:3103

Engelking, P. C. 1987 J. Chem. Phys. 87:936

Farges, J.; de Peraudy, M. F.; Raoult, B.; Torchet, G. 1981 Surf. Sci. 106:95

Gellene, G. I.; Porter, R. F. 1984 J. Phys. Chem. 88:6680

Glownia, J. H.; Riley, S. J.; Colson, S. D.; Nieman, G. C. 1980 J. Chem. Phys. 72:5998

Gutmann, M.; Willberg, D. M.; Zewail, A. H. 1992a J. Chem. Phys. 97:8037

Gutmann, M.; Willberg, D. M.; Zewail, A. R. 1992b J. Chem. Phys. 97:8048

Herzberg, G. 1960 Molecular Spectra and Molecular Structures, Vol. 3, Electronic Spectra and Electronic Structure of Polyatomic Molecules, Van Nostrand Reinhold, New York, pp. 463–466

Hineman, M. F.; Brucker, G. A.; Kelley, D. F.; Benstein, E. R. 1992 J. Chem. Phys. 97:3341

Hineman, M. F.; Kelley, D. F.; Bernstein, E. R. 1993 J. Chem. Phys. 99:4533

Kay, B. D.; Castleman, A. W., Jr. 1983 J. Chem. Phys. 78:4297

Keesee, R. G.; Castleman, A. W., Jr. 1986 J. Phys. Chem. Ref. Data 15:1011

Khundkar, L. R.; Zewail, A. H. 1990 Annu. Rev. Phys. Chem. 41:15

Klots, C. E. 1985 J. Chem. Phys. 83:5854

Klots, C. E. 1987a Nature 327:222

Klots, C. E. 1987b Z. Phys. D 5:83

Klots, C. E. 1991a Z. Phys. D 20:105

Klots, C. E. 1991b Z. Phys. D 21:335

Lee, N.; Keesee, R. G.; Castleman, A. W., Jr. 1980 J. Colloid Interface Sci. 75:555

Levine, R. D.; Bernstein, R. B. 1987 *Molecular Reaction Dynamics and Chemical Reactivity*, Oxford University Press, Oxford

Lifshitz, C.; Iraqi, M.; Peres, T. 1990 Rap. Comm. Mass Spectrom. 4:485

Lifshitz, C.; Louage, F. 1989 J. Phys. Chem. 93:5633

Lifshitz, C.; Louage, F. 1990 Int. J. Mass Spectrom. Ion Proc. 101:101

Liu, Q.; Wang, J. K.; Zewail, A. H. 1993 Nature 364:427

Märk, T. D. 1987 Int. J. Mass Spectrom. Ion Proc. 79:1

Märk, T. D.; Castleman, A. W., Jr. 1984 Adv. At. Mol. Phys. 20:65

Märk, T. D.; Scheier, P.; Leiter, K.; Ritter, W.; Stephan, K.; Stamatovic, A. 1986 Int. J. Mass Spectrom. Ion Proc. 74:281

Misaizu, F.; Houston, P. L.; Nishi, N.; Shinohara, H.; Kondow, T.; Kinoshita, M. 1993 J. Chem. Phys. 98:336

Morgan, S.; Castleman, A. W., Jr. 1989 J. Phys. Chem. 93:4544

Papanikolas, J. M.; Gord, J. R.; Levinger, N. E.; Ray, D.; Vorsa, V.; Lineberger, W. C. 1991 J. Phys. Chem. 95:8028

Papanikolas, J. M.; Vorsa, V.; Nadal, M. E.; Campagnola, P. J.; Gord, J. R.; Lineberger, W. C. 1992 J. Chem. Phys. 97:7002

Potter, E. D.; Liu, Q.; Zewail, A. H. 1992 Chem. Phys. Lett. 200:605

Purnell, J.; Wei, S.; Buzza, S. A.; Castleman, A. W., Jr. 1993 J. Phys. Chem. 97:12530

Quinton, A. M.; Simons, J. P. 1982 J. Chem. Soc. Faraday Trans. 2 78:1261

Ray, D.; Levinger, N. E.; Papanikolas, J. M.; Lineberger, W. C. 1989 J. Chem. Phys. 91:6533

Rosker, M. J.; Dantus, M.; Zewail, A. H. 1988 J. Chem. Phys. 89:6113

Scherer, N. F.; Khundkar, L. R.; Bernstein, R. B.; Zewail, A. H. 1987 J. Chem. Phys. 87:1451

Schreiber, E.; Kühling, H.; Kobe, K.; Rutz, S.; Wöste, L. 1992 Ber. Bunsenges. Phys. Chem. 96:1301

Shinohara, H.; Nishi, N. 1987 Chem. Phys. Lett. 141:292

Stace, A. J. 1986 J. Chem. Phys. 85:5774

Stace, A. J.; Moore, C. 1983 Chem. Phys. Lett. 96:80

Steadman, J.; Syage, J. A. 1990 J. Chem. Phys. 92:4630

Steadman, J.; Syage, J. A. 1991 J. Phys. Chem. 95:10326

Syage, J. A.; Steadman, J. 1991 J. Chem. Phys. 95:2497

Tomoda, S. 1986 Chem. Phys. 110:431

Torchet, G.; Schwartz, P.; Farges, J.; de Feraudy, M. F.; Raoult, B. 1983 J. Chem. Phys. 79:6196

Valente, E. J.; Bartell, L. S. 1983 J. Chem. Phys. 79:2683

Valente, E. J.; Bartell, L. S. 1984 J. Chem. Phys. 80:1451, 1458

Wei, S.; Purnell, J.; Buzza, S. A.; Castleman, A. W., Jr. 1993 J. Chem. Phys. 99:755

Wei, S.; Purnell, J.; Buzza, S. A.; Stanley, R. J.; Castleman, A. W., Jr. 1992 J. Chem. Phys. 97:9480

Wei, S.; Tzeng, W. B.; Castleman, A. W., Jr. 1990a J. Chem. Phys. 92:332

Wei, S.; Tzeng, W. B.; Castleman, A. W., Jr. 1990b J. Chem. Phys. 93:2506

Willberg, D. M.; Gutmann, M.; Breen, J. J.; Zewail, A. H. 1992 J. Chem. Phys. 96:198

Ziegler, L. D. 1985 J. Chem. Phys. 82:664

7

Magic Numbers, Reactivity, and Ionization Mechanisms in Ar$_n$X$_m$ Heteroclusters

GOPALAKRISHNAN VAIDYANATHAN AND JAMES F. GARVEY

7.1. INTRODUCTION

During the past decade, there has been an enormous increase in experimental and theoretical studies directed toward obtaining a fundamental understanding of the properties of van der Waals (vdW) clusters (Castleman 1990; Castleman and Keesee 1986b, 1988b; Garvey et al. 1991; Goyal et al. 1993; Janda 1985; Jortner 1984; Levy 1981; Märk 1987; Märk and Castleman 1985; Ng 1983; Stace 1992). As the term "cluster" is frequently used in the scientific community with different connations, it would be appropriate to define "cluster" in the context of this field. Gas phase clusters are defined as finite gas phase aggregates composed of two to several million components (i.e., atoms or molecules). These species are held together by different types of forces, ranging from the weak van der Waals forces all the way up to strong electrostatic forces. A techique of classifying clusters based on the type of the binding forces has been developed. Another convenient basis for the classification of clusters is according to the size, such that clusters with $n = 2$–10 or 13 have been termed microclusters, clusters with $n = 10$–10^2 have been referred to as small clusters, and aggregates with $n \geq 10^2$ are called large clusters (Jortner 1984).

Clusters composed of a few molecules can be treated, to a first approximation, as isolated gas phase species, while clusters with sizes $n > 10^3$ begin to exhibit properties resembling those of condensed or bulk materials. Between these two extremes lies the regime where cluster systems cannot be adequately treated by either the molecular concepts or the conventional solid-state approach. Thus, clusters have previously been described as the conceptual bridge linking the gas and the condensed phases (bulk liquids or solids) (Castleman 1990; Castleman and Keesee 1986b, 1988b; Garvey et al. 1991; Goyal et al. 1993; Janda 1985; Jortner 1984; Levy 1981; Märk 1987; Märk and Castleman 1985; Ng 1983; Stace 1992).

Traditionally, researchers have concentrated on understanding these two forms of matter as separate entities and it is only in the last few years that fundamental questions have been posed [i.e., what is the minimum number of atoms required to exhibit an electric conduction band or the number of atoms or molecules necessary to show crystal structure similar to the bulk phase? (Recknagel 1984)]. Gas phase clusters offer the physicist an ideal model system to investigate the evolution of electronic properties from single particles to the bulk materials. The atomic clusters of rare gases are being investigated to understand the transition from localized single-particle excitation to excitons (Moller and Zimmerman 1989). The development of metallic behavior with nonmetallic atoms has also received considerable attention (Rademann et al. 1987).

For the chemist, clusters offer a unique medium to investigate the effects of solvation on chemical reactivity (Brutschy 1992; Coolbaugh and Garvey 1992, 1994; Stace 1992). Traditional gas phase studies have concentrated on under-standing the intrinsic structural factors, in order to rationalize the variations in the reaction pathways for molecules. Bimolecular gas phase reactions have been found to be significantly different from the reactions in condensed phase, and this has been typically attributed to "solvation effects" (Braumann et al. 1971). Thus, studies aimed at investigating the variations in the chemical reactivity with increasing cluster size enable, at least in principle, observation of the transition from bimolecular reactivity to reactions in the bulk medium (Brutschy 1992; Coolbaugh and Garvey 1992; Stace 1992).

However, chemical reactions observed in vdW clusters can be significantly different from those of two isolated molecules, as several new effects may be introduced (Naaman 1988). In clusters, atoms or molecules other than the reactant can act as a catalyst by introducing additional terms in the reactive potential. The presence of the solvent molecules may lead to new reaction pathways, or to an increase in the density of states for the reacting complex, which will lengthen its lifetime and thus enable complicated multistep processes to occur. The reactivity in vdW clusters has been shown to exhibit a pronounced dependence on both the number of solvent molecules and their nature, as well as the amount of energy deposited into the cluster. One area of intense study has been that of energy deposition and its randomization within the cluster (Brutschy et al. 1987; Dao and Castleman 1986; Dao et al. 1985; Fund et al. 1981; Märk et al. 1992; Stace 1987b). Last, it is also possible that endothermic reactions which are not normally observed in the gas phase may be observed in the solvent environment. That is, endothermic reactions may be initiated by other exothermic reactions occurring within the same cluster (Coolbaugh et al. 1990). The study of factors which govern the structures of finite clusters may provide an understanding of the microscopic structure of bulk solvents. Thus, clusters can be used as simple models for understanding energetics and dynamics in condensed phases from a microscopic point of view (Brutschy 1990) and may eventually provide an understanding in areas as diverse as atmospheric sciences, catalysis, combustion engineering, and materials research (Castleman 1990; Takagi 1986).

Although gas phase clusters can be generated by a number of techniques, adiabatic expansions are the most widely utilized method for the generation of vdW clusters. A wealth of information regarding the energetics, dynamics, and structures of clusters has been recently obtained due to the availability of many new and improved experimental techniques. Although a number of spectroscopic techniques have been utilized in investigating vdW clusters, mass spectrometry (MS) is extensively employed for the study of clusters, as it enables size selective studies of cluster ions to be conducted with very high sensitivity. Recent developments in tandem MS techniques enable truly size selective studies to be performed on cluster ions, and a considerable amount of information regarding the stability, structure, and reactivity of cluster ions has been obtained. The study of "intracluster ion–molecule reactions" is one of the areas that is being intensively pursued by a number of groups (Castleman and Keesee 1986a; Coolbaugh et al. 1993; Garvey and Bernstein 1986a; Wei et al. 1991b,c).

Several aspects of the binary cluster ion system $Ar_n X_m^+$, where X is either an atom or a polyatomic molecule, will be discussed in this chapter. These studies are of interest for the following reasons.

(1) The study of vdW cluster ions consisting of an atomic or a molecular ion surrounded by an increasing number of polarizable species, such as rare gas atoms, offers an elegant approach to the physical description of the solvation phenomenon of ions in the gas phase. The study of solvated homogeneous cluster ions is complicated by the fact that the charge may be delocalized over the entire cluster. However, when the ionization potential of the dopant atom or molecule (X) is lower than that of Ar/Ar_n, then one can expect the positive charge to reside on the X_m^+ ion. The atomic or molecular ion will either be surrounded by the argon atoms in certain arrangements or lie on, or close to, the surface of the argon cluster. Extensive studies of rare gas cluster (RGC) ions, including Ar_n^+, has led to the conclusion that they have icosahedral structures. The enhanced stability of certain sizes of RGC ions arises from the formation of completed solvent shells (Echt et al. 1981, 1990). The effects of the nature number, size, and symmetry of X on the stability and structural arrangements can therefore be investigated in binary cluster ions (Stace 1983, 1984, 1985a,b, 1987c; Ozaki and Fukuyama 1985).

(2) An additional motivation for investigations of binary cluster ions is the potential to understand the influence of the solvating component on the ion–molecule chemistry, that is, observe whether the product cluster ion sequences observed in mixed expansions are affected by the identity of the carrier gas.

(3) Mass spectrometric investigations of clusters begin by the ionization of the neutral clusters. The mechanisms of charge and energy transfer are of fundamental importance to obtain a better understanding of the ionization and fragmentation processes within the cluster. Although ionization is primarily an electronic process, the subsequent ion–molecule and fragmentation reactions, which ultimately give rise to the cluster mass spectrum, require that this electronic energy be converted to nuclear motion. Mass spectrometric investigations alone are not sufficient for the isolation and identification of the charge and energy transfer steps in homoclusters (Stace 1983, 1984, 1985a,b, 1987c). However, in heteroclusters, it may be possible to isolate and identify at least some of the individual steps involved in charge and energy transfer (Stace 1983, 1984, 1985a,b, 1987c; Ozaki and Fukuyama 1985).

(4) The measurement of the ionization efficiency (IE) curves of cluster ions has been useful in observing the change in the appearance energy (AE) of cluster ions as a function of cluster size. The IE curves of cluster ions also show resonance-like structures, as well as distinct thresholds at various ionization energies. These features have been attributed to various charge and energy transfer processes and provide insight into the dominant ionization mechanisms in clusters

of various sizes (Brutschy 1990; Brutschy et al. 1987; Märk et al. 1992; Vaidyanathan et al. 1991a).

7.2. METHODOLOGY

7.2.1. Generation of van der Waals Clusters

The method of choice for the generation of vdW clusters utilizes supersonic expansions. In this technique, the species to be clustered are allowed to expand from a high pressure to a low pressure region through a molecular beam nozzle. The basic principles of adiabatic expansion have been the focus of a number of reviews (Hagena 1974, 1987; Scoles 1988) and only the pertinent aspects will be described here.

Adiabatic expansions normally produce internally cold molecules and clusters of atoms and/or molecules. In the early part of the expansion, a large number of many-body collisions occur, leading to the formation of aggregates. The terminal temperature attained depends on a number of parameters, which include stagnation pressure, stagnation temperature, the nozzle-to-skimmer distance, the nozzle geometry, and the nature and concentration of the carrier gas. The main advantage of utilizing a supersonic expansion is the ease with which an intense cluster beam can be generated; the inherent drawback is that a wide range of cluster sizes are produced. The width of the cluster distribution is strongly dependent on the experimental conditions employed for the expansion (Scoles 1988). The overall cluster size distribution can be qualitatively manipulated by a judicious choice of the nozzle diameter, stagnation pressure, and expansion temperature, as well as the mixing ratio of the various components for the generation of heteroclusters. Large nozzle diameters, high stagnation pressures, and low stagnation temperatures shift the cluster distrubution towards larger cluster sizes. It has been observed that conditions favoring the formation of larger clusters also lead to a broader distribution of cluster sizes (Scoles 1988).

Heteroclusters can be generated by using seeded expansions or by employing a pickup cell. Seeded expansions involve expanding a dilute mixture of one or more of the components in the carrier gas, whereas in the pickup experiments the neat cluster beam is passed through a gas cell where the other component is present in controlled amounts. In the case of seeded expansions, the gas samples may be premixed or mixed in real time using flowmeters. Flow control allows for greater control of the cluster size distribution that is generated. When one of the components is a liquid, a carrier gas can be bubbled through a reservoir containing the liquid. Controlling the temperature of the liquid in the reservoir allows the vapor pressure to be regulated, and thus the overall composition of the cluster may be manipulated more effectively.

7.2.2. Detection of Clusters

The study of neutral clusters has posed several technological difficulties which have hindered the development of this field. The main problem has been to identify and isolate neutral clusters of a single size from the cluster size distribution produced in supersonic expansions. Several methods have been employed to solve this problem partially. One of the techniques of physically separating the neutral clusters was based on the ability to focus or defocus molecules with a dipole moment or high polarizability using an electric field gradient. This technique proved useful in separating the monomer from the dimer and, in exceptional cases, the trimer. The dipole moment decreases rapidly with size due to dipole–dipole interactions within the cluster, and may result in the cancelation of the net dipole moment in larger clusters (Novick et al. 1973). The other technique is to physically separate clusters of different sizes via ionization, followed by charge exchange, but this method is hindered by the fragmentation that accompanies the neutralization (Fayet et al. 1985).

In spectroscopic studies of clusters, it is possible in many cases to identify a particular optical spectrum with a particular cluster size, especially for the smaller cluster sizes. However, when reactions within clusters are also being investigated, there are usually no "fingerprint regions" in the spectrum of the product that would allow the unique identification of the size of the parent cluster from which the product was formed.

Ionized clusters are easier, in this respect, in that identification and isolation of the cluster ion is possible, prior to the reaction, by mass selecting the cluster ion of interest. As a result, a large number of research groups have employed various mass spectrometric techniques to investigate cluster ions. These techniques offer the advantages of high sensitivity and size selection up to very large cluster sizes (following the initial ionization step which creates the cluster ion). Electron impact, single-photon (synchrotron radiation), and multiphoton (laser) ionization are the most commonly employed methods of ionization. It has been recognized that the ionization of a distribution of neutral cluster sizes leads to substantial evaporation of monomers from the initially generated cluster ion. This process is referred to as "van der Waals fragmentation" and is usually regarded as the size determining step (Buck and Meyer 1986; Klots 1988). The size distribution of the cluster ions is narrower than the neutral cluster size distribution. The fragmentation channels reflect the nature of the intermolecular bond, that is, hydrogen-bonded clusters usually fragment via a proton transfer reaction, whereas vdW clusters fragment by cleavage of the intermolecular bond. It is also possible that the generated cation may react chemically with the surrounding neutrals within the cluster to give novel product ions. In a few cases, the products observed cannot be rationalized in terms of conventional bimolecular or condensed phase reactions (Garvey and Bernstein 1986b; Romanowski and Wanczek 1984; Stace 1993). The qualitative changes in the reactivity of cluster ions, as a function of the degree of solvation, are therefore directly reflected in the cluster mass spectrum (CMS).

7.2.3. Experimental Setup

The experimental setup used by our group is shown in Figure 7-1. This apparatus consists of a molecular beam source coupled to a chamber housing the quadrupole mass spectrometer. The continuous beam source consists of a Campargue-type nozzle, an expansion chamber, and a collimation chamber. The nozzle assembly itself is mounted on a micrometer and is fitted with a gas handling line which

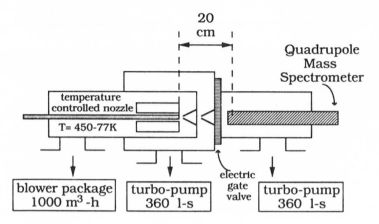

Figure 7-1. Schematic side view of differentially pumped cluster beam apparatus and quadrupole mass spectrometer. The temperature of the nozzle in the stagnation region is regulated by a circulating chiller. Reprinted with permission from Vaidyanathan et al. 1991b. Copyright 1991 American Institute of Physics.

allows the stagnation volume (immediately behind the nozzle) to be pressurized. The stagnation pressure, P_0, is measured by a diaphragm pressure gauge. The nozzle is fitted with a sheath through which a fluid may be circulated to control the nozzle temperature. The temperature of the nozzle, or the expansion temperature, T_0, is measured by a thermocouple situated behind the nozzle. A cooling fluid is circulated through the sheath using a circulating chiller when the nozzle is to be cooled, whereas when the nozzle has to be heated, heated water is circulated around the nozzle. The nozzle used for all of the experiments reported in this chapter has a diameter, $d = 250\ \mu m$.

The central portion of the neutral cluster beam is skimmed by a 0.5 mm skimmer. The skimmed portion of the beam is passed into a collimation chamber where it is skimmed by a second 0.5 mm skimmer before entering the mass spectrometer chamber. The expansion and collimation chambers are differentially pumped by a roots blower stack ($1000\ m^3\ h^{-1}$) and a turbomolecular pump ($360\ l\ s^{-1}$), respectively. The pressures in the expansion chamber (P_1) and collimation chamber (P_2) are monitored using a Pirani gauge and ion gauge, respectively. During operation, the operating pressures are usually in the following ranges: $P_0 \leq 5$ atm; $P_1 = 10^{-2}$–10^0 torr, and $P_2 = 10^{-5}$–10^{-4} torr.

The skimmed and collimated cluster beam passes into the quadrupole mass spectrometer (Extrel, C-50), which has a mass range of 0–1200 amu with unit mass resolution. The mass spectrometer chamber is pumped by a turbomolecular pump ($360\ l\ s^{-1}$). The pressure in the mass spectrometer chamber (P_3), when the beam is in operation, is always less than 1×10^{-6} torr. This is necessary to ensure that the contributions from reactions of the cluster ions with the background gas are not significant. The distance of the nozzle from the ion source varies in the range 20.5–22.5 cm, depending on the nozzle to skimmer distance.

The cluster beam enters the open-design ionizer section of the mass spectrometer, collinear to the axis of the ion optics. Electrons generated from an iridium filament are accelerated towards the ion source by the variable potential difference. A small fraction of the neutral cluster distribution entering the ion source is ionized by the energetic electrons. The electron energy (E_{el}) can be varied continuously over the nominal energy range 5.0–100.0 eV. The modular nature of the mass spectrometer allows it to be operated in either the mass scan or the energy scan modes.

a. *Mass scan mode*

In this mode, the electron energy is maintained constant while a mass spectrum is acquired. The ionizer section is usually set to a more positive voltage than the quadrupole rods to ensure that the cations are repelled towards the mass filter. The ion energy is fixed to ensure maximum sensitivity without distortion of the individual peak shapes. As ions of different masses would have different velocities, the sensitivity is expected to change with the mass of the ion. The programmed ion optics (PIO) developed by Extranuclear Laboratories offers a convenient solution: the ion energy changes as a function of the mass, such that all ions enter the mass filter with optimum velocities. The ions are focused, by an Einzel lens array, into the aperture of a quadrupole mass filter (0.95 × 20 cm, 200 W RF/DC, 1.2 MHz). The ions exiting the quadrupole are accelerated onto a conversion dynode (CD) maintained at −3.0 kV. The resulting secondary electrons are accelerated towards a channeltron particle detector. The use of the CD increases the secondary emission ratio for the heavier ions relative to the lighter ions. This reduces the mass discrimination against the higher masses as compared with the setups where the ions are detected directly by the channeltron particle multiplier. The emission current is usually maintained at 0.65 mA when acquiring the cluster mass spectra but this can also be varied over the range 0.10–30.0 mA.

The data acquisition and analysis were for the most part done by using a PC-based data system. The mass scale is typically calibrated with argon cluster ions being set to their nominal masses, such that the peaks correspond to integral values. The peak shapes and widths were adjusted to be similar over the entire mass range under study to ensure uniform resolution across the mass range. The ion energy, the low and high mass resolution adjusts, the voltage on the extractor lens, and the Einzel lens system all have profound effects on the resolution, as well as the sensitivity. Continuous spectra were acquired, whenever necessary, using a digital oscilloscope (LeCroy 9400) and these spectra are presented as line traces.

b. *Energy scan mode*

Several modifications were made to the commercial ionizer circuitry to enable scanning of the electron impact energy. In this mode, the quadrupole mass filter is set to a particular mass: that is, constant m/z ratio, while the electron energy is scanned externally. The C50 is manually set to pass the ion of interest with a very high resolution, leading to a considerable decrease in the signal intensity of the ions. The measurement of ionization efficiency (IE) curves also requires low emission currents, which again reduces the ion current considerably. Hence, the IE curves can be acquired only for ions that are at least relatively abundant. The emission currents were maintained at 0.35 mA or lower for all the measurements. Emission currents ranging from 0.10 to 0.50 mA did not lead to any significant changes in the appearance energy (AE) values. The scanning of the electron energy is accomplished by applying a constant DC voltage to the ionizer filament in conjunction with a ramped DC acceleration voltage. The constant voltage (E_0) is supplied by a regulated DC power supply with the external ramped voltage (E_r) being supplied by a ramp generator. Thus the nominal energy, E_{el}, of the electron beam can be represented by $E_{el} = E_0 + E_r + CF$, where the last term is a correction factor arising primarily from the spatial nonuniformity of the potential gradient. The value of CF was determined by acquiring the IE curves of argon and helium. The ionization potential (IP) and the AE of the various cluster ions were determined using the "zero current" method (i.e., the point at which the ion current rises above the baseline was taken to be the IP or the AE for the ion). Usually, the lowest ionization energy required for observing an atomic ion or a molecular ion is termed as IP, whereas the onset of a fragment ion is referred to as AE. The analog ion current was collected, averaged, and processed using the boxcar averager and signal processor.

7.3. MAGIC NUMBERS

7.3.1. Cluster Mass Spectrum

In subsection 7.2.1, it was discussed that the neutral clusters generated from the beam source have a wide range of cluster sizes. Although there is no single, easily available technique which allows the size specific detection of neutral clusters, it is now well accepted that the cluster size distribution can be manipulated, to a certain extent, by careful control of the various experimental parameters. Mass spectrometric detection of clusters in general and tandem mass spectrometry in particular offer the inherent advantage of enabling size specific studies of the cluster ions.

Mass spectrometry involves the detection of charged particles, and, in the present case, a portion of the neutral cluster beam is ionized. Ionization essentially involves electronic excitation and occurs on the time scale of the order of 10^{-16} s (Haberland 1985; Märk 1987). The mass spectrometric detection of the ions is usually achieved on a microsecond time scale after the ionization event. As a result, the ionization process is taken to be time zero in the discussion of the processes which occur following the actual ionization of the neutral clusters, yet before the mass selection of the cluster ions. That is, the resulting cluster ion will "incubate" in the ionizer for microseconds before being accelerated into the mass filter. On that time scale, the cluster ion may lose monomer units, and the cation within the cluster may fragment or react chemically with the adjacent molecules.

In the early years, there was some controversy regarding whether ionization of neutral clusters was accompanied by substantial evaporation of monomers from the cluster ions or if the cluster ion distribution reflects the neutral cluster distribution. Some groups concluded from their data that the CMS was a reflection of the neutral cluster distribution, that is, ionization of neutral clusters did not lead to significant fragmentation (Ding and Hesslich 1983; Echt et al. 1982; Sattler 1985). The size selective ionization studies conducted by Buck and Meyer (1984a, 1986) provided conclusive evidence to suggest that ionization of RGCs leads to substantial evaporation. For example, ionization of Ar_5 did not yield Ar_5^+ and Ar_4^+, yet produced monomer and dimer ions.

It has now been established, for the case of RGCs, that following the ionization event, the positive charge is localized on either a highly vibrationally excited dimer (Haberland 1985; Kunz and Valldorf 1988) or trimer ion (Bohmer and Peyerimhoff 1986; Hesslich and Kunz 1981). The localization of charge on a dimer or a trimer is commonly referred to as charge trapping, and this process is thought to occur within 10^{-12} s after the ionization of RGCs. The vibrational relaxation of the binding energies of the dimer ion into the cluster is completed within 10^{-8} s after ionization, and it is this process which leads to further evaporation of rare gas atoms (Haberland 1985).

Depending on the technique used for mass detection, it is possible to detect metastable decompositions, that is, delayed loss of monomers or fragments from less stable cluster ions on a microsecond time scale. Metastable decompositions of singly and multiply charged cluster ions have been studied widely. On this time scale, loss of a single monomer was observed for most cases (Märk 1987). Scheier and Märk (1987c) investigated the metastable decompositions and reported that the loss of two monomers from Ar_n^+ was a sequential loss of individual monomers, when $n \geq 7$. On the other hand, the Ar_4^+ yielded the Ar_2^+ in a single-step fission process. Recently, Märk and coworkers (Foltin and Märk 1991; Foltin et al. 1991a,b; Märk et al. 1992) reported an unusual metastable decomposition channel leading to the loss of anywhere from 2 to 10 rare gas atoms from argon cluster ions following electron impact ionization, and this process had a threshold of 27.0 eV. A similar process has also been reported by Hertel and coworkers (Stager et al. 1991) during photoionization of argon clusters using synchrotron radiation. Thus, the evaporation sequences appear to extend from the initial at around 10^{-10} s after the initial ionization into the microsecond time regime. These studies strengthened the belief that the sequence of monomer evaporations plays an important role in determining the overall appearance of the resultant CMS.

Thus, the CMS acquired is a reflection of: (1) the neutral cluster size distribution generated, (2) the energy employed in ionizing the neutral clusters, (3) the various processes that occur after the ionization event, and (4) the time elapsed between the ionization event and the detection of the cluster ions.

7.3.2. Origin of Magic Numbers

The mass spectrometric studies of clusters composed of atoms and/or molecules have shown that cluster ions of certain sizes have enhanced intenstities, that is, the smoothly decreasing intensity pattern of cluster-ion distributions exhibit marked discontinuities. The sizes at which cluster ions exhibit enhanced intensity are referred to as "magic numbers," a term coined by Herzog et al. (1973) to describe especially abundant species. This term was originally used in nuclear physics, where the extra stability of certain isotopes was correlated with proton or neutron shell closings. In the case of cluster ions that are strongly hydrogen bonded, the link between the observation of magic numbers and the formation of stable, closed shell cluster ions has been developed (Castleman and Keesee 1986a; Coolbaugh et al. 1990; Peifer et al. 1989; Stace and Moore 1982; Wei et al. 1991b,c). Magic numbers observed in rare gas, atomic, and even some molecular cluster ions have been attributed to especially stable icosahedral structures with completed shells (Bohmer and Peyerimhoff 1989; van der Waal 1983). Martin et al. (1993) recently reported that the magic numbers observed in the CMS of $(C_{60})_n^+$ and $(C_{70})_n^+$ are similar, but not identical, to those observed for small and medium sized rare gas cluster ions and they proposed that the cluster ions of fullerene molecules may prefer the icosahedral symmetry.

As discussed in the previous section, the CMS is affected by a number of factors, such as neutral cluster size distribution, cluster size dependent ionization or fragmentation, fragmentation rates, and the technique used for acquiring the resultant mass spectrum. Thus, more strongly bound structures may decay at substantially slower rates than weakly bound structures. These stable structures act as a kinetic bottleneck and, as a result, appear with greater intensity in the CMS. Although the thermodynamic stability of the cluster ions may contribute to the observation of magic numbers in the ion intensity distribution, this alone may not be directly responsible for the observation of magic numbers. The direct origin of magic numbers has been traced to the kinetics of the unimolecular decompositions following the ionization (Casero and Soler 1991; Castleman and Keesee 1988a). After ionization, the cluster ions have a high internal energy and therefore the rates of evaporation are quite high. However, as successive evaporations occur, the internal energy of the cluster ion drops, leading to a decrease in the rates of evaporation. This has been suggested to lead to the development of magic numbers in the microsecond time regime (metastable decompositions have been detected at up to ca. 100 μs after the initial ionization event). Märk et al. (1986) found that the metastable decay rates for the various argon cluster ions exhibited a size dependence and speculated that the anomalies in the intensity distribution of cluster ions may be due to the differencs in the metastable decay rates of cluster ions of various sizes. The dependence of the metastable decomposition rates of large neon cluster ions on the time elapsed since the ionization event was experimentally confirmed and the cluster ions with the highest intensities were reported to have the smallest metastable decomposition rates (Märk and Scheier 1987b). Even though it has been assumed that a qualitative relationship exists between the cluster ion intensities and their stabilities, there was no direct correlation between the evaporation energy and the magic numbers. Castleman and coworkers (Wei et al. 1991a) have shown that the magic numbers observed in the CMS of xenon are identical to the local maxima observed in the relative binding energies.

Thus, the apparent stability of a cluster ion in any specific CMS may arise from a combination of one or more of the following factors, such as structural stability, electronic bonding pattern, and dynamics of unimolecular dissociation.

7.3.3. Magic Numbers in $Ar_nX_m^+$

The first experimental evidence for the existence of magic numbers was reported by Recknagel and coworkers (Echt et al. 1981). Since then, a number of experimental and theoretical studies of magic numbers in rare gas clusters have been published (Buck and Meyer 1986; Carnovale et al. 1989; Castleman et al.

1981; Ding and Hesslich 1983; Echt et al. 1984, 1987; 1990; Fieber et al. 1991; Haberland 1985; Harris et al. 1984; Kunz and Valldorf 1988; Lethbridge and Stace 1988; Levinger et al. 1988; Märk and Scheier 1987a; Stace and Moore 1983; Stevens and King 1984). The magic numbers observed in the inert gas cluster ions coincide with the most compact structures calculated for the neutral clusters, and the magic numbers observed in rare gas cluster-ion distributions were suggested to arise on account of especially stable neutral clusters (Echt et al. 1981). Although, in the early years, the stability of the rare gas cluster ions generated considerable debate centering on whether the intensity fluctuations found in the mass spectra reflect the size distributions in the neutral or the atomic clusters, it has now been firmly established that these magic numbers are due to the stability of the cluster ions (Buck and Meyer 1984a, 1986; Haberland 1985; Märk and Scheier 1987b, Märk et al. 1986; Wei et al. 1991a).

A number of cluster mass spectra obtained by ionization of neat argon expansions have been published over the years. Ding and Hesslich (1983) observed magic numbers at $n = 3, 14, 16, 19, 21, 23$ and 27; whereas Northby and coworkers (Harris et al. 1984) reported magic numbers at $n = 13, 19, 23, 26, 29, 32$, and 34 for the same region of the CMS. Although the sets of magic numbers reported differ in several positions, the observation of a magic number at $n = 13$ in one case and $n = 14$ in the other has drawn considerable attention in the literature over the years. The study by Ding and Hesslich (1983) involved the generation of neutral clusters, followed by electron-induced ionization, whereas Northby and coworkers (Harris et al. 1984), employed an experimental setup where the argon ions generated by a corona discharge source act as the nucleating agent for the formation of larger cluster ions. The discrepancy in the size at which the magic number was observed has been rationalized by suggesting that the exact size of the ion core (Ar^+, Ar_2^+, Ar_3^+, or Ar_4^+) has an effect on the magic number sequence (Bohmer and Peyerimhoff 1988; Haberland 1985; Saenz et al. 1985). The Ar_{13}^+ has been found to be especially stable in theoretical calculations when an Ar^+ or Ar_3^+ is assumed to form the ion core, whereas studies that use an Ar_2^+ as the ion core found the Ar_{14}^+ ion to be especially stable.

Many of the magic number combinations observed in the CMS of inert gas atoms have been identified with stable structures having an icosahedral symmetry (Echt et al. 1981). The Mackay icosahedra series (Hoare 1979; Mackay 1962) exhibits completion of the first three solvation shells as $n = 13, 55$, and 147, respectively, such that the completion of solvation shells at $n = 13, 55$, etc., can arise from structures with a cuboctahedron symmetry (Hoare 1979). However, theoretical studies indicate that the icosahedral structures are more stable than those with cuboctahedral symmetry (Hoare 1979). The theoretical studies of Farges et al. (1986) and Northby (1987) provide insight into the growth of icosahedral structures.

There have also been a number of experimental and theoretical studies on mixed clusters of the type $Ar_n X_m^+$ where X is either an atom or a polyatomic molecule. Extensive experimental investigations of the heterocluster ions of argon with various polyatomic molecules were conducted by Stace (1983, 1984, 1985a,b, 1987c). For mixed expansions of argon with dimethyl ether, mixed cluster ions of the type $Ar_n CHO^+$ and $Ar_n CH_3 OCH_2^+$ were observed. The intensity distribution of Ar_nCHO^+ cluster ions showed strong magic numbers at $n = 12$ and 18, whereas the intensity distribution of the $Ar_nCH_3OCH_2^+$ showed a weak magic number only at $n = 18$. Based on this data, it was concluded that the $Ar_{12,18}CHO^+$ cluster ions form structures with icosahedral symmetry, with the oxygen atom providing the extra atom, that is, $Ar_{12}O$ forms the 13-atom cluster. Stace and coworkers suggested that the linear geometry of the CHO^+ would enable its incorporation into cluster ions without significant steric hindrance. In the case of the $CH_3OCH_2^+$ cation, its large size and the steric hindrance due to the methyl and the methylene groups were stated to be the reason for not observing a magic number at $n = 12$ in the case of the $Ar_nCH_3OCH_2^+$. Stace also reported that Ar_nI^+ has enhanced stability at $n = 12$ and 18 whereas $Ar_nI_2^+$ has a local maxima only at $n = 17$. These results were interpreted in terms of icosahedral packing with the I^+ and I_2^+ forming the ion core that was solvated by the argon atoms.

Ozaki and Fukuyama (1985) also investigated mixed cluster ions Ar_nM^+, where M = Xe, N_2, CO_2, and CH_3OH. They observed an enhanced stability for $n = 12$ in all the cases and interpreted this as the solvation of the M^+ ion by 12 argon atoms to form a stable shell, regardless of the chemical identity of the M^+ ion. They also report that the magnitude of the intensity drop between $n = 12$ and 13 becomes larger as the difference in the IPs between Ar and M increases (all the molecules investigated had an IP lower than that of Ar). This was taken to be a measure of the degree of localization of the positive charge on M.

A Monte-Carlo simulation of Ar_nXe^+ clusters by Bohmer and Peyerimhoff (1988) predicted that clusters with $n = 9$, 12, and 18 should show pronounced stability, with $n = 15$ being weakly magic. An icosahedron and a double icosahedron structure were suggested for $Ar_{12}Xe^+$ and $Ar_{18}Xe^+$, respectively. The $Ar_{15}Xe^+$ was proposed to have a structure intermediate to the icosahedron and the double icosahedron, with the three extra atoms bound not only to the atoms in the first shell but also among themselves. The enhanced stability of Ar_9Xe^+ was attributed to a structure where all the argon atoms are at the same optimum equivalent distance from the central Xe^+ ion, as well as from each other. It was also demonstrated that nine argon atoms is the maximum number that can be placed around a central xenon atom without stretching or shrinking any of the Ar–Xe bonds, that is, all the argon atoms occupy the most favorable position with respect to the Xe^+ ion.

In our own studies of $Ar_nH_2O^+$ cluster ions, a prominent magic number was observed for $n = 9$, and weakly magic numbers were observed at $n = 3, 5, 7$, and 15 (Vaidyanathan et al. 1991c). The plot of the intensity of various $Ar_nH_2O^+$ ions as a function of electron energy and stagnation pressures of argon are shown in Figure 7-2 and 7-3 respectively. The IP of both the Xe atom and the H_2O molecule are lower than that of the Ar atom by 3.63 and 3.14 eV, respectively. As a result, the charge would eventually be localized on the dopant atom/molecule in both cases. Thus, it is possible that central H_2O^+ ion may be solvated by nine equivalent argon atoms, similar to that suggested in the case of the Ar_9Xe^+ cluster ion. However, the variation from the expected magic number sequence in the case of $Ar_{12}H_2O^+$ is contrary to the results obtained for $Ar_{12}Xe^+$. A simple rationale for this difference could be the distortion of the argon atoms from the expected icosahedral packing by the nature of the ion–neutral cluster interactions. Neutron diffraction patterns of pure and water-doped vapor-deposited rare gas solids have been reported to show significant differences. These changes have been attributed to a change from a face-center cubic (fcc) structure for pure rare gas crystals to a distorted hexagonal close-packing (hcp) structure when doped with water (Langel et al. 1988). In the case of pure Ar films, the fcc and hcp structures differ only in the stacking sequence, with the fcc being slightly favored energetically. An important factor involved in the transformation observed in water-doped solid argon was conjectured to be the nature of the interaction of water with the rare gas environment (Knozinger et al. 1988). Such interactions may also be of considerable importance for small water-doped argon cluster ions and could account for the deviations we observe from the stable structures with an icosahedral symmetry.

Cluster ions of the type $Ar_n(NO)_m^+$ ($n \leq 22$, $m \leq 4$) have also been investigated (Desai et al. 1992) and magic numbers were found in the intensity distribution of $Ar_n NO^+$ at $n = 12$ and 18, suggesting formation of stable icosahedral structures. The intensity distribution of $Ar_n(NO)_2^+$ exhibited a magic number at $n = 17$, which was rationalized by suggesting that the $(NO)_2^+$ is incorporated into the double icosahedron since this species is too large to be accommodated within a single icosahedron without causing significant distortion to the structure. Märk and coworkers (Lezius et al. 1992) have also investigated the magic numbers in the $Ar_n(NO)^+$ by attaching NO^+ to neutral argon clusters, as well as directly ionizing a beam containing $Ar_n(NO)_m$ heteroclusters using electron impact ionization. The $Ar_n NO^+$ was reported to have enhanced intensity at $n = 12$, 18, 22, 25, and 54 in both cases. The magic numbers were in agreement with expected icosahedral structures.

Stace and coworkers also investigated mixed expansions of argon with methyl formate (Bernard et al. 1990). A number of heterocluster ions, including $Ar_n CH_3 OH^+$, $Ar_n CH_2 OH^+$, and $Ar_n CO^+$ were observed to be more intense than their neighbors for $n = 12$, 18, 54, 70, and 73. As it is not very likely that 54 argon atoms can completely solvate ions of varying sizes such as $CH_3 OH^+$, $CH_2 OH^+$, and CO^+ to generate a stable 55-atom cluster and exhibit a magic number, it was speculated that these molecular ions occupy surface sites or lie close to thr surface of the cluster ion. Likewise, magic numbers have also been observed for $Ar_n N_2^+$ and $Ar_n O_2^+$ at $n = 54$ (Stace 1993). Although Northby and coworkers (Harris et al. 1984) observed Ar_{55}^+ to be especially stable when using preformed ions to generate clusters, the Ar_{54}^+ and Ar_{55}^+ cluster ions have not been reported to have enhanced stability in neat argon expansions (Echt et al. 1990). On the other hand, Whetten et al. (1990) observed pronounced magic numbers at $n = 12$, 18, 54, and 146 for the $Ar_n Al^+$ cluster ions and in this case it was suggested that the Al^+ ion may form the central core of the icosahedral shell structure.

Stamatovic et al. (1988) studied mixed expansions of SF_6 with argon and reported that the intensity distribution of both $Ar_n SF_6^+$ and $Ar_n SF_5^+$ shows magic numbers at $n = 13$. Ding et al. (1985) studied mixed expansions of argon with N_2 and CO and magic numbers were observed in the intensity distribution of $Ar_n(N_2)_m^+$ and $Ar_n(CO)_m^+$ for $n = 13$ ($m = 1$–4 for N_2 and $m = 1$ and 2 for CO). These results differ from the other resports discussed in this section in that the magic number is observed at $n = 13$ rather than at $n = 12$ for heterocluster ions

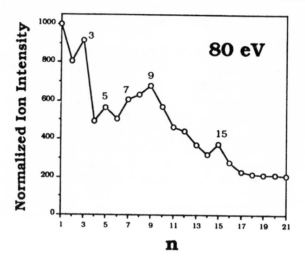

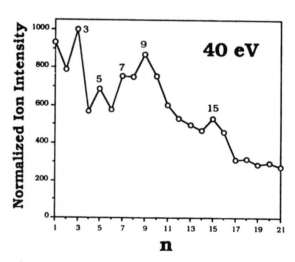

Figure 7-2. Plot of normalized ion intensities as a function of n, for the cluster ion $Ar_nH_2O^+$, as a function of electron energy ($P_0 = 3.8$ atm). Magic numbers are noted by numbering of individual data points. Note that the magic number structure becomes more pronounced at lower electron energies where monomer evaporation is expected to occur to a smaller extent. Reprinted with permission from Vaidyanathan et al. 1991c. Copyright 1991 American Chemical Society.

of the type $Ar_nX_m^+$. The results of the study conducted by Ding et al. (1985) are not in agreement with the observations of Ozaki and Fukuyama (1985) for $Ar_n(N_2)_m^+$ cluster ions. The enhanced stability of $Ar_{13}X_m^+$ independent of the number of N_2 or CO molecules attached to the heterocluster ions may suggest that the polyatomic molecules are on, or close to, the surface of the cluster and it is the packing of the argon atoms in the heteroclusters that determines the magic numbers.

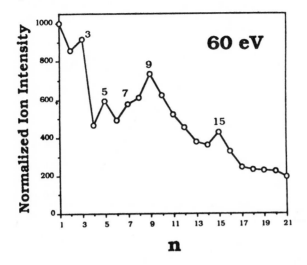

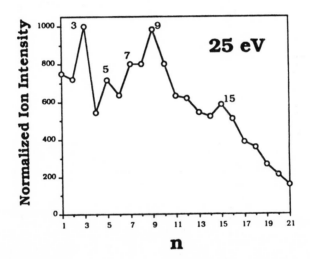

Figure 7-2 (*continued*)

From the above discussion, it is evident that mixed cluster ions of the type $Ar_n M_m^+$ exhibit strong magic numbers at values of $(n + m) = 13$, 19, 55, 71, and 147 in a variety of different studies. These values correspond to the completion of the first, second, and third icosahedral shells occurring at 13, 55, and 147; whereas 19 and 71 correspond to especially stable subshells formed by interpenetrating double icosahedron structures. The size and symmetry of the dopant moiety appear to be the most important factors in observing magic numbers that can be rationalized on the basis of icosahedral-like structures. The inability to observe magic numbers has been attributed to the distortion of the icosahedral structure due to size and steric factors associated with the dopant ion which destroys the delicate balance between the monomer interactions. One of the issues that has been interpreted differently involves the location of the dopant atomic/molecular

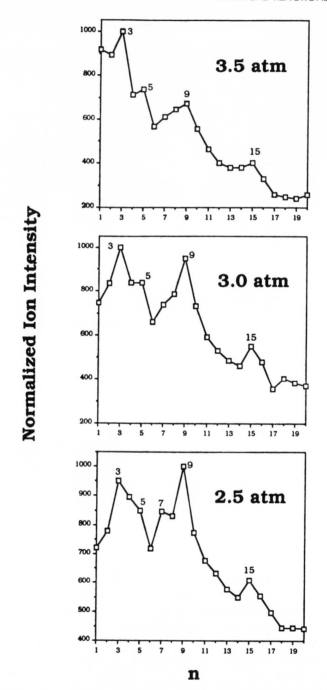

Figure 7-3. Plot of normalized ion intensities as a function of n, for the cluster ion $Ar_nH_2O^+$, as a function of expansion pressure (electron energy = 25 eV). Magic numbers are noted by numbering of individual data points. Reprinted with permission from Vaidyanathan et al. 1991c. Copyright 1991 American Chemical Society.

ion in the heteroclusters. Some studies conclude that the dopant ion was solvated by the rare gas atoms while others suggested that the molecular ion was located on, or close to, the surface of the rare gas cluster. Thus, the nature, number size, and symmetry of the dopant ion, as well as the total number of argon atoms in the heterocluster ions, seem to play an important role in the location of the dopant ion.

7.4. REACTIONS WITHIN CLUSTERS

The CMS obtained by the ionization of an inert gas expansion consists of a series of peaks separated from each other by the monomer mass. In the case of neat expansions of molecules, or when helium is used as the carrier gas, the CMS very often contains a dominant cluster ion sequence composed of protonated cluster ions (i.e., $\{M\}_n H^+$). Additional cluster ion sequences arise from the solvation of a fragment ion (arising from the fragmentation of a cation within the cluster). (Coolbaugh et al. 1993; Garvey et al. 1991). Certain molecular systems have exhibited "new" fragment ions in the CMS which cannot be rationalized on the basis of the known unimolecular or bimolecular chemistry of the molecule which have been suggested to be due to unique chemical reactions characteristic of the cluster environment (Coolbaugh et al. 1990; Garvey and Bernstein 1986b; Romanowski and Wanczek 1984; Stace 1993; Vaidyanathan et al. 1993; Wei et al. 1991a). Since the study of clusters can conceptually bridge the gap between gas phase and condensed phase chemistry (see, e.g., Jena et al. 1986), it is important to understand the changes in reactivity of molecules with varying degrees of solvation, as well as the nature of the solvating molecules (Garvey and Bernstein 1986b; Romanowski and Wanczek 1984; Stace 1993; Wei et al. 1991a).

We will be presenting the results of various studies which suggest that the reactions of molecules are influenced by the choice of carrier gas, as well as the energy utilized in ionizing the neutral clusters.

7.4.1. Effect of Carrier Gas on Cluster Mass Spectrum

In the mass spectrometric investigations of most hydrogen-bonded clusters, the protonated clusters form the dominant cluster ions as a result of a rapid proton transfer reaction (Märk and Castleman 1985). Castleman and coworkers have extensively investigated the neat expansions of methanol (Morgan and Castleman 1987, 1989; Morgan et al. 1989; Zhang et al. 1991) and observed the protonated cluster ions to form the most intense cluster series. In the case of methanol, the exoergic proton transfer reaction may be written as:

$$CH_3OH^+ + CH_3OH \rightarrow CH_3OH_2^+ + CH_3O, \tag{1}$$

Within the cluster environment, the reaction may proceed as:

$$(CH_3OH)_n^+ \rightarrow (CH_3OH)_{n-1}H^+ + CH_3O \tag{2}$$

$$(CH_3OH)_{n-1}H^+ \rightarrow (CH_3OH)_{n-k-1}H^+ + kCH_3OH, \tag{3}$$

where $k \geq 1$. The energy released by the exothermic reaction (2) leads to the evaporation of additional methanol molecules, as shown in reaction (3). Very weak signals were observed for the unprotonated methanol dimer ions but unprotonated cluster ions were not observed for larger cluster sizes (Herron et al. 1992; Morgan and Castleman 1987, 1989; Morgan et al. 1989; Zhang et al. 1991). The protonated cluster ions were also reported to undergo secondary reactions to give a protonated dimethyl ether:

$$(CH_3OH)_{n-2}(CH_3OH + CH_3OH_2^+) \rightarrow (CH_3OH)_{n-2}(CH_3)_2OH^+ + H_2O. \quad (4)$$

This reaction was observed to occur as a metastable decomposition process for $n = 4$–9. For clusters with $n > 9$, retention of water was observed as a prompt reaction:

$$(CH_3OH)_nH^+ \rightarrow (CH_3OH)_{n-3}H_3O^+ + (CH_3)_2O + CH_3OH. \quad (5)$$

The protonated cluster ions were observed to be the major product ions in the CMS of ammonia, water, etc., although unprotonated cluster ions have also been observed in these cases, depending on the choice of carrier gas, method of ionization, and electron energy utilized in the investigation (Shinohara et al. 1985, 1986). The inability to observe unprotonated cluster ions is usually attributed to poor Franck–Condon factors for the vertical ionization transitions. These poor Franck–Condon factors arise from the large differences in the configuration of the neutral and ionic clusters (Stace 1987a).

Our group at SUNY has recently examined the CMS of pure methanol expansion, helium/methanol seeded expansion, and argon/methanol coexpansion obtained by electron impact ionization (Vaidyanathan et al. 1991b). In pure methanol expansions, neat methanol clusters were generated by allowing methanol vapors to expand directly through the nozzle. Helium/methanol seeded expansion and argon/methanol coexpansions were obtained by using a split-flow system which allows the amount of methanol entrained in the carrier gas to be controlled. The concentration of methanol in argon/methanol coexpansions was adjusted such that cluster ions of the type $Ar_n(CH_3OH)_m^+$ with $m \geq 2$ were not observed in significant amounts.

In the case of the helium/methanol expansion, the CMS obtained using 70.0 eV electrons [Figure 7-4(b)] exhibited the $(CH_3OH)_mH^+$ cluster ions as the major series. With an argon/methanol coexpansion, the 70.0 eV CMS showed the Ar_n^+ and $(CH_3OH)_mH^+$ as the most intense cluster ions [Figure 4(a)]. Very weak signals corresponding to $(CH_3OH)_m^+$, $(CH_3OH)_mCH_3O^+$, $(CH_3OH)_mCH_2O^+$, and $Ar_nCH_xO^+$ ($x = 0$–5) were also observed.

Argon is usually not the best carrier gas for generating neat clusters in an adiabatic expansion as heteroclusters will also be generated. As has been discussed in subsection 7.2.1, the size distribution, as well as the composition of the neutral clusters (neat as well as mixed), is affected by the ratio of the two components and the expansion conditions. Stace and coworkers (Bernard et al. 1990; Stace 1983, 1984, 1985a,b, 1987b,c) investigated a large number of heterocluster systems where the expansion conditions were adjusted to generate the Ar_n^+ and $Ar_nX_m^+$ cluster ions as the major products. Stamatovic et al. (1988) also demonstrated that by varying the concentration of SF_6 in an argon expansion, they were able to produce, at will, expansions containing either SF_6 cluster ions or the

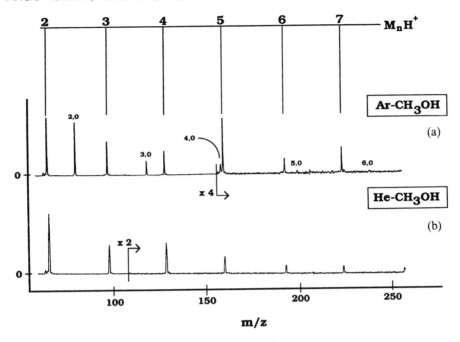

Figure 7-4. (a) Raw 70 eV mass spectra of Ar/MeOH expansion, (b) raw 70 eV mass spectra of He/MeOH expansion. Number above bar indicates value of n (i.e., number of methanol unit in M_nH^+ where $M = CH_3OH$). Peaks labeled (n, m) correspond to ions of the type $Ar_n(CH_3OH)_m^+$ where the n and m are listed, respectively. Reprinted with permission from Vaidyanathan et al. 1991b. Copyright 1991 American Institute of Physics.

argon/SF_6 heterocluster ions. The relative intensities of the various fragment ions, as well as the fragmentation patterns, have been found to be significantly different for argon/molecule expansions as compared with neat expansions of the molecule (Stace 1983, 1984, 1985a,b, 1987c; Stace and Bernard 1988; Stamatovic et al. 1988; Isenor and Qi 1989; Vaidyanathan et al. 1991b). SF_6^+ was either a weak peak or not observed at all in the mass spectrum of the isolated molecule, and cluster ions of the type $(SF_6)_n^+$ were also not observed in the neat expansions of SF_6. However, the cluster ions $Ar_n(SF_6)_m^+$ were observed for $m \geq 3$ in the argon/SF_6 expansions. Isenor and Qi (1989) also observed anomalous fragmentation patterns for the electron impact ionization of SF_6 and SiF_4 seeded in argon expansions. They proposed that the Ar_nX_m ($X = SF_6$ and SiF_4) neutral clusters were the precursors for the ions observed in the CMS. Stace and Bernard (1988) also noted that, in mixed expansions of argon with traces of ethylene, cluster ions such as $Ar_nCH_2^+$ and Ar_nCH^+ were not observed and suggested that the absence of these ions could be due to energetic considerations.

  Nishi and coworkers (Shinohara et al. 1985, 1986) have reported the production of unprotonated water and ammonia cluster ions, as well as $Ar_nX_m^+$ ($X = H_2O$ and NH_3) cluster ions when argon/molecule mixed expansions were ionized using UV lines from an argon lamp. The "quenching" of the proton transfer reaction was attributed to the cooling of the heterocluster ions via

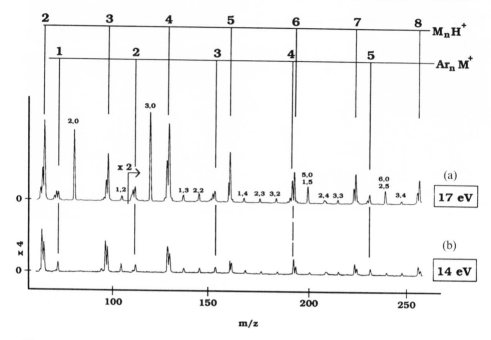

Figure 7-5. (a) Raw 17 eV mass spectra of Ar/MeOH expansion, (b) raw 14 eV mass spectra of Ar/MeOH expansion. Number above bar indicates value of n (i.e., number of methanol unit in M_nH^+ where $M = CH_3OH$). Peaks labeled (n, m) correspond to ions of the type $Ar_n(CH_3OH)_m^+$ where the n and m are listed, respectively. Reprinted with permission from Vaidyanathan et al. 1991b. Copyright 1991 American Institute of Physics.

evaporation of argon atoms. The observation of unprotonated cluster ions was of considerable interest because these ions are usually not produced in significant quantities from the neat clusters of molecules that can hydrogen bond extensively. The initial ionization event in such molecules would lead to the formation of the protonated cluster ions as the major product, via an exoergic proton transfer reaction similar to the one suggested for methanol [reaction (2)]. Thus, it seemed that the ionization of Ar_nX_m ($X = CH_3OH$, H_2O, NH_3) heteroclusters could be used as a general method of generating unprotonated cluster ions from molecules that can hydrogen bond extensively. In fact, Huisken and coworkers (Ehbrecht et al. 1993) have reported that the electron impact ionization of Ar_nX_m, where X = methanol and methyl fluoride, is also an efficient method of producing unprotonated methanol and methyl fluoride cluster ions.

7.4.2. Effect of Ionization Energy on Cluster Mass Spectrum

We observed that when the energy of the ionizing electrons is reduced from 70.0 eV to 17.0 eV [slightly above the IP of the argon atom (15.76 eV)], a number of interesting changes can be observed in the CMS, as shown in Figure 7-5(a).

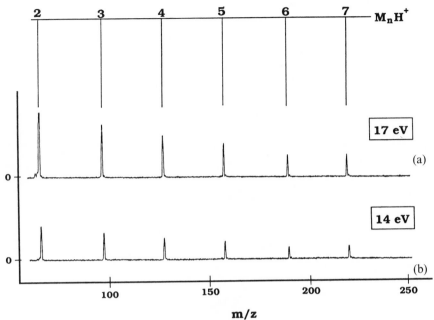

Figure 7-6. (a) Raw 17 eV mass spectra of He/MeOH expansion, (b) raw 14 eV mass spectra of He/MeOH expansion. Number above bar indicates value of n (i.e., number of methanol unit in M_nH^+ where $M = CH_3OH$). Reprinted with permission from Vaidyanathan et al. 1991b. Copyright 1991 American Institute of Physics.

The intensity of the heterocluster ions $Ar_nCH_xO^+$ increased substantially, relative to the neat methanol cluster ions, and the cluster ions of the type $Ar_n(CH_3OH)_m^+$ were observed with significant intensity over the entire mass range. At electron energies below the IP of argon—for example, 14.0 eV—the intensity of the unprotonated methanol cluster ions was observed to be greater than that of the pronated cluster ions, as shown in Figure 7-5(b).

However, the CMS for the helium/methanol system is qualitatively the same for 70.0, 17.0, and 14.0 eV ionizing energies [Figures 7-4(a), 7-6(a) and 7-6(b)] The only changes observed were a decrease in the overall signal intensity and a shift of the cluster-ion size distribution toward a larger n. There can be very little doubt that the observed cluster chemistry of argon/methanol heteroclusters following ionization is considerably different from that of the neat methanol expansions, especially at low electron energies. In the ionization of heteroclusters such as $Ar_n(CH_3OH)_m$, the difference in the IPs of the two components [IP $(CH_3OH) = 10.85$ eV vs. IP (Ar) $= 15.76$ eV] might lead one to assume that any ionization below 15.76 eV would proceed via the direct ionization of the methanol component within the argon/methanol heteroclusters. Although the IP values are useful in predicting the final location of the charge in a heterocluster ion, they offer no conclusive indications of the site at which the initial excitation or ionization will occur.

7.5. IONIZATION MECHANISMS

Extensive photoionization studies by Ding et al. (1987) have shown that there are essentially three different mechanisms responsible for the ionization of neutral heteroclusters of the type R_nX_m (where R = rare gas atoms and X = polyatomic molecule). Similar ionization mechanisms are also possible using electron-induced ionization studies. The various possibilities are summarized, as follows.

(1) The molecular component of the heteroclusters with the lower IP may be ionized directly. In this case, the ionization threshold of the heteroclusters would correspond to the IP of the polyatomic molecule and the amount of internal energy deposited would be relatively small at the ionization threshold, leading to negligible fragmentation (Gover et al. 1991).

$$R_nX_m + e^- \rightarrow R_nX_m^+ + 2e^-. \tag{6}$$

(2) The rare gas atom (R) with the higher IP is ionized first, followed by a charge transfer to the polyatomic molecule. This can either involve a direct charge transer:

$$R_nX_m + e^- \rightarrow R_n^+X_m + 2e^-,$$
$$R_n^+X_m \rightarrow (R_nX_m^+)^* \tag{7a}$$

or self-trapping, followed by charge transfer:

$$R_n^+X_m \rightarrow R_{n-x}R_x^+X_m \rightarrow (R_nX_m^+)^* \tag{7b}$$

$$(R_nX_m^+)^* \rightarrow R_iX_k^+ + (n-i)R + (m-k)X, \tag{7c}$$

where $x = 2$ or 3 depending on whether the charge is eventually localized on a dimer or a trimer are gas ion, and $k, i \geq 1$. A study by Carnovale et al. (1991) indicates that the size of the rare gas ion core is highly dependent on the size of the cluster ion. Direct charge transfer [reaction (7a)] would result in a considerable amount of energy being deposited at the ionization threshold. Hence, substantial evaporation of rare gas atoms is possible, as well as fragmentation of the polyatomic molecule as shown in reaction (7c). Charge transfer from a trapped hole [reaction (7b)] would lead to the dopant atom/molecule receiving less energy than that available in the direct charge transfer from the inert gas atom (ca. 2 eV less in the case of argon).

(3) The third possibility arises when the rare gas atom(s) are initially *electronically* excited, and this excitation energy can then be transferred to the polyatomic molecule. If the IP of the molecule is lower than the excitation energy (E_{ex}) of the rare gas atoms, then ionization of the polyatomic molecule may occur. This process has been termed "intramolecular Penning ionization" (Kamke et al. 1985a). It is also possible for the E_{ex} to result in the vibrational excitation of the neutral cluster. The extent of fragmentation will then depend on the relative efficiencies of the two channels and the excess energy within the cluster ion $[E_{ex} - IP(X)]$. The appearance energy of the heterocluster ions in this case would be E_{ex}, where $IP(X) < E_{ex} < IP(R)$.

$$R_nX_m + e^- \rightarrow R_n^*X_m + e^- \tag{8a}$$

$$R_n^*X_m \rightarrow R_nX_m^+ + e^- \qquad (8b)$$

$$\rightarrow R_iX_k^+ + (n-i)R + (m-k)X + e^- \qquad (8c)$$

$$\rightarrow X_m^+ + nR + e^- \qquad (8d)$$

The appearance energy (AE) curves for the various cluster ions, such as X_m^+, $R_iX_k^+$, and $R_nX_m^+$, will reflect the dominant mechanism for the ionization of the heteroclusters. Thus, the AE for the heterocluster will correspond to the IP of the polyatomic molecule when direct ionization is the dominant ionization mechanism. However, when the ionization of the polyatomic molecule is indirect (i.e., involves either the ionization or the excitation of the inert gas component as the first step, followed by charge or excitation energy transfer to produce X^+ ions in heteroclusters), then the AE will correspond to either the IP or the excitation energy of the rage gas atom or atoms.

In addition to single ionization, clusters can also undergo multiple ionization/excitation processes. The measurement of the double ionization energies of isolated gas phase atoms and molecules has predominantly utilized synchrotron radiation (Dujardin et al. 1984) or electron-induced ionization (Märk 1975). The double ionization energy for isolated atoms or molecules was found to the approximately 2.8 times the single ionization energy (Tsai and Eland 1980). In the case of solids, the double ionization energy equals twice the single ionization energy (Biester et al. 1987; Scheier and Märk 1987a). The currently accepted model for double ionization in condensed phase involves, at threshold, a primary ionization event, followed by a secondary ionization event of another surrounding atom/molecule by the emitted electron, such that two electrons are emitted. Thus, the IE curves of mixed clusters can also show structures at higher energies, indicating that double ionization/excitation occurs at larger cluster sizes. The double ionization/excitation process would occur when the incident electron (e_i) has enough energy that either: (1) the ejected electron (e_e) or (2) the scattered electron (e_{sc}) has enough energy to excite/ionize an adjacent site. Thus, thresholds could be expected at $2E_{ex}$, $[IP(R/R_n) + E_{ex}]$, and $2IP(R/R_n)$ in the IE curves. A threshold corresponding to $2IP(R)$ has been reported for the following process (Scheier and Märk 1987a):

$$R_n + e^- \rightarrow R_n^{2+} + 3e^-. \qquad (9)$$

It has been shown, by electron impact ionization, that the double ionization thresholds of stable doubly charged argon cluster ions are considerably lowered (32.0 eV) (Scheier and Märk 1987a) compared with the atomic dication (43.4 eV) (Holland et al. 1979).

We will now review the studies of intramolecular Penning ionization and multiple scattering/ionization in vdW clusters, followed by our own electron impact studies on argon/methanol heteroclusters.

7.5.1 Ionization Efficiency Curves

The energetics of ion formation is of considerable importance in understanding the ion thermochemistry and hence, a considerable amount of work has been done to determine the IP and AE of ions by mass spectrometric techniques (Lias et al. 1988). A number of experimental difficulties are associated with the meaurement of IE curves, including accurate and precise calibration of the energy scale, the energy spread on the electron or photon beams, and the effective detection of small currents. These potential sources of errors may be minimized by taking proper precautions, and this topic has been covered by Rosenstock (1976).

The measurement of the IP and AE values for cluster ions and their fragments has enabled the determination of thermochemical data, such as their heats of formation, proton affinities, and binding energies (Brutschy et al. 1987). From these data, it has been possible to obtain insight into the effects of solvation on the structures of cluster ions. In addition to observing the change in the IPs as a function of cluster size, the various thresholds observed in the ion yield curves have provided evidence of the ionization mechanism, that is, direct ionization, ionization following transfer of excitation energy, or ionization arising from multiple excitation/ionization (Brutschy et al. 1987; Scheier and Märk 1987a; Vaidyanathan et al. 1991a).

7.5.2   Intramolecular Penning Ionization

The term "Penning ionization" was originally applied to the spontaneous ionization of a ground state atom M when it collides with a metastable atom A*. (Lampe 1972; Niehaus 1981). In more recent years, this term has been extended to include reactions of other electronic states, as well as reactions involving molecules. An analogous process has been observed within vdW clusters and is referred to as "intramolecular Penning ionization" (Kamke et al. 1985a). The term Penning ionization may be a little misleading in that two different ionization mechanisms can be envisaged for the overall reaction:

$$A^* + M \rightarrow A + M^+ + e^-. \tag{10}$$

The fundamental understanding of Penning ionization is based on the classical exchange model proposed by Hotop and Niehaus (1969), where excitation of A* occurs such that an electron in the outer shell of A* is ejected into the ionization continuum, concomitant with the transfer of an electron in one of the orbitals of M into the vacant inner shell of A*. It is also possible for reaction (10) to proceed via an excitation energy transfer mechanism (Miller and Morgner 1977):

$$A^* + M \rightarrow A + M^* \rightarrow A + M^+ + e^-. \tag{11}$$

It is therfore important to bear in mind that intramolecular Penning ionization is a general description of an ionization process in which the excited state of the component with the higher IP mediates the eventual ionization of the component with the lower IP.

Dehmer and Pratt were one of the earliest groups to observe resonance structures in the photoionization efficiency (PIE) curves of small argon ions and mixed rare gas cluster ions (Dehmer and Pratt 1982; Pratt and Dehmer 1982). The structures in the PIE curves of small argon cluster ions were attributed to interband transitions while the features observed in the PIE curves of mixed rare gas cluster ions were suggested to arise from autoionizing states. Walters et al. (1985) also observed similar resonances in the PIE curves of C_6H_6/HCl

heteroclusters. These spectral features were attributed to an autoionization process where the photon is initially absorbed by the HCl, followed by an energy transfer across the weak intermolecular bond.

Kamke et al. (1985a,b) studied the resonances in the PIE curves of a number of aromatic compounds seeded in various rare gases, and they rationalized the resonance features in terms of intramolecular Penning ionization (IPI). These studies show that IPI is a general process in mixed clusters of aromatic compounds with rare gas atoms, in that it is independent of the functional group attached to the ring or the nature of the ring itself and the resonances were blue-shifted relative to the monomer absorption lines (Kamke et al. 1985a,b). Ruhl et al. (1986) studied the PIE curves of mixed clusters obtained in argon/benzene and argon/fluorobenzene expansions and reported two sharp lines, red-shifted with respect to the monomer absorption lines. As these shifts cannot be explained by autoionization, they also rationalized their results in terms of IPI, providing strong support for the earlier interpretation of Kamke and coworkers. However, the position and shape of the resonance lines reported by Kamke et al. (185a) and Ruhl et al. (1986) were not similar. The discrepancy between the blue shifts in one study and red shifts in the other was resolved in a later study (Kamke et al. 1987) when they showed that the position and the shape of the resonance lines were dependent on the stagnation pressure of the rare gas. By studying the PIE curves of a number of ions in an argon/benzene expansion, they concluded that the broad resonances which are blue-shifted are observed under conditions that favor the formation of large heteroclusters, whereas narrow, red-shifted lines were attributed to the excitation of a single argon atom bound to the benzene molecule (Kamke et al. 1987). Hertel and coworkers (Kamke et al. 1989a) have also reported exciton absorption and Penning ionization in mixed molecular clusters, and Castleman and coworkers have reported similar processes in paraxylene/trimethylamine complexes (Dao and Castleman 1986) and phenylacetylene/ammonia clusters (Breen et al. 1989). Examples of indirect ionization in heteroclusters of rare gas atoms with diatomic and polyatomic molecules have also been reported by Ding (1989).

As discussed in subsection 7.4.2, electron-induced ionization studies of argon/methanol expansion conducted by our group also showed dramatic changes in the CMS at low electron energies (Vaidyanathan et al. 1991b). In order to gain an insight into the origin of the ionization mechanisms at low electron energies, we acquired the IE curves for pure methanol (no carrier gas) expansions, helium/methanol seeded expansions and argon/methanol coexpansions. The IE curves of CH_3OH^+ and $(CH_3OH)_2H^+$ in argon/methanol and helium/methanol expansions are shown in Figures 7-7 and 7-8, respectively. Figure 7-9 shows the IE curves of $ArCH_3OH^+$ and $ArCH_3O^+$ heterocluster ions. Table 7-1 lists the AEs of various ions in pure methanol, helium/methanol, and argon/methanol expansions.

Examination of Figures 7-7–7-9 and the data listed in Table 7-1 reveals that the AE of a number of ions found in the pure methanol and helium/methanol expansions are very similar, and are in excellent agreement with the values reported in the literature (Cook et al. 1980; Momigny et al. 1980). However, the AEs of the very same ions obtained in an argon/methanol coexpansion differ significantly from the reported values and all lie in the energy range of 11.3–11.8 eV. We

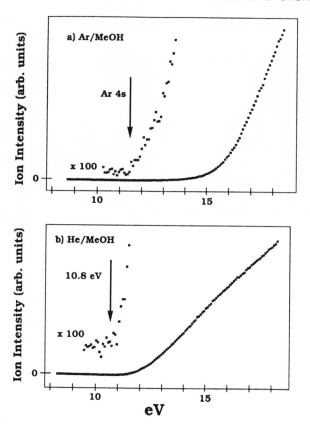

Figure 7-7. (a) Electron ionization efficiency curve for CH_3OH^+ ion ($m/z = 32$) from an Ar/MeOH expansion. Arrow indicates energy corresponding to the first excited 4s state of Ar (11.55 eV). (b) Electron ionization efficiency curve for CH_3OH^+ ion ($m/z = 32$) from an He/MeOH expansion. Arrow indicates onset of ionization (10.8 eV). Reprinted with permission from Vaidyanathan et al. 1991b. Copyright 1991 American Institute of Physics.

interpret this observation as an indication that all these ions arise from the ionization of similar precursors, that is, $Ar_n(CH_3OH)_m$ (Ding et al. 1985). Isenor and Qi (1989) had also reported that the anomalous fragmentation patterns observed for SF_6 and SiF_4 seeded in argon may be due to the Ar_nSF_6 and Ar_nSiF_4 being the precursors for the majority of the ions observed in the CMS. Resonance structures observed in the PIE curves of various ions in mixed expansions of rare gases with atomic, diatomic and organic molecules have also been attributed to arise from indirect ionization involving heteroclusters of the type Ar_nX_m.

The AEs measured by our group are consistent with an ionization mechanism which involves the creation of the excited metastable argon atom $Ar(^3P_{2,0})$ via electron impact. After generation of this Ar 4s state, its excitation energy may then be transferred to the methanol component of the heterocluster. The excitation energies for the formation of Ar 4s $^3P_{2,0}$ states are 11.55 and 11.72 eV, respectively.

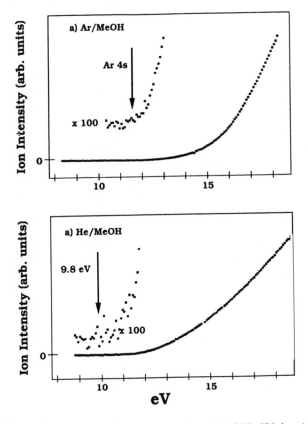

Figure 7-8. (a) Electron ionization efficiency curve for $(CH_3OH)_2H^+$ ion ($m/z = 65$) from an Ar/MeOH expansion. Arrow indicates energy corresponding to the first excited 4s state of Ar (11.55 eV). (b) Electron ionization efficiency curve for $(CH_3OH)_2H^+$ ion ($m/z = 65$) from an He/MeOH expansion. Arrow indicates onset of ionization (9.8 eV). Reprinted with permission from Vaidyanathan et al. 1991b. Copyright 1991 American Institute of Physics.

Our speculation is further strengthened by the fact that energy transfer from the excited metastable argon atom to neutral methanol is quite facile. Tao et al. (1987) have reported such reactions:

$$Ar\,(4s\,^3P_{2,0}) + CH_3OH \rightarrow Ar + CH_2O + 2H \qquad (12)$$

$$\rightarrow Ar + CH_2 + O + H, \qquad (13)$$

where the thermodynamic thresholds for the reactions are 5.4 and 8.7 eV, respectively. Thus, a near resonant excitation energy transfer within the cluster between the Ar 4s and an adjacent methanol molecule seems to be the dominant ionization at the lowest energy threshold (Vaidyanathan et al. 1991b).

7.5.3 Multiple Scattering/Ionization

A particularly interesting area involves the study of multiple scattering and ionization in vdW clusters. The mechanism of production and stability of multiply

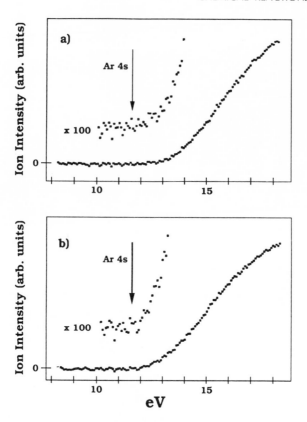

Figure 7-9. (a) Electron ionization efficiency curve for $ArCH_3OH^+$ ion ($m/z = 72$) from an Ar/MeOH expansion. Arrow indicates energy corresponding to the first excited 4s state of Ar (11.55 eV). (b) Electron ionization efficiency curve for $ArCH_3O^+$ ion ($m/z = 71$) from an Ar/MeOH expansion. Arrow indicates energy corresponding to the first excited 4s state of Ar (11.55 eV). Reprinted with permission from Vaidyanathan et al. 1991b. Copyright 1991 American Institute of Physics.

charged cluster-ion vdW clusters has received considerable attention (Echt 1988; Echt et al. 1988; Lezius and Märk 1989; Lezius et al. 1989; Mark 1988; Märk and Castleman 1985; Märk et al. 1989; Sattler et al. 1981; Scheier and Märk 1987a,b; Scheier et al. 1988, 1989). Multiply charged cluster ions are observed only above a certain critical appearance size (n_c), and the determination of this critical size for a large number of vdW clusters has led to an understanding of the properties necessary for the formation of stable multiply charged cluster ions (Echt 1988). The absence of small doubly charged cluster ions gave rise to the speculation that the M_n^{2+} ions must consist of two M^+ ions and $(n - 2)$ M components. It was suggested that when the coulombic repulsion between the cationic centers exceeds the cohesive energy of the cluster, then the cluster ion would decompose in a process termed a "Coulomb explosion." The measurement of the ionization cross sections as a function of electron energy showed that the AEs of the multiply charged cluster ions were far below any of the onsets for the formation of multiply

Table 7-1. List of Appearance Energies for Different Ions

| | (A) Pure Methanol and Helium/Methanol Seeded Expansion | | |
Ions	Pure Methanol (eV)	He/Methanol (eV)	Reported AEs (eV)
$CH_3OH_2^+$	10.5 ± 0.2	10.5 ± 0.2	10.4 ± 0.1[a]
CH_3OH^+	10.9 ± 0.2	10.8 ± 0.2	10.8 ± 0.1[b]
CH_3O^+	11.7 ± 0.2	11.8 ± 0.2	11.7 ± 0.2[b]
CH_2O^+	12.1 ± 0.3	12.2 ± 0.1	12.8 ± 0.3[b]
CHO^+	14.1 ± 0.3	14.1 ± 0.3	14.4 ± 0.5[b]
$(CH_3OH)_2H^+$	9.8 ± 0.2	9.8 ± 0.2	9.8 ± 0.5[a]

| | (B) Argon/Methanol Coexpansion |
Ions	Ar/Methanol (eV)
$CH_3OH_2^+$	11.4 ± 0.2
CH_3OH^+	11.5 ± 0.2
CH_3O^+	11.7 ± 0.2
CH_2O^+	11.5 ± 0.2
$(CH_3OH)_2H^+$	118 ± 0.2
Ar^+	15.8 ± 0.2
$Ar(CH_3OH)^+$	11.7 ± 0.2
$Ar(CH_3O)^+$	11.7 ± 0.2
$Ar(CH_2O)^+$	11.8 ± 0.2

[a] Cook et al. 1980
[b] Momigny et al. 1980.

charged monomer ions. As these large red shifts could not be attributed to solution, the formation of multiply charged cluster ions was suggested to arise from multiple sequential single ionization events of one incoming electron at different monomer sites within the same cluster (Lezius et al. 1989; Märk et al. 1989; Scheier and Märk 1987a; Scheier et al. 1988, 1989).

In spite of the strong experimental and theoretical evidence that a doubly charged cluster ion (M_n^{2+}) consists of two distinct cationic monomers (M^+) along with $(n - 2)$ monomers, and that doubly charged cluster ions below a certain cluster size are not observed (in a microsecond time domain) due to Coulomb explosion, there was no direct evidence for the occurrence of Coulomb explosion in doubly charged cluster ions. Lezius and Märk (1989) provided this direct experimental evidence by measuring the IE curves of singly and doubly charged argon cluster ions. A steep increase in the ion yield of Ar_{60}^+ around 30.0 eV was attributed to the formation of Ar_{120}^{2+} (which cannot be uniquely distinguished from Ar_{60}^+). The IE curves of Ar_n^+ ions, with $n = 10, 20, 30$, and 40, were similar to that of Ar_{60}^+ in that they also displayed an increase in ion yields above 30.0 eV. The high energy feature was rationalized as the contribution to the ion signal from the Coulomb explosion of an unstable doubly charged cluster ion (size $< n_2$ for Ar_n^{2+}) to yield two single charged cluster fragments.

Recently, Foltin et al. (1991a,b) have reported an unusual metastable decomposition channel in singly charged argon cluster ions involving the loss of a number of argon atoms. The number of argon atoms lost increases from two for Ar_4^+ to ten for Ar_{30}^+. This process has a well defined onset energy of ca. 27.0 eV, whereas the metastable decomposition channel involving the loss of a single argon occurs at all energies. This unusual metastable decay process also showed an exponential decrease in the metastable decay fraction with increasing time after the ionization. However, the metastable decay channel involving the loss of a single argon showed a nonexponential behavior. The delayed release of energy in the metastable time regime and the exponential decrease in the metastable decay fraction with time were suggested to be indicative of formation of a localized long-lived exicted state with a definite lifetime within the cluster. The lowest metastable electronically excited states of the Ar^+ ion are at least 14.0 eV above the ground ionic state, whereas the lowest metastable excited states of argon atoms, that is, Ar 4s $^3P_{2,0}$, are 11.55–11.83 eV above the ground state. Thus, the metastable decay channel was attributed to the formation of both a charged and an excitation center (inside the same cluster) by a single incoming electron which undergoes multiple scattering. The metastable lifetimes of these ions are in agreement with the radiative lifetimes measured for $Ar_2^*(^3\Sigma_u^+)$ in solid and liquid argon (Schwenter et al. 1985). The radiative decay of this excimer to the ground state $Ar_2(^1\Sigma_g)$ would lead to the repulsion of the argon atoms and cause the decomposition of the parent cluster ions. A similar metastable decomposition channel arising from multiple scattering has also been observed in neon cluster ions (Foltin and Märk 1991). Hertel and coworkers (Steger et al. 1991) have noted a similar high energy threshold for the metastable decomposition within relatively large argon clusters (n about 20) following photoionization. This study shows the existence of a second state, 1.5 eV above the first one. The first state is attributed to the $n = 1$ exciton state, while the second is suggested to arise from the excitation of the $n = 2$ exciton state in the cluster. The observation of two exciton states separated by 1.5 eV is consistent with the studies conducted in solid argon (Schwenter et al. 1985).

In our labs at SUNY, we have measured the IE curves of Ar_3^+ ion (neat argon expansion) and $CH_3OH_2^+$ ion (argon/methanol coexpansion) with the stagnation pressure of argon varying over the range 1.2–3.5 atm (Vaidyanathan et al. 1992). Figure 7-10 shows the IE curve of the Ar_3^+ ion from a neat argon expansion at a number of stagnation pressures. The AE of the Ar_3^+ ion was determined to be around 15.0 eV, which is consistent with the reported values (Kamke et al. 1989b) The Ar_3^+ was selected because it has been shown that this ion arises predominantly from larger clusters (Buck and Meyer 1984b); hence, the IE curve of the Ar_3^+ ion might be expected to reflect the ionization cross sections of larger clusters from which it arises. Examination of Figure 7-10 reveals that the ion signal increases rapidly beyond 15.5 eV and displays a linear increase in signal intensity between 16.5 and 22.6 eV. However, it is only at higher stagnation pressures that a high energy feature becomes observable at ca. 27.0 eV, followed by an even greater increase in ion yield beyond 30.0 eV.

The near linear increase in ion signal between 16.5 and 22.5 eV is in the energy regime where Ding et al. (1985) observed broad maxima in PIE curves of ions generated by the ionization of $Ar_n(CO)_x$ clusters. This feature was attributed to interband transitions by comparison with the energy loss spectrum of condensed argon (Nuttal et al. 1975). The strong pressure dependence of the structure around 27.0 eV was taken to be an indication that this feature arises from the ionization

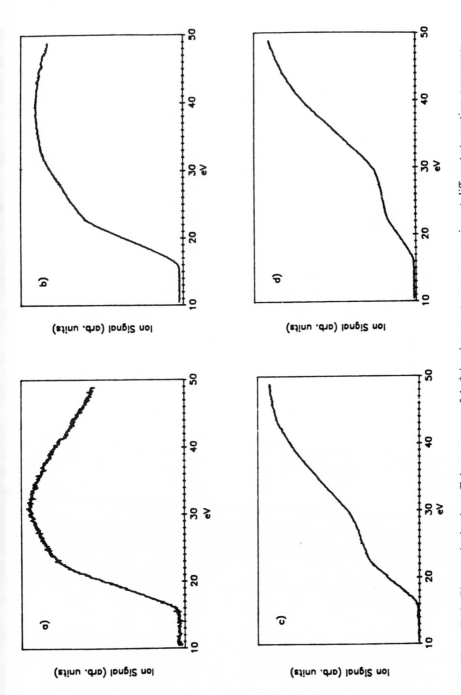

Figure 7-10. Electron ionization efficiency curves of Ar_3^+ ion in a neat argon expansion at different stagnation pressures: (a) 1.2, (b) 2.0, (c) 2.5, and (d) 3.0 atm. Reprinted with permission from Vaidyanathan et al. 1992. Copyright 1992 American Chemical Society.

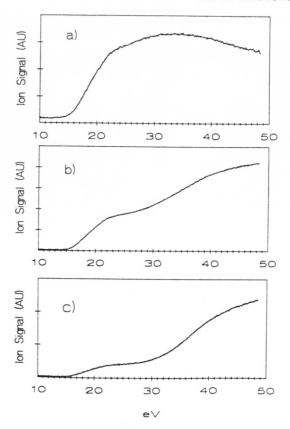

Figure 7-11. Electron ionization efficiency curves of $CH_3OH_2^+$ ion in a neat argon expansion at different stagnation pressures: (a) 1.5, (b) 2.5, and (c) 3.0 atm. Reprinted with permission from Vaidyanathan et al. 1992. Copyright 1992 American Chemical Society.

of larger neutral clusters (higher stagnation pressures favor the formation of larger neutral clusters with a broader size distribution). One of the processes within clusters which has already been shown to display such size effects is multiple ionization. As discussed earlier, multiple ionization occurs as a result of sequential single ionizations, at separate locations within the same cluster, by a single incoming electron (Lezius et al. 1989; Märk et al. 1989; Scheier and Märk 1987a; Scheier et al. 1988, 1989) Lezius and Märk (1989) have reported a significant increase in ion yield of even small cluster ions at around 32.0 eV and they interpreted this as contributions to ion signal from Coulomb explosion of unstable doubly charged cluster ions. Foltin and coworkers (Foltin and Märk 1991; Foltin et al. 1991a,b) have also shown that metastable decomposition leading to multiple monomer loss occurs above 27.0 eV in argon clusters. Hertel and coworkers observed a similar metastable process only in relatively large argon clusters (Steger et al. 1991). Coulomb explosion is prompt while metastable decay, because of the radiative decay of the excimer, has a lifetime of 1.2 μs. Ion trajectory simulations suggest that the ions produced by either process should be detected in our experimental setup as the cluster ions spend several tens of microseconds within the ion lens system before entering the quadrupole mass filter. Thus, we concluded that the higher energy features we observed in the IE curve of the Ar_3^+ ion arise from multiple scattering, that is, the formation of ion and exciton within the same cluster or Coulomb explosion.

The IE curves of $CH_3OH_2^+$ generated in an argon/methanol expansion for various stagnation pressures are displayed in Figure 7-11. The AE of this ion at various stagnation pressures is again consistent with intramolecular Penning ionization within the cluster. The similarity of between the IE curves of the Ar_3^+ ion and the $CH_3OH_3^+$ ion, even at higher electron energies, provides additional support to our contention that the argon component of the cluster mediates the ionization of the $Ar_n(CH_3OH)_m$ heteroclusters. Thus, the high energy feature observed in the IE curves of $CH_3OH_2^+$ should also arise from processes resulting from multiple scattering/multiple ionization.

7.6. SUMMARY AND CONCLUDING REMARKS

Mass spectrometric investigation of gas phase clusters has come to play a significant role in cluster research because it is presently the only technique that allows size-dependent analysis of cluster beams. We anticipate that, in the near future, CMS studies utilizing tandem mass spectrometric techniques will be invaluable in confirming the importance of metastable decompositions and in determining the factors responsible for the enhanced intensities of various ions in the CMS. Such an instrument is currently under construction in our own labs, and we should soon have the ability to perform collision-induced dissociation studies on large vdW clusters.

The study of rare gas clusters has attracted great interest because they are the prototype of vdW clusters. Various studies have probed their structure, stability, energetics, and dynamics. A natural extension to these studies is to probe the behaviour of various atomic, diatomic, and polyatomic components when solvated by rare gas atoms. In addition to acting as a carrier gas, argon also readily forms both homoclusters and heteroclusters. As discussed in this chapter, the influence of the dopant on the stability and structure of the mixed cluster ions has been the subject of a large number of studies. Although there is a considerable body of evidence to suggest that the heterocluster ions of the type $Ar_n X_m^+$ form structures with an icosahedron symmetry, more studies need to be performed to obtain a better understanding of the actual location of the dopant atom/molecule in the heterocluster ions. Two important questions which need to be addressed are: is the dopant ion inside the cluster or is it on, or close to, the surface of the cluster? And does the location of the dopant ion change with the number of rare gas atoms in the heterocluster?

Some of the early studies reported abnormal fragmentation patterns when polyatomic molecules were seeded in argon. It was speculated that heteroclusters, such as $Ar_n X_m$, are the neutral precursors for the various ions observed in the cluster mass spectrum. Our study of the argon/methanol coexpansions showed that the CMS is definitely affected by the choice of carrier gas, as well as the energy employed in ionizing the neutral clusters. Over the years, we have found that a considerable amount of unprotonated cluster ions can be produced when molecules that hydrogen bond extensively are expanded with argon and ionized by using low energy electrons (≤ 20.0 eV). It is therefore an important caveat to realize that the "inert" gas component of the heterocluster can, in fact, mediate the ionization event.

The mechanisms of charge and energy transfer are of fundamental importance in obtaining insight into the ionization and fragmentation processes. These processes are not easy to identify in a homocluster using only mass spectrometric methods. However, as shown in this chapter, argon initially absorbs the energy. That is, in both electron impact ionization and photoionization studies, the inert gas atom *mediates* the ionization event by transferring either excitation energy or the charge to the dopant molecule when the IP (X) < IP (Ar/Ar_n). The pressure dependence of the intramolecular Penning ionization seems to reflect the transition from excitation of a single atom to a group of atoms; thus, reflecting properties that are more characteristic of bulk material.

Although considerable progress has been made in the study of gas phase clusters, the future prospects seem to be equally promising. The impetus for a better understanding of gas phase clusters arises from the novel ability to generate new materials uniquely within the cluster environment. With increasing emphasis being given to nanofabrication and materials fabrication, cluster research may play a crucial role not only in synthesizing new materials, but also in investigating the chemical reactivities of these unique "species." These considerations make it more important to achieve a fundamental understanding of the properties of these clusters, which truly constitute the fifth phase of matter.

ACKNOWLEDGMENTS: The Office of Naval Research is gratefully acknowledged for financial support of this work. We also acknowledge the work of M. Todd Coolbaugh on the studies of magic numbers and ionization mechanisms of heteroclusters.

REFERENCES

Bernard, D. M.; Gotts, N. G.; Stace, A. J. 1990 Int. J. Mass. Spectrom. Ion Proc. 95:327

Biester, H. W.; Besnard, M. J.; Dujardin, G.; Hellner, L.; Koch, E. E. 1987 Phys. Rev. Lett. 59:1277

Bohmer, H. U.; Peyerimhoff, S. D. 1986 Z. Phys. D 4:195

Bohmer, H. U.; Peyerimhoff, S. D. 1988 Z. Phys. D. 8:91

Bohmer, H. U.; Peyerimhoff, S. D. 1989 Z. Phys. D 11:239

Brauman, J. I.; Riveros, J. M.; Blair, L. K. 1971 J. Amer. Chem. Soc. 93:3914

Breen, J. J.; Kilgore, K.; Tzeng, W. B.; Wei, S.; Keesee, R. G.; Castleman, A. W., Jr. 1989 J. Chem. Phys. 90:11

Brutschy, B. 1990 J. Phys. Chem. 94:8637

Brutschy, B. 1992 Chem. Rev. 92:1567

Brutschy, B.; Bisling, P.; Ruhl, E.; Bäumgartel, H. 1987 Z. Phys. D 5:217

Buck, U.; Meyer, H. 1984a Ber. Bunsenges, Phys. Chem. 88:254

Buck, U.; Meyer, H. 1984b Phys. Rev. Lett. 52:109

Buck, U.; Meyer, H. 1986 J. Chem. Phys. 84:4854

Carnovale, F.; Barrie-Peel, J.; Rothwell, R. G.; Vallforf, J.; Kunz, P. J. 1989 J. Chem. Phys. 90:1452

Carnovale, F.; Peel, J. B.; Rothwell, R. G. 1991 J. Chem. Phys. 95:6473

Casero, R.; Soler, J. M. 1991 J. Chem. Phys. 95:2927

Castleman, A. W., Jr. 1990 J. Clus. Sci. 1:3

Castleman, A. W.; Jr.; Kay, B. D.; Herman, V.; Holland, P. M.; Märk, T. D. 1981 Surf. Sci. 106:179

Castleman, A. W., Jr.; Keesee, R. G. 1986a Annu. Rev. Phys. Chem. 37:525

Castleman, A. W., Jr.; Keesee, R. G. 1986b Chem. Rev. 86:589

Castleman, A. W., Jr.; Keesee, R. G. 198a In

Elemental and Molecular Clusters, G. Benedek, T. P. Martin, and G. Pacchioni, eds., Springer-Verlag, Berlin, pp. 307–328

Castelman, A. W., Jr.; Keesee, R. G. 1988b Science 241:36

Cook, K. D.; Jones, G. G.; Taylor, J. W. 1980 Int. J. Mass Spectrom. Ion Phys. 35:273

Coolbaugh, M. T.; Garvey, J. F. 1992 Chem. Soc. Rev. 21:163

Coolbaugh, M. T.; Pfeifer, W. R.; Garvey, J. F. 1990 J. Amer. Chem. Soc. 112:3692

Coolbaugh, M. T.; Vaidyanathan, G.; Garvey, J. F. 1994 Int. Rev. Phys. Chem. 13:1

Dao, P. D.; Castleman, A. W., Jr. 1986 J. Chem. Phys. 84:1435

Dao, P. D.; Morgan, S.; Castleman, A. W., Jr. 1985 Chem. Phys. Lett. 113:219

Dehmer, P. M.; Pratt, S. T. 1982 J. Chem. Phys. 76:843

Desai, S. R.; Feigerle, C. S.; Miller, J. C. 1992 J. Chem. Phys. 97:1793

Ding, A. 1989 Z. Phys. D. 12:253

Ding, A.; Cassidy, R. A.; Futrell, J. H.; Cordis, L. 1987 J. Phys. Chem. 91:2562

Ding, A.; Futrell, J. H.; Cassidy, R. A.; Cordis, L.; Hesslich, J. 1985 Surf. Sci. 156:282

Ding, A.; Hesslich, J. 1983 Chem. Phys. Lett. 94:54

Dujardin, G.; Leach, S.; Dutuit, O.; Guyon, P. M.; Richard-Viard, M. 1984 Chem. Phys. 88:339

Echt, O. 1988 In Elemental and Molecular Clusters, G. Benedek, T. P. Martin, and G. Pacchioni, eds. Springer-Verlag, Berlin, pp. 263–284

Echt, O.; Cook, M. C.; Castelman, A. W.; Jr. 1987 Chem. Phys. Lett. 135:229

Echt, O.; Kandler, O.; Leisner, T.; Miehle, W.; Recknagel, E. 1990 J. Chem. Soc. Farday Trans. 2 36:2411

Echt, O.; Kreisle, D.; Knapp, M.; Recknagel, E. 1984 Chem. Phys. Lett. 108:401

Echt, O.; Kreisle, D.; Recjnagel, E.; Saenz, J. J.; Casero, R.; Soler, J. M. 1988 Phys. Rev. A 38:3236

Echt, O.; Reyes Flotte, A.; Knapp, M.; Sattler, K.; Recknagal, K. 1982 Ber. Bunsenges. Phys. Chem. 86:54

Echt, O.; Sattler, K.; Recknagel, E. 1981 Phys. Rev. Lett. 47:1121

Ehbrecht, M.; Stemmler, M.; Huisken, F. 1993 Int. J. Mass Spectom. Ion Phys. 123:R1–R5

Farges, J.; de Feraudy, M. F.; Raoult, B.; Torchet, G. 1986 J. Chem. Phys. 84:3491

Fayet, P.; Granzer, G.; Hegenbart, G.; Moisar, E.; Pische, B.; Woste, L. 1985 Phys. Rev. Lett. 55:3002

Fieber, M.; Broker, G.; Holub-Krappe, E.; Ding, A. 1991 Z. Phys. D 20:21

Foltin, M.; Märk, T. D. 1991 Chem. Phys. Lett. 180:317

Foltin, M.; Walder, G.; Castelman, A. W., Jr.; Märk, T. D. 1991a J. Chem. Phys. 94:810

Foltin, M.; Walder, G.; Mohr, S.; Scheier, P.;

Castelman, A. W., Jr.; Märk, T. D. 1991b Z. Phys. D 20:157

Fund, K. M.; Henke, W. E.; Hays, T. R.; Selzle, H. L.; Schlag, E. W. 1981 J. Chem. Phys. 85:3560

Garvey, J. F.; Bernstein, R. B. 1986a J. Amer. Chem. Soc. 86:589

Garvey, J. F.; Bernstein, R. B. 1986b J. Phys. Chem. 90:3577

Garvey, J. F.; Herron, H. J.; Vaidyanathan, G.; Coolbaugh, M. T. 1994 Chem. Rev. 94:1999

Garvey, J. F.; Peifer, W. R.; Coolbaugh, M. T. 1991 Acc. Chem. Res. 24:48

Goyal, S.; Schutt, D. L.; Scoles, G. 1993 Acc. Chem. Res. 26:123

Grover, J. R.; Herron, W. J.; Coolbaugh, M. T.; Peifer, W. R.; Garvey, J. F. 1991 J. Phys. Chem. 95:6473

Haberland, H. 1985 Surf. Sci. 156:305

Hagena, O. 1974 In Molecular Beams and Low Density Gas Dynamics, P. P. Wegener, ed., Marcel Dekker, New York

Hagena, O. F. 1987 Z. Phys. D 4:291

Harris, I. A.; Kidwell, R. S.; Northby, J. A. 1984 Phys. Rev. Lett. 53:2390

Herron, W. J.; Coolbaugh, M. T.; Vaidyanathan, G.; Garvey, J. F. 1992 J. Amer. Chem. Soc. 114:3684

Herzog, R. F. K.; Poschenrider, W. P.; Satkiewicz, R. G. 1973 Radiat. Effects 18:199

Hesslich, J.; Kuntz, P. J. 1981 Z. Phys. D 2:251

Hoare, M. R. 1979 Adv. Chem. Phys. 40:49

Holland, D. M. P.; Codling, K.; West, J. B.; Marr, G. V. 1979 J. Phys. B 12:2465

Hotop, H.; Niehaus, A. 1969 Z. Phys. 228:68

Isenor, N. R.; Qi, J. 1989 Chem. Phys. Lett. 155:283

Janda, K. C. 1985 Adv. Chem. Phys. 60:201

Jena, P.; Rao, B. K.; Khanna, S. N. (eds.) 1986 Physics and Chemistry of Small Clusters, Plenum Press, New York

Jortner, J. 1984 Ber. Bunsenges, Phys. Chem. 88:188

Kamke, B.; Kamke, W.; Herrmann, R.; Hertel, I. V. 1989a Z. Phys. D 11:153

Kamke, B.; Kamke, W.; Wang, Z.; Rohl, E.; Brutschy, B. 1987 J. Chem. Phys. 86:2525

Kamke, W.; de Vries, J.; Krauss, J.; Kaiser, E.; Kamke, B.; Hertel, I. V. 1989b Z. Phys. D 14:339

Kamke, W.; Kamke, B.; Keifl, H. U.; Hertel, I. V. 1985a Chem. Phys. Lett. 122:356

Kamke, W.; Kamke, B.; Kiefl, H. U.; Wang, Z.; Hertel, I. V. 1985b Chem. Phys. Lett. 128:399

Klots, C. E. 1988 J. Phys. Chem. 92:5864

Knozinger, E.; Schuller, W.; Langel, W. 1988 Faraday Discuss. Chem. Soc. 86:285

Kuntz, P. J.; Valldorf, J. 1988 Z. Phys. D 8:195

Lampe, F. W. 1972 In Ion–Molecule Reactions, J. L. Franklin, ed., Plenum Press, New York, pp. 601–646

Langel, W.; Schuller, W.; Knozinger, E.; Fleger, H. W.; Fleger, J. 1988 J. Chem. Phys. 89:1741

Lethbridge, P. G.; Stace, A. J. 1988 J. Chem. Phys. 89:4062

Levinger, N. E.; Ray, D.; Alexander, M. L.; Lineberger, W. C. 1988 J Chem. Phys. 89:5654

Levy, D. H. 1981 Adv. Chem. Phys. 47:323

Lezius, M.; Scheier, P.; Märk, T. D. 1992 Chem. Phys. Lett. 196:118

Lezius, P.; Märk, T. D. 1989 Chem. Phys. Lett. 155:496

Lezius, P.; Scheier, P.; Stamatovic, A.; Märk, T. D. 1989 J. Chem. Phys. 91:3240

Lias, S. G.; Bartmess, J. E.; Liebman, J. F.; Holmes, J. L.; Levin, R. D.; Mallard, W. G. 1988 J. Phys. Chem. Ref. Data

Mackay, A. L. 1962 Acta Crystallogr, 15:916

Märk, T. D. 1975 J. Chem. Phys. 63:3731

Märk, T. D. 1987 Int. J. Mass Spectrom. Ion Proc. 79:1

Märk, T. D. 1988 In Electronic and Atomic Collisions, H. B. Gilbody, W. R. Newell, F. H. Read, and A. C. H. Smith, eds. Elsevier, Amsterdam, pp. 705–717

Märk, T. D.; Castleman, A. W.; Jr. 1985 Adv. At. Mol. Phys. 20:65

Märk, T. D.; Foltin, M.; Grill, V.; Rauth, T.; Wlader, G. 1992 Ber. Bunsenges. Phys. Chem. 96:1125

Märk, T. D.; Scheier, P. 1987a Chem. Phys. Lett. 137:245

Märk, T. D.; Scheier, P. 1987b J. Chem. Phys. 87:1456

Märk, T. D.; Scheier, P.; Leiter, K.; Ritter, W.; Stephan, K.; Stanatovic, A. 1986 Int. J. Mass Spectrom. Ion Proc. 74:281

Märk, T. D.; Scheier, P.; Lezius, P.; Walder, G.; Stamatovic, A. 1989 Z. Phys. D 12:279

Martin, T. P.; Naher, U.; Schaber, H.; Zimmerman, U. 1993 Phys. Rev. Lett. 70:3079

Miller, W. H.; Morganer, H. 1977 J. Chem. Phys. 67:4923

Moller, T.; Zimmerman, G. 1989 J. Opt. Soc. Am. B 6:1062

Momigny, J.; Wankenne, H.; Krier, C. 1980 Int. J. Mass Spectrom. Ion Phys. 35:151

Morgan, S.; Castelman, A. W.; Jr. 1987 J. Amer. Chem. Soc. 109:2867

Morgan, S.; Castelman, A. W., Jr. 1989 J. Phys. Chem. 93:4544

Morgan, S.; Keesee, R. G.; Castelman, A. W., Jr. 1989 J. Amer. Chem. Soc. 111:3841

Naaman, R. 1988 Adv. Chem. Phys. 70:181

Ng, C. Y. 1983 Adv. Chem. Phys. 52:263

Niehaus, A. 1981 Adv. Chem. Phys. 45:399

Northby, J. A. 1987 J. Chem. Phys. 87:6166

Novick, S. E.; Davies, P. B.; Dyke, T. R.; Klemperer, W. 1973 J. Amer. Chem. Soc. 95:8547

Nuttal, J. D.; Gallon, T. E.; Devey, M. G.; Mathews, J. A. D. 1975 J. Phys. C 8:445

Ozaki, Y.; Fukuyama, T. 1985 Int. J. Mass Spectrom, Ion Proc. 88:227

Peifer, W. R.; Coolbaugh, M. T.; Garvey, J. F. 1989 J. Chem. Phys. 91:6684

Pratt, S. T.; Dehmer, P. M. 1982 J. Chem. Phys. 76:4865

Rademann, K.; Kaiser, B.; Even, U.; Hensel, F. 1987 Phys. Rev. Lett. 59:2319

Recknagel, E. 1984 Ber. Bunsenges. Phys. Chem. 88:201

Romanowski, G.; Wanczek, K. P. 1984 Int. Mass Spectrom. Ion Proc. 62:277

Rosenstock, H. M. 1976 Int. Mass Spectrom. Ion Phys. 20:139

Ruhl, E.; Bisling, P. G. F.; Brutschy, B.; Beckman, K.; Leisin, O.; Morganer, H. 1986 Chem. Phys. Lett. 128:512

Saenz, J. J.; Soler, J. M.; Garcia, N. 1985 Surf. Sci. 156:121

Sattler, K. 1985 Surf. Sci. 156:292

Sattler, K.; Muhlbach, O.; Echt, O.; Pfau, P.; Recknagel, E. 1981 Phys. Rev. Lett. 47:160

Scheier, P.; Märk, T. D. 1987a Chem. Phys. Lett. 136:423

Scheier, P.; Märk, T. D. 1987b J. Chem. Phys. 86:3056

Scheier, P.; Märk, T. D. 1987c Phys. Rev. Lett. 59:1813

Scheier, P.; Walder, G.; Stamatovic, A.; Märk, T. D. 1988 Chem. Phys. Lett. 150:222

Scheier, P.; Walder, G.; Stanatovic, A.; Märk, T. D.1989 J. Chem. Phys. 90:4091

Schwenter, N.; Koch, E. E.; Jortner, J. (eds.) 1985 Elecgronic Excitation in Condensed Rare Gases, Springer, Berlin

Scoles, G. (ed.) 1988 Atomic and Molecular Beam Methods, Vol. I, Oxford University Press, New York

Shinohara, H.; Nishi,; Washida, N. 1985 J. Chem. Phys. 83:1939

Shinohara, H.; Nishi, N.; Washida, N. 1986 J. Chem. Phys. 84:5561

Stace, A. J. 1983 J. Phys. Chem. 87:2286

Stace, A. J. 1984 J. Amer. Chem. Soc. 106:4380

Stace, A. J. 1985a Chem. Phys. Lett. 113:355

Stace, A. J. 1985b J. Amer. Chem. Soc. 107:755

Stace, A. J. 1987a In Mass Spectrometry, Specialist Periodical Report, Vol. 9, M. E. Rose, ed., Royal Society of Chemistry, London, p. 96

Stace, A. J. 1987b J. Chem. Soc. Faraday Trans. 2 83:29

Stace, A. J. 1987c J. Phys. Chem. 91:2286

Stace, A. J. 1992 Ber. Bunsenges. Phys. Chem. 96:1136

Stace, A. J. 1993 Org. Mass Spectrom. 28:3

Stace, A. J.; Bernard, D. M. 1988 Chem. Phys. Lett. 146:531

Stace, A. J.; Moore, C. B. 1983 Chem. Phys. Lett. 96:80

Stace, A. J.; Moore, C. J. 1982 J. Phys. Chem. 86:3681

Stamatovic, A.; Scheier, P.; Märk, T. D. 1988 J. Chem. Phys. 88:6884

Steger, H.; de Vries, J.; Kamke, W.; Hertel, I. V. 1991 Z. Phys. D 21:85

Stephens, P. W.; King, J. G. 1984 Phys. Rev. Lett. 53:2390

Takagi, I. 1986 Z. Phys. D 3:170

Tao, W.; Golde, M. F.; Ho, G. H.; Moyle, A. M. 1987 J. Chem. Phys. 87:1045

Tsai, B. P.; Eland, J. H. D. 1980 Int. J. Mass Spectrom. Ion Phys. 36:143

Vaidyanathan, G.; Coolbaugh, M. T.; Garvey, J. F. 1991a J. Clus, Sci. 2:183

Vaidyanathan, G.; Coolbaugh, M. T.; Garvey, J. F. 1992 J. Phys. Chem. 96:1589

Vaidyanathan, G.; Coolbaugh, M. T.; Peifer, W. R.; Garvey, J. F. 1991b J. Chem. Phys. 94:1850

Vaidyanathan, G.; Coolbaugh, M. T.; Peifer, W. R.; Garvey, J. F. 1991c J. Phys. Chem. 95:4193

Vaidyanathan, G.; Herron, W. J.; Garvey, J. F. 1993 J. Phys. Chem. 97:78870

van der Waal, B. W. 1983 J. Chem. Phys. 79:3948

Walters, E. A.; Grover, J. R.; White, M. G.; Hui, E. T. 1985 J. Phys. Chem. 89:3914

Wei, S.; Shi, Z.; Castelman, A. W., Jr. 1991a J. Chem. Phys. 94:8604

Wei, S.; Tzeng, W. B.; Castleman, A. W., Jr. 1991b J. Phys. Chem. 95:5080

Wei, S.; Tzeng, W. B.; Keesee, R. G., Castelman, A. W., Jr. 1991c J. Amer. Chem. Soc. 113:1960

Whetten, R. L.; Schriver, K. E.; Persson, J. L.; Hahn, M. Y. 1990 J. Chem. Soc. Farday Trans. 2 86:2375

Zhang, X.; Yang, X.; Castleman, A. W., Jr. 1991 Chem. Phys. Lett. 185:298

Index